# Recent Advances and Applications of Machine Learning in Metal Forming Processes

# Recent Advances and Applications of Machine Learning in Metal Forming Processes

Editors

**Pedro Prates**
**André Pereira**

MDPI • Basel • Beijing • Wuhan • Barcelona • Belgrade • Manchester • Tokyo • Cluj • Tianjin

*Editors*
Pedro Prates
University of Aveiro
Portugal

André Pereira
University of Coimbra
Portugal

*Editorial Office*
MDPI
St. Alban-Anlage 66
4052 Basel, Switzerland

This is a reprint of articles from the Special Issue published online in the open access journal *Metals* (ISSN 2075-4701) (available at: https://www.mdpi.com/journal/metals/special_issues/machine_learning_forming).

For citation purposes, cite each article independently as indicated on the article page online and as indicated below:

LastName, A.A.; LastName, B.B.; LastName, C.C. Article Title. *Journal Name* **Year**, *Volume Number*, Page Range.

**ISBN 978-3-0365-5771-7 (Hbk)**
**ISBN 978-3-0365-5772-4 (PDF)**

# Contents

# About the Editors

**Pedro Prates**

Pedro Prates is an Assistant Professor at the University of Aveiro, Portugal, in the subarea of applied and computational mechanics. His research interests are focused on the numerical modelling and simulation of metal forming processes, calibration of elastoplastic constitutive models (using FEMU and machine learning approaches), robustness and sensitivity analyses, elastoplastic fatigue and fracture mechanics and machine learning. He has been involved in several externally funded research projects in the areas mentioned above, one of them as a principal investigator. His research work has led to over 50 publications in international journals and conference proceedings.

**André Pereira**

André Pereira is a researcher at the Centre for Mechanical Engineering, Materials and Processes (CEMMPRE), and Invited Assistant Professor at the University de Coimbra, Portugal. His main areas of research are the numerical simulation of forming processes and nanotubes, constitutive parameter identification, uncertainty, and sensitivity analysis. He has participated in 6 R&D projects in the abovementioned areas, being the principal investigator in one of them. His cooperation with more than 50 international researchers has resulted in the publication of 36 works in international journals and conference proceedings.

metals MDPI

*Editorial*

# Recent Advances and Applications of Machine Learning in Metal Forming Processes

**Pedro A. Prates [1,2,3,*] and André F. G. Pereira [3,4,*]**

1 TEMA—Centre for Mechanical Technology and Automation, Department of Mechanical Engineering, University of Aveiro, 3810-193 Aveiro, Portugal
2 LASI—Intelligent Systems Associate Laboratory, 4800-058 Guimarães, Portugal
3 CEMMPRE, Department of Mechanical Engineering, University of Coimbra, 3030-788 Coimbra, Portugal
4 ARISE—Advanced Production and Intelligent Systems Associated Laboratory, 4200-465 Porto, Portugal
* Correspondence: prates@ua.pt (P.A.P.); andre.pereira@uc.pt (A.F.G.P.)

## 1. Introduction

Machine Learning (ML) is a subfield of artificial intelligence, focusing on computational algorithms that are designed to learn and improve themselves, without the need to be explicitly programmed. ML algorithms have been applied in several fields, being particularly useful for solving complex tasks that would normally require the understanding and building of either impossible or complex first-principle models, i.e., based on fundamental physical laws.

ML approaches are emerging in the area of metal forming processes, driven by the increasing availability of large datasets, coupled with the exponential growth of computer performance. In fact, there has been a growing interest in evaluating the capabilities of ML algorithms in studying topics related to metal forming processes, such as: classification, detection and prediction of forming defects; material modelling and parameters identification; process classification and selection; process design and optimization. The purpose of this Special Issue is to disseminate state-of-the-art ML applications in metal forming processes.

## 2. Contributions

The Special Issue is comprised of a total of ten research articles related to ML applications for metal forming processes, including: prediction of forming results [1] and their energy consumption [2]; constitutive modelling [3] and parameters identification [4]; process parameters optimization [4,5]; prediction, detection and classification of defects [6–8]; prediction of mechanical properties [9,10]. The following paragraphs summarize the contributions of these works.

Several Machine Learning (ML) algorithms can be found in the literature, each with their advantages and disadvantages. In this Special Issue, two papers [1,2] compared the predictive performance of various ML algorithms in two different applications. Marques et al. [1] studied the performance of different parametric and non-parametric metamodels in predicting the forming results of the U-Channel and the Square Cup forming process. For the non-parametric techniques (ML-based), the metamodels trained with Multi-Layer Perceptron, Gaussian Processes, Kernel Ridge and Support Vector Regression algorithms were more accurate than those trained with Decision Trees, Random Forest and k-Nearest Neighbors algorithms. Additionally, the parametric metamodeling techniques, Response Surface Method and Polynomial Chaos Expansion, also showed themselves to be competitive alternatives to the best ML-based metamodels. Mirandola et al. [2] also compared the performance of different ML algorithms, but for the prediction of the energy consumption in radial axial-ring rolling forming process. Eight different ML algorithms (Random Forest, Gradient Boosting, Artificial Neural Network, Ridge, Lasso,

**Citation:** Prates, P.A.; Pereira, A.F.G. Recent Advances and Applications of Machine Learning in Metal Forming Processes. *Metals* **2022**, *12*, 1342. https://doi.org/10.3390/met12081342

Received: 19 July 2022
Accepted: 3 August 2022
Published: 12 August 2022

Elastic Net, Kernel Ridge and Support Vector Regression) were used to predict energy consumption based on material, geometrical, and process parameters. The trained ML models proved to be reliable, even for extrapolation predictions (i.e., outside the training data range). The best prediction accuracy was reached by the model trained with the Gradient Boosting algorithm.

Numerical simulation is nowadays an essential tool for the development and optimization of metal forming processes. The accuracy of the numerical simulations requires constitutive models capable of describing the mechanical behavior of the material. On this topic, two papers of the special issue [3,4] investigated the application of ML to constitutive modeling and material parameters identification. Lourenço et al. [3], explored and discussed the potential contributions of ML in terms of elastoplasticity constitutive modelling. This work discusses and analyses four recent advances and applications of ML: parameters identification; the enhancement of traditional constitutive models; the development of data-driven constitutive models with embedding physical and empirical knowledge; the development of constitutive models fully based on data-driven approaches. In summary, the authors demonstrated the potential of ML-based approaches to solve diverse and complex constitutive modelling problems. Cruz et al. [4] explored the performance of ML models, using shallow artificial neural networks, in two applications: (i) identification of constitutive parameters (isotropic hardening) with a three-point bending test; (ii) optimization of a process parameter (punch displacement), to obtain a desired bending angle (after springback) in press-brake air bending process. In both applications, the trained ANN models were able to solve the identification and optimization problem with reliable results.

The optimization of forming process parameters was also investigated by Palmieri et al. [5]. This work proposes a method to perform real-time control of the blank holder force during a deep-drawing process. Metamodels were initially built with kriging technique to establish a relationship between the process parameters and quality indices. Afterwards, a multi-objective optimization was performed to obtain the blank holder force that guarantees a component without defects in the presence of variability in the yield stress values and lubrication conditions. The result was a regulation curve that is useful for real-time control of the blank holder force to avoid defects in the deep-drawing component.

The application of ML algorithms to predict, detect, and classify defects was investigated by three works of this Special Issue [6–8]. Hao et al. [6] proposed a method to classify hot rolling strip surface defects based on the Wasserstein Generative Adversarial Network (WGAN) and an attention mechanism. A dataset of defects images was collected, and then the WGAN model was used to generate additional images (data augmentation). The image classification was performed with the model (SE-ResNet34), coupled with an attention mechanism that enables SE-ResNet34 model to focus on the most valuable information, in order to improve the classification accuracy. The proposed method revealed an excellent classification accuracy of hot rolling strip steel surface defects. Wang et al. [7] also investigated an intelligent recognition model for hot rolling strip surface defects based on convolutional neural networks (CNN), to improve the detection accuracy. The most common defects were classified in five categories (Upwarp, Black Line, Crack, Slag Inclusion and Gas Hole), and a database of defect images was built. Using this data, defect recognition models were established using CNN. The results showed that, depending on the type of the CNN, it was possible to obtain a high recognition accuracy in a short period of time. Lee et al. [8], also proposed a methodology based on CNN to predict the buckling instability of automotive sheet metal panels. A CNN model was used to establish relations between image results on indentation points, evaluated in several localizations of the panels, and the magnitude of the buckling instability. The developed method was able to accurately predict the buckling instability magnitude for automotive sheet metal panels.

Besides forming defects, the mechanical properties of the final product are also an essential aspect to control during metal forming processes. Two works from this Special Issue focused on the application of ML to predict mechanical properties [9,10]. Wu et al. [9]

proposed a new prediction model based on Multidimensional Support Vector Regression, combined with a feature selection method, which involves maximum information coefficient correlation characterization and complex network clustering. This method allows for the selection of the most representative input variables to reduce the input dimensionality. The proposed model was used to predict the steel mechanical properties based on the conditions of four main processes (smelting, continuous casting, hot rolling, and cold rolling). Compared to other models, the proposed model had, simultaneously, the highest prediction accuracy and the lowest computational complexity. On the same subject, Merayo et al. [10] developed a methodology to optimize the topology of a multilayer artificial neural network, to predict the ultimate tensile strength of aluminum alloys based on the information of their chemical composition and tempering process. The methodology consists in optimizing the number of nodes of two hidden layers, to maximize the accuracy of the predictions without compromising the computational cost. The optimized artificial neural network was able to give accurate predictions.

## 3. Conclusions

The Special Issue covers 10 papers about the application of Machine Learning (ML) approaches to Metal Forming Processes. Based on these works, the application of ML approaches revealed itself to be a success, reaching accurate predictions and classification tasks. As Guest Editors, we are confident that the quality of the methods and results presented in this Special Issue, represent a significant contribution for the dissemination and advancement of future research of Machine Learning in metal forming processes.

**Funding:** This book was sponsored by FEDER funds through the program COMPETE (Programa Operacional Factores de Competitividade), by national funds through FCT (Fundação para a Ciência e a Tecnologia) under the projects UIDB/00285/2020, UIDB/00481/2020, UIDP/00481/2020, CENTRO-01-0145-FEDER-022083, LA/P/0104/2020 and LA/P/0112/2020; this book was also co-funded by POCI under the projects PTDC/EME-EME/31243/2017 (RDFORMING) and PTDC/EME-EME/31216/2017 (EZ-SHEET).

**Acknowledgments:** The guest editors would like to thank the authors, the reviewers, and the editorial team of Metals for their valuable contributions to this Special Issue.

**Conflicts of Interest:** The authors declare no conflict of interest.

## References

1. Marques, A.E.; Prates, P.A.; Pereira, A.F.G.; Oliveira, M.C.; Fernandes, J.V.; Ribeiro, B.M. Performance Comparison of Parametric and Non-Parametric Regression Models for Uncertainty Analysis of Sheet Metal Forming Processes. *Metals* **2020**, *10*, 457. [CrossRef]
2. Mirandola, I.; Berti, G.A.; Caracciolo, R.; Lee, S.; Kim, N.; Quagliato, L. Machine Learning-Based Models for the Estimation of the Energy Consumption in Metal Forming Processes. *Metals* **2021**, *11*, 833. [CrossRef]
3. Lourenço, R.; Andrade-Campos, A.; Georgieva, P. The Use of Machine-Learning Techniques in Material Constitutive Modelling for Metal Forming Processes. *Metals* **2022**, *12*, 427. [CrossRef]
4. Cruz, D.J.; Barbosa, M.R.; Santos, A.D.; Miranda, S.S.; Amaral, R.L. Application of Machine Learning to Bending Processes and Material Identification. *Metals* **2021**, *11*, 1418. [CrossRef]
5. Palmieri, M.E.; Lorusso, V.D.; Tricarico, L. Robust Optimization and Kriging Metamodeling of Deep-Drawing Process to Obtain a Regulation Curve of Blank Holder Force. *Metals* **2021**, *11*, 319. [CrossRef]
6. Hao, Z.; Li, Z.; Ren, F.; Lv, S.; Ni, H. Strip Steel Surface Defects Classification Based on Generative Adversarial Network and Attention Mechanism. *Metals* **2022**, *12*, 311. [CrossRef]
7. Wang, D.; Xu, Y.; Duan, B.; Wang, Y.; Song, M.; Yu, H.; Liu, H. Intelligent Recognition Model of Hot Rolling Strip Edge Defects Based on Deep Learning. *Metals* **2021**, *11*, 223. [CrossRef]
8. Lee, S.; Quagliato, L.; Park, D.; Berti, G.A.; Kim, N. A Buckling Instability Prediction Model for the Reliable Design of Sheet Metal Panels Based on an Artificial Intelligent Self-Learning Algorithm. *Metals* **2021**, *11*, 1533. [CrossRef]
9. Wu, Y.; Yan, Y.; Lv, Z. Novel Prediction Model for Steel Mechanical Properties with MSVR Based on MIC and Complex Network Clustering. *Metals* **2021**, *11*, 747. [CrossRef]
10. Merayo, D.; Rodríguez-Prieto, A.; Camacho, A.M. Topological Optimization of Artificial Neural Networks to Estimate Mechanical Properties in Metal Forming Using Machine Learning. *Metals* **2021**, *11*, 1289. [CrossRef]

metals

MDPI

Article

# Performance Comparison of Parametric and Non-Parametric Regression Models for Uncertainty Analysis of Sheet Metal Forming Processes

**Armando E. Marques [1,*], Pedro A. Prates [1], André F. G. Pereira [1], Marta C. Oliveira [1], José V. Fernandes [1] and Bernardete M. Ribeiro [2]**

1 CEMMPRE, Department of Mechanical Engineering, Univ Coimbra, 3030-788 Coimbra, Portugal; pedro.prates@dem.uc.pt (P.A.P.); andre.pereira@dem.uc.pt (A.F.G.P.); marta.oliveira@dem.uc.pt (M.C.O.); valdemar.fernandes@dem.uc.pt (J.V.F.)

2 CISUC, Department of Informatics Engineering, Univ Coimbra, 3030-290 Coimbra, Portugal; bribeiro@dei.uc.pt

* Correspondence: armando.marques@uc.pt

Received: 2 March 2020; Accepted: 30 March 2020; Published: 1 April 2020

**Abstract:** This work aims to compare the performance of various parametric and non-parametric metamodeling techniques when applied to sheet metal forming processes. For this, the U-Channel and the Square Cup forming processes were studied. In both cases, three steel grades were considered, and numerical simulations were performed, in order to establish a database for each combination of forming process and material. Each database was used to train and test the various metamodels, and their predictive performances were evaluated. The best performing metamodeling techniques were Gaussian processes, multi-layer perceptron, support vector machines, kernel ridge regression and polynomial chaos expansion.

**Keywords:** sheet metal forming; uncertainty analysis; metamodeling; machine learning

## 1. Introduction

Sheet metal forming is a widely used manufacturing technique in the automotive and aerospace industries. As the standards of modern industry become more demanding, the traditional trial-and-error approach to process design is too costly to be viable, both in scrap losses and in time spent. As such, researchers look for ways to make process design more efficient. Since sheet metal forming problems present high non-linearity with regard to material properties, boundary conditions, and geometry, the creation of analytical models is unfeasible. As a result of this, researchers began to focus on the use of the finite element method (FEM) to model forming processes. However, the FEM simulation of complex forming processes can be computationally expensive, and a large number of simulations can be required in order to find a good design solution, due to the high number of variables. Alternatively, metamodeling techniques can be used to create predictive models based on the data obtained from a set of numerical simulations, limiting the amount of simulations required during the design process and, as such, reducing the computational cost. Parametric metamodeling techniques, such as the response surface method (RSM), have been substantially used in sheet metal forming problems. Wei et al. [1] applied this method to reduce the amount of FEM simulations required to optimize the forming process of a deck-lid outer panel, while Naceur et al. [2] used a moving least squares iterative adaptation of the method in two different problems: the minimization of springback in the deep drawing of a cylindrical cup and the optimization of the initial blank shape for a forming process. These models can achieve good prediction accuracy; however, they may struggle in cases with high non-linearity. As such, in recent years much attention has been given to machine learning (ML) metamodels. In particular, the artificial neural network (ANN), support vector machine (SVM), and Gaussian process (GPs) have been applied to various sheet metal forming processes. Sun et al. [3] applied SVM, alongside RSM and Kriging (a particular case of GP), in the optimization of the forming process of an automobile inner panel. Teimouri et al. [4] explored various ANN algorithms in a springback optimization problem, and compared them with the RSM, concluding that the ANN algorithms showed better performance. Wessing et al. [5] compared the application of ANN and Kriging in predicting the final sheet thickness of a B-pillar and concluded that Kriging performed better. Similarly, Ambrogio et al. [6] obtained better results from the Kriging method when compared to ANNs and RSM, when applied to the prediction of the final sheet thickness in an incremental sheet metal forming problem. Feng et al. [7] used SVM in an optimization problem related to variable blank-holder force and Jingdong et al. [8] used GP in the prediction of forming defects, namely, the occurrence of fractures and appearance of wrinkles. Despite the growing interest in the application of these techniques, researchers usually select just one or a few based on subjective criteria. To the authors' knowledge, no in-depth study has been conducted to determine the relative performance of the many available regression metamodeling techniques when applied to sheet metal forming processes, as the existing studies only focus on a small number of techniques.

The present work consists of a performance evaluation of various regression metamodeling techniques when applied to the prediction of results of sheet metal forming processes. The parametric metamodeling techniques evaluated are the response surface method (RSM) and polynomial chaos expansion (PCE), while the non-parametric metamodeling techniques evaluated include Gaussian processes (GPs), artificial neural networks (multi-layer perceptron or MLP), decision trees (DTs), random forest (RF), k-nearest neighbors (kNN), support vector regression (SVR), and kernel ridge regression (KRR). All the non-parametric techniques considered can be classified as machine learning (ML) techniques. The forming processes considered were the U-Channel and the Square Cup. For each forming process, three steel grades were considered to cover a wide range of hardening behavior. For each grade, it was assumed that the elastic and plastic properties present some variability, described by a normal distribution [9]. The same type of distribution was also used to describe the initial thickness of the sheet, the contact with friction conditions, and one process parameter. The values of maximum thinning were evaluated for both processes, as well as springback for the U-Channel case and maximum equivalent plastic strain for the Square Cup case.

The rest of the paper is arranged as follows: first, a brief theoretical introduction of each of the metamodeling techniques is given, followed by a description of the FEM models built for each forming process, including the material properties. Then, the dataset generation process is described, followed by the results obtained and respective discussion. The final section contains the general conclusions taken from this work.

## 2. Metamodeling

Metamodeling techniques allow mathematical relationships to be established between the design variables (i.e., sources of variability) and the simulated outputs (i.e., responses) of forming processes. The vector of design variables is defined as $\mathbf{x} = x_i,\ i = 1, \ldots, p$, where $p$ is the total number of sources of variability (inputs). In order to train the metamodel, it is necessary to evaluate the metamodel response $\mathrm{y}^*(\mathbf{x})$ for a predefined set of training points, $\mathbf{x}^{\mathrm{t}}$, to ensure that at those points the simulation outputs $\mathrm{y}(\mathbf{x})$ are well represented. In this context, it is possible to define a training matrix $\mathbf{X} = x_{im}$, with $i = 1, \ldots, p$ and $m = 1, \ldots, q$, where $q$ is the total number of training points.

### 2.1. Response Surface Method (RSM)

RSM is a regression model that fits a polynomial function to a set of training points [3]. In this work, a quadratic function is used, as follows:

$$\mathrm{y}^*(\mathbf{x}) = \beta_0 + \sum_{i=1}^{p} \beta_i x_i + \sum_{i=1}^{p} \sum_{j>i}^{p} \beta_{ij} x_i x_j + \sum_{i=1}^{p} \beta_{ii} x_i^2 \quad (1)$$

where $y^*(\mathbf{x})$ is the estimated response for a given set of inputs $\mathbf{x}$ and $\beta_0$, $\beta_i$, $\beta_{ij}$. and $\beta_{ii}$ are the set of RSM coefficients, which can be organized in the vector of unknowns $\boldsymbol{\beta}$, with a dimension equal to the total number of RSM coefficients: $B = 0.5p^2 + 1.5p + 1$. Note that for $q < B$ the system of equations is underdetermined while for $q > B$ it is overdetermined (i.e., there is a unique solution only when $q = B$). Thus, for $q \neq B$, the least squares method is used. This means that for $q < B$ the Euclidean norm $||\boldsymbol{\beta}||$ is minimized, imposing that $\mathbf{y} = \mathbf{H}\boldsymbol{\beta}$; where $\mathbf{H}$ is the linear system matrix and $\mathbf{y}$ is the vector of simulation responses. For $q > B$ it is the Euclidean norm $||\mathbf{y} - \mathbf{H}\boldsymbol{\beta}||$ that is minimized.

### *2.2. Polynomial Chaos Expansion (PCE)*

The polynomial chaos expansion (PCE) is a metamodel that estimates the response, $y^*(\mathbf{x})$, for a given vector of probabilistic input variables, $\mathbf{x}$, through a basis of orthogonal stochastic polynomials. Assuming that the input variables $x_i$ are independent, the model response, $y^*(\mathbf{x})$, is given by:

$$y^*(\mathbf{x}) = \sum\nolimits_{\alpha \in A} \beta_\alpha \Psi_\alpha(\mathbf{x}), \tag{2}$$

where $\Psi_\alpha$ is an orthogonal polynomial basis, $\beta_\alpha$ are the associated coefficients, and A is a set of pre-selected multi-index $\boldsymbol{\alpha}$, which represents the input variables. In order to avoid a high number of response evaluations, only the multi-indexes that consider input variables up to a degree of 4 and low-order interactions between those variables, following a hyperbolic truncation scheme [10], are assumed. Hermite polynomials are used to construct the polynomial basis, $\Psi_\alpha$, since the input variables are Gaussian. The coefficients $\beta_\alpha$ are calculated with the ordinary least squares method by minimizing the difference between the model responses $y^*(\mathbf{x}^t)$ and the simulated outputs $y(\mathbf{x}^t)$.

### *2.3. Gaussian Process (GP)*

A Gaussian process (GP) corresponds to a collection of random variables, which have a Gaussian distribution [8]. The properties of these variables can be specified by the mean and covariance functions of the GP. In practice, the mean function is often considered to be zero, which means that the GP is completely defined by the covariance function. The GP regression model is represented as follows:

$$y(\mathbf{x}) = f(\mathbf{x}) + \epsilon, \tag{3}$$

where $y(\mathbf{x})$ is an observed response, $f(\mathbf{x})$ is the corresponding random GP variable, and $\epsilon$ is the noise. The joint probability of the normal distribution of the training outputs $\mathbf{y}(\mathbf{x}^t)$ and the test outputs $\mathbf{y}(\mathbf{x}^e)$ is given by:

$$\begin{bmatrix} \mathbf{y}(\mathbf{x}^t) \\ \mathbf{y}(\mathbf{x}^e) \end{bmatrix} \sim N\left(0, \begin{bmatrix} \mathbf{K}(\mathbf{X},\mathbf{X}) + \sigma_\epsilon^2 \mathbf{I} & \mathbf{K}(\mathbf{X},\mathbf{X}_*) \\ \mathbf{K}(\mathbf{X}_*,\mathbf{X}) & \mathbf{K}(\mathbf{X}_*,\mathbf{X}_*) \end{bmatrix}\right), \tag{4}$$

where $\sigma_\epsilon^2$ represents the noise variance, $\mathbf{I}$ is the identity matrix, and each $\mathbf{K}$ matrix is a covariance matrix evaluated for all pairs of points considered, with $\mathbf{X}$ representing training points and $\mathbf{X}_*$ representing test points. The GP prediction for the group of testing points can be obtained through the following equations:

$$\overline{\mathbf{f}_*} = \mathbf{K}(\mathbf{X}_*,\mathbf{X})\left[\mathbf{K}(\mathbf{X},\mathbf{X}) + \sigma_\epsilon^2 \mathbf{I}\right]^{-1} \mathbf{y}\left(\mathbf{x}^t\right), \tag{5}$$

$$\mathbf{cov}(\mathbf{f}_*) = \mathbf{K}(\mathbf{X}_*,\mathbf{X}_*) - \mathbf{K}(\mathbf{X}_*,\mathbf{X})\left[\mathbf{K}(\mathbf{X},\mathbf{X}) + \sigma_\epsilon^2 \mathbf{I}\right]^{-1} \mathbf{K}(\mathbf{X},\mathbf{X}_*), \tag{6}$$

where $\overline{\mathbf{f}_*}$ is the vector of predicted results (mean) and $\mathbf{cov}(\mathbf{f}_*)$ represents the covariance of model outputs, which acts as a measure of the predictions uncertainty.

### *2.4. Multi-Layer Perceptron (MLP)*

A multi-layer perceptron is a type of feed forward neural network, which can be used for both classification and regression. It is formed by a series of nodes (neurons) grouped into layers [5]. Each node is connected to the nodes in the next layer, but there are no interconnections between nodes in the same layer. The first layer, called the input layer, is formed by a number of nodes equal to the number of inputs in the data, $p$. The output layer receives information from the previous layer to make a prediction. Between the input and output layer, the model has one or more hidden layers. Each node in a hidden layer has a nonlinear activation function. The output of a node in a hidden layer can be described by the following equation:

$$z_i = \varnothing\left(\sum{}_j w_{ij} z'_j + b_i\right), \tag{7}$$

where $z_i$ is the output of the current node, $i$; $z'_j$ is the value obtained from node $j$ of the prior layer; $w_{ij}$ is the weight associated to $z'_j$; $b_i$ is the bias term; and $\varnothing$ represents the activation function. For regression, the output layer nodes have a similar formulation, the only difference being the lack of an activation function.

The weights are adjusted when the model is fitted to the training data, through a process called backpropagation. This algorithm consists of assessing how each weight should be changed (increased or decreased) in order to obtain a better prediction, and then updating all weights in the network accordingly, in small increments, until a minimum error estimate for the prediction is achieved.

### 2.5. Decision Trees (DTs) and Random Forest (RF)

Decision trees are models that split data continuously, based on simple decision rules. During training, the choice of how to split the data at each node is made so that an error metric is minimized [11]. The most common metric in this case is the MSE (mean squared error). This process is repeated until each of the final nodes (leaf nodes) has a value of the MSE associated to its data under a certain threshold, defined a priori. The prediction value for each leaf node becomes the average of the values for the dependent variable associated with the training points in the node.

The Random forest model is an extension of the decision tree model. It consists of training multiple decision trees, each with a different sample of the training data. The predictions made by this model are the average of the prediction obtained by each of the trees [12].

### 2.6. k-Nearest Neighbors (kNN)

The k-nearest neighbors method does not create a model with the training data. Each time it makes a prediction, it calculates the distance between the point for which the prediction will be made and each of the training points. Then, the $k$ training points that are closest to the prediction point are selected. The result of the prediction will be either the average of the result values associated to these $k$ training points, or a weighted average based on the distance, so that between these $k$ training points, more influence is given to the ones closer to the prediction point [13].

### 2.7. Support Vector Regression (SVR)

The support vector regression model consists of finding a function that fits the training data, and is as flat as possible, while under the assumption that error values under a certain value of $\gamma$ are accepted without penalty [14]. This means finding the function that can include the most training points in the area around it, with distance less or equal to $\gamma$. Since sometimes this restriction can be unfeasible, slack variables $\xi_i$ and $\xi_i^*$ can also be defined, which work as soft margins. Values with errors between $\gamma$ and these variables still affect the functions shape, but under a penalty.

When applied to a linear case, this problem can be represented by:

$$\begin{cases} \min\left(\frac{1}{2}\|\mathbf{w}\|^2 + V\sum_i(\xi_i + \xi_i^*)\right) \\ \text{with :} \\ y_i - wx_i - \beta_0 \le \gamma + \xi_i \\ wx_i + \beta_0 - y_i \ \le \gamma + \xi_i^* \end{cases}, \quad (8)$$

where $\mathbf{w}$ is the normal weight vector to the surface that is being approximated, and $V$ is a constant that represents the trade-off between function flatness and tolerance for deviations above $\gamma$.

This problem can be generalized for non-linear cases by applying a kernel trick. A kernel is a similarity function between the training inputs and the unlabeled inputs for which the model will make a prediction. The kernel trick is used to transform the data into a higher dimensional space, allowing a linear learning model to learn non-linear functions without explicit mapping.

### 2.8. Kernel Ridge Regression (KRR)

Ridge regression creates a model of similar form to the one obtained by support vector regression, the main difference being the loss function used, with ridge regression using squared error loss. For a linear case, training this model consists in minimizing the cost function $J$:

$$J = \frac{1}{2}\sum_i\left(y_i - \mathbf{w}^{\mathrm{T}}x_i\right)^2 + \frac{1}{2}\lambda\|\mathbf{w}\|^2, \quad (9)$$

where $\lambda$ is the regularization term. Once again, in order to generalize this model to non-linear cases, a kernel trick is applied, mapping the data into a higher dimensional space [15].

## 3. Forming Simulations and Metamodeling Procedure

This section presents the details of the numerical models of the U-Channel and the Square Cup forming processes, including the materials considered and the relevant input variables. The procedure adopted for the generation and evaluation of the metamodels is also described.

### 3.1. Numerical Models

The numerical models of the U-Channel and Square Cup forming processes are represented in Figure 1. Both processes comprise three main elements: the blank holder, the die, and the punch. The first stage of the forming process consists of reducing the distance between the die and the blank holder, until an imposed force is attained (blank holder force (BHF)). Then, the punch moves to promote the material flow into the die cavity, while the BHF remains constant. The U-Channel forming process ends after a total punch displacement of 30 mm, while the Square Cup forming process ends after a total punch displacement of 40 mm. The last stage consists of the tools removal, which promotes the recovery of the elastic energy stored in the part (springback). The initial dimensions of the blank of the U-Channel and the Square Cup forming processes are, respectively, $150 \times 35 \times 0.78$ mm$^3$ and $75 \times 75 \times 0.78$ mm$^3$. The material is considered orthotropic. Due to material and geometry symmetries, only one fourth of the blank is simulated for the Square Cup deep-drawing process, considering a finite element mesh with 1800, eight-node hexahedral solid elements. For the U-Channel, only half of the blank is considered, and boundary conditions are set to guarantee a plain strain state along the width of the blank, which enables the use of a total of 450, eight-node hexahedral solid elements.

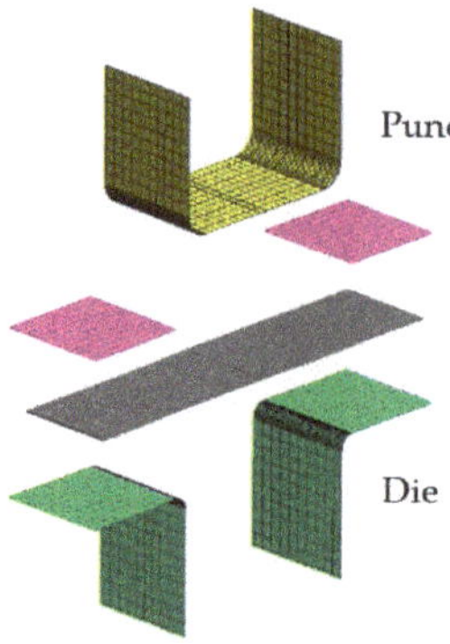

(a)

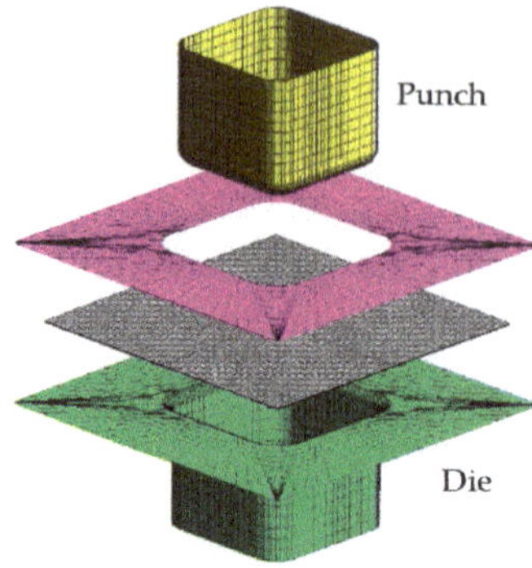

(b)

**Figure 1.** Representation of the finite element models for the (**a**) U-Channel; (**b**) Square Cup.

The numerical simulations were carried out with the in-house finite element code DD3IMP, developed and optimized for simulating sheet metal forming processes [16]. The forming tool geometry was modeled using Nagata patches [17]. The contact with friction is described by Coulomb's law with a constant value for the friction coefficient, $\mu$, between the sheet and the tools. The constitutive model adopted in this study assumes (i) the isotropic elastic behavior described by the generalized Hooke's law and (ii) the plastic behavior is anisotropic, as generally observed in metallic sheets, and as described by the orthotropic Hill'48 yield criterion combined with Swift isotropic hardening law. The Hill'48 yield criterion is described as follows:

$$F\left(\sigma_{yy}-\sigma_{zz}\right)^2+G(\sigma_{zz}-\sigma_{xx})^2+H\left(\sigma_{xx}-\sigma_{yy}\right)^2+2L\tau_{yz}^2+2M\tau_{xz}^2+2N\tau_{xy}^2=Y^2, \quad (10)$$

where $\sigma_{xx}$, $\sigma_{yy}$, $\sigma_{zz}$, $\tau_{xy}$, $\tau_{yz}$ and $\tau_{xz}$ are the components of the Cauchy stress tensor defined in the orthotropic coordinate system of the material; $F$, $G$, $H$, $L$, $M$ and $N$ are the anisotropy parameters and $Y$. is the flow stress. The condition $G+H=1.$ is assumed and so $Y$. is represented by the uniaxial tensile stress along the rolling direction of the sheet. The parameters $L$ and $M$ are assumed equal to 1.5, as in isotropy (von Mises). The parameters $F$, $G$, $H$ and $N$ can be related with the anisotropy coefficients $r_0$, $r_{45}$ and $r_{90}$, as follows:

$$F=\frac{r_0}{r_{90}(r_0+1)},\ G=\frac{1}{r_0+1},\ H=\frac{r_0}{r_0+1},\ N=\frac{1}{2}\frac{(r_0+r_{90})(2r_{45}+1)}{r_{90}(r_0+1)}. \quad (11)$$

The Swift hardening law is expressed by:

$$Y=C\left[(Y_0/C)^{(1/n)}+\bar{\varepsilon}^p\right]^n, \quad (12)$$

where $\bar{\varepsilon}^p$ is the equivalent plastic strain and $C$, $Y_0$ and $n$ are material parameters. Two types of numerical simulation outputs were considered for each forming process: (i) springback and maximum thinning for the U-Channel process; and (ii) maximum equivalent plastic strain and maximum thinning for the Square Cup deep-drawing.

### 3.2. Parameter Variability

Three different steel grades were considered for each forming process: DC06, DP600, and HSLA340. For each of these materials, a normal distribution was assumed for describing the variability of the following inputs: $C$, $Y_0$ and $n$ of the Swift hardening law; Young's modulus, $E$ and Poisson coefficient

$\nu$ of the generalized Hooke's law; anisotropy coefficients $r_0$, $r_{45}$ and $r_{90}$; initial sheet thickness, $t_0$; and friction coefficient $\mu$. The mean and standard deviation (SD) values of each parameter are detailed in Table 1. In addition to material parameters, the value of the BHF was also considered to introduce some variability in the process conditions; the mean and standard deviation values of the BHF for the U-Channel are 4900 N and 245 N, respectively; in the case of the Square Cup, the mean and standard deviation values of the BHF are 2450 N and 122.5 N, respectively.

**Table 1.** Mean and standard deviation (SD) values for each input of the three materials considered [18].

| Materials | | C (MPa) | $n$ | $Y_0$ (MPa) | E (GPa) | $\nu$ | $\mu$ | $r_0$ | $r_{45}$ | $r_{90}$ | $t_0$ (mm) |
|---|---|---|---|---|---|---|---|---|---|---|---|
| DC06 | Mean | 565.32 | 0.259 | 157.12 | 206 | 0.3 | 0.144 | 1.790 | 1.510 | 2.270 | 0.780 |
| | SD | 26.85 | 0.018 | 7.16 | 3.85 | 0.015 | 0.029 | 0.051 | 0.037 | 0.121 | 0.013 |
| DP600 | Mean | 1093.0 | 0.187 | 330.30 | 210 | 0.3 | 0.144 | 1.010 | 0.760 | 0.980 | 0.780 |
| | SD | 52.46 | 0.02 | 9.64 | 7.35 | 0.015 | 0.029 | 0.04 | 0.03 | 0.06 | 0.01 |
| HSLA340 | Mean | 673.0 | 0.131 | 365.30 | 210 | 0.3 | 0.144 | 0.820 | 1.070 | 1.040 | 0.780 |
| | SD | 32.30 | 0.011 | 10.67 | 7.35 | 0.015 | 0.029 | 0.033 | 0.039 | 0.061 | 0.005 |

### 3.3. Metamodel Generation and Evaluation

Based on the normal distribution of each input shown in Table 1, 1000 sets of inputs were randomly generated for each material. Numerical simulations of the U-Channel and Square Cup forming processes were performed for each of these randomly generated inputs, $\mathbf{x}$, with a total of 3 (materials) × 1000 (sets of inputs) = 3000 simulations for each forming process. For each material, the numerical simulations of each forming process were grouped into two sets: one training set, $\mathbf{x}^t$, with 700 simulations used to generate the metamodels, and one testing set, $\mathbf{x}^e$, with 300 simulations to evaluate the performance of the generated metamodels, by comparing the estimated/predicted output values with those obtained by numerical simulation. In addition to these sets, an extra training set and test set that includes simulations from all three materials was considered for each forming process. This was done to evaluate the impact of considering multiple materials on the predictive performance of the metamodels. The root mean square relative error (RMSRE) was used to evaluate the performance of each metamodel:

$$RMSRE = \sqrt{\frac{1}{l}\sum_{i=1}^{l}\left(\frac{y_i(\mathbf{x}^e) - y_i^*(\mathbf{x}^e)}{y_i(\mathbf{x}^e)}\right)}, \tag{13}$$

where $y(\mathbf{x}^e)$ and $y^*(\mathbf{x}^e)$ are the simulated and predicted response values for the set of testing inputs $\mathbf{x}^e$, respectively, and $l$ is the number of testing points.

The parametric metamodels (RSM and PCE) where generated in Excel, while the ML metamodels where generated with python libraries, specifically, GPy [19] for the GP metamodels and Scikit-learn [20] for the remaining models.

## 4. Results and Discussion

Table 2 presents the RMSRE values of the metamodels generated for each forming process ("U-Channel" and "Square Cup"), material ("DC06", "DP600", and "HSLA340"), and simulation output ("Springback", "Maximum Thinning", and "Maximum Equivalent Plastic Strain"); the results for the cases labeled as "Mixed" correspond to the metamodels generated from a training set that includes all three materials. The lowest value of RMSRE for each case, which corresponds to the best predictive performance, is highlighted.

**Table 2.** Values of root mean square relative error (RMSRE, %) obtained for the metamodel prediction for the U-Channel and Square Cup processes.

| U-Channel | Springback | | | | Maximum Thinning | | | |
|---|---|---|---|---|---|---|---|---|
| | DC06 | DP600 | HSLA340 | Mixed | DC06 | DP600 | HSLA340 | Mixed |
| RSM | 2.223 | 1.084 | 1.117 | 1.574 | 0.905 | 1.046 | 0.970 | 2.146 |
| PCE | 2.111 | 1.084 | 1.184 | 2.098 | 0.861 | 1.058 | 1.144 | 1.563 |
| GP | 2.132 | 1.060 | 1.023 | 1.572 | 0.849 | 1.025 | 0.875 | 1.017 |
| MLP | 1.986 | 1.113 | 1.019 | 1.476 | 0.772 | 1.242 | 0.782 | 1.097 |
| SVR | 2.230 | 1.056 | 1.090 | 1.538 | 0.883 | 1.048 | 0.864 | 1.083 |
| DT | 4.620 | 4.004 | 3.448 | 4.776 | 5.742 | 4.290 | 3.992 | 6.047 |
| RF | 2.876 | 2.726 | 2.298 | 3.545 | 4.022 | 2.815 | 2.796 | 4.104 |
| kNN | 4.068 | 3.777 | 2.950 | 4.542 | 10.846 | 5.599 | 5.585 | 6.877 |
| KRR | 2.168 | 1.057 | 1.072 | 1.546 | 0.851 | 1.035 | 0.865 | 1.060 |
| **Square Cup** | **Maximum Equivalent Plastic Strain** | | | | **Maximum Thinning** | | | |
| | DC06 | DP600 | HSLA340 | Mixed | DC06 | DP600 | HSLA340 | Mixed |
| RSM | 0.519 | 1.741 | 0.744 | 1.523 | 0.688 | 1.607 | 2.258 | 1.687 |
| PCE | 0.514 | 1.741 | 0.750 | 1.201 | 0.688 | 1.577 | 2.242 | 1.748 |
| GP | 0.493 | 1.618 | 0.666 | 1.230 | 0.672 | 1.538 | 2.195 | 1.529 |
| MLP | 0.607 | 1.653 | 0.952 | 1.416 | 0.661 | 1.553 | 2.233 | 1.633 |
| SVR | 0.533 | 1.606 | 0.836 | 1.231 | 0.688 | 1.585 | 2.221 | 1.556 |
| DT | 1.210 | 2.415 | 1.688 | 2.009 | 2.499 | 2.582 | 3.612 | 3.550 |
| RF | 0.823 | 1.830 | 1.308 | 1.490 | 1.734 | 1.850 | 2.682 | 2.527 |
| kNN | 1.322 | 2.754 | 2.253 | 2.899 | 2.481 | 2.419 | 3.084 | 4.188 |
| KRR | 0.518 | 1.689 | 0.665 | 1.321 | 0.673 | 1.545 | 2.147 | 1.532 |

Abbreviations: RSM, response surface method; PCE, polynomial chaos expansion; GP, Gaussian process; MLP, multi-layer perceptron, SVR, support vector regression; DT, decision tree; RF, random forest; kNN, k-nearest neighbors; KRR, kernel ridge regression.

The MLP model achieved the best performance in 6 of the 16 cases presented. For the remaining cases, the best performances were achieved by the GP (5), SVR (2), KRR (2), and PCE (1) models. It should be noted that the differences in performance between these five models were generally small. The RSM metamodels showed performances that were, in general, very similar to the PCE metamodels, and as such, can be considered as competitive with the models that achieved the best performances. On the other hand, the DT, RF, and kNN models performed clearly worse than the remaining metamodels for all cases. The few comparative studies found in the literature, namely Wessing et al. [5] and Ambrogio et al. [6], favored the use of the GP technique instead of ANN and RSM for thickness prediction, as it achieved significantly better results. In the current study, the GP technique also tended to present the best performance for the prediction of maximum thinning, but this was not valid for the other responses.

The inclusion of all three materials in the training and testing of the metamodels did not lead to significantly worse results, when compared to the performances obtained for the single material cases. In fact, in certain cases, the performance obtained for the models trained with the three materials surpassed the performance of models trained with just one material. For example, in the springback prediction for the U-Channel case, more than half of the metamodels tested achieved better performance when trained with the three materials than when trained specifically for the DC06 material. Thus, when training metamodels to predict forming process results considering various materials, it is worth considering the usage of just one dataset containing training data representative of all materials available, instead of training a different model for each material.

As an example, Figure 2 represents the comparison between the simulated values and the values predicted by the MLP and kNN algorithms for the testing dataset for the maximum thinning in the U-Channel process using DC06. The algorithms MLP and kNN were chosen for this comparison because they achieved the best and poorest performances, respectively, in this case.

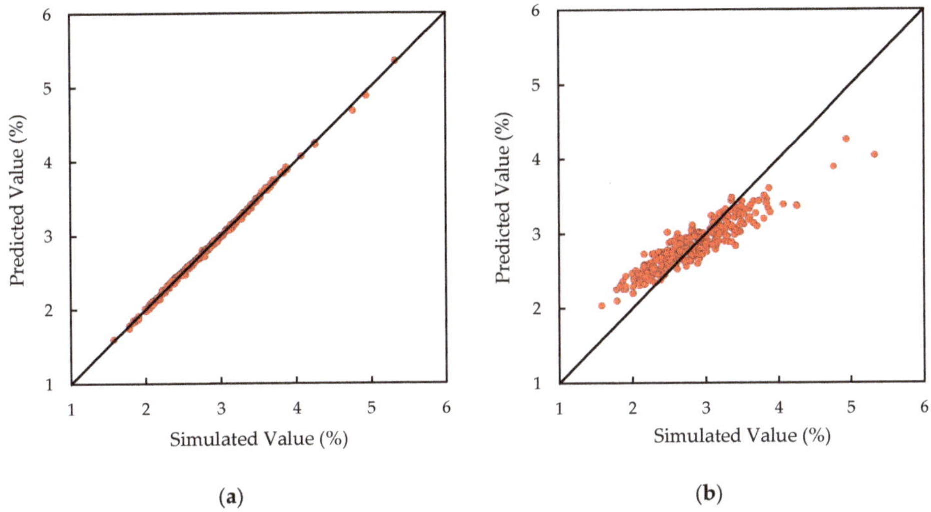


**Figure 2.** Comparison between simulated and predicted values of maximum thinning in the U-Channel case with the DC06 material by the algorithms: (**a**) MLP; (**b**) kNN.

Figure 3 presents the frequency distributions corresponding to the previous example, generated from 1000 new random points, according to the variability described in Table 1. The frequency distribution of the numerical simulation results is also represented and taken as a reference. The distribution obtained for the MLP metamodel closely resembles the distribution of the simulated results, however, the kNN shows more predictions in the range between 2.6% and 3.2%. In fact, the average value of the simulated results is 2.87%. This is in agreement with Figure 2b, where it is clear that the difference between predicted values and the corresponding simulated values is, in general, larger when the simulated values are further from the average.

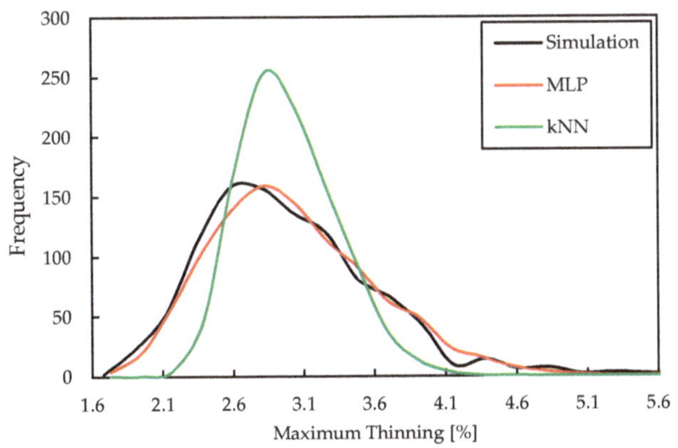


**Figure 3.** Frequency distributions obtained for the numerical simulations and predictions by the MLP and kNN algorithms, considering the maximum thinning in the U-Channel case with the DC06 material.

## 5. Conclusions

In this work, parametric and non-parametric regression models were applied to predict the results of sheet metal forming processes, with the goal of evaluating their performance, and establish which metamodels offer the best results. It was concluded:

- The ML techniques can be divided into two groups in terms of performance:
  - The first group consists of the DT, RF, and kNN metamodeling techniques, which generally showed poor performances, with kNN in particular producing the poorest predictions.
  - The second group consists of the MLP, GP, SVR, and KRR techniques. For almost all cases studied, the best predictive performance corresponded to one of these techniques, with MLP showing the best performance in more cases than any other. It is also of note that the performance of these techniques is comparable and, as such, the usage of any of them can be recommended.
- The parametric modeling techniques, RSM and PCE, showed competitive performances when compared with the second group of ML techniques and should be considered as valid alternatives.
- The performance of both types of modeling techniques depends on the response under analysis. For the particular case of the maximum thinning, GP shows better performance compared to the other techniques.
- When training metamodels to predict forming process results for different materials, the usage of just one dataset containing the training data of all materials should be considered, instead of training a different model for each material.

**Author Contributions:** Formal analysis, A.E.M., P.A.P., A.F.G.P., and M.C.O.; funding acquisition, P.A.P. and J.V.F.; investigation, A.E.M.; software, M.C.O. and B.M.R.; supervision, P.A.P., J.V.F., and B.M.R.; writing—original draft, A.E.M.; writing—review and editing, A.E.M., P.A.P., A.F.G.P., M.C.O., J.V.F., and B.M.R. All authors have read and agreed to the published version of the manuscript.

**Funding:** This research is sponsored by FEDER funds through the program COMPETE–Programa Operacional Factores de Competitividade and by national funds through FCT–Fundação para a Ciência e a Tecnologia, under the projects UID/EMS/00285/2020 and UID/CEC/00326/2020. It was also supported by projects: SAFEFORMING, co-funded by the Portuguese National Innovation Agency, by FEDER, through the program Portugal-2020 (PT2020), and by POCI, with ref. POCI-01-0247-FEDER-017762; RDFORMING (reference PTDC/EME-EME/31243/2017), co-funded by Portuguese Foundation for Science and Technology, by FEDER, through the program Portugal-2020 (PT2020), and by POCI, with reference POCI-01-0145-FEDER-031243; EZ-SHEET (reference PTDC/EME-EME/31216/2017), co-funded by Portuguese Foundation for Science and Technology, by FEDER, through the program Portugal-2020 (PT2020), and by POCI, with reference POCI-01-0145-FEDER-031216. All supports are gratefully acknowledged.

**Conflicts of Interest:** The authors declare no conflict of interest.

## Notations

| Symbol | Description |
|---|---|
| $\mathbf{A}$ | Set of pre-selected multi-index |
| $B$ | Number of RSM coefficients |
| $C, Y_0, n$ | Swift hardening law parameters |
| $E$ | Young's modulus |
| $F, G, H, L, M, N$ | Anisotropy parameters |
| $f(\mathbf{x})$ | GP random variable |
| $\overline{\mathbf{f}}_*, \mathbf{cov}(\mathbf{f}_*)$ | Vector of GP predictions, covariance of GP outputs |
| $\mathbf{H}$ | Linear system matrix |
| $i, j, m$ | Index variables |
| $\mathbf{I}$ | Identity matrix |
| $J$ | KRR cost function |
| $\mathbf{K}$ | Covariance matrix |
| $k$ | Selected number of nearest training points in the kNN model |

| | |
|---|---|
| $p$ | Number of sources of variability |
| $q, l$ | Number of training and testing points, respectively |
| $r_0, r_{45}, r_{90}$ | Anisotropy coefficients |
| $t_0$ | Initial sheet thickness |
| $V$ | SVR trade-off variable |
| $\mathbf{w}$ | Normal weight vector |
| $w, b$ | Weight parameter, bias |
| $\mathbf{x}$ | Vector of design variables (Inputs) |
| $\mathbf{X}, \mathbf{X}_*$ | Training matrix, testing matrix |
| $\mathbf{x}^t, \mathbf{x}^e$ | Vector of design variables of a training and a testing point, respectively |
| $Y$ | Flow stress |
| $\mathbf{y}$ | Vector of simulation responses |
| $\mathrm{y}(\mathbf{x})$ | Simulation response for set of inputs $\mathbf{x}$ |
| $\mathrm{y}^*(\mathbf{x})$ | Metamodel response for set of inputs $\mathbf{x}$ |
| $z, z\prime$ | Output of the current MLP node, output of a node at the prior layer |
| $\boldsymbol{\alpha}$ | Multi-index |
| $\boldsymbol{\beta}$ | Vector of unknowns for the RSM model |
| $\beta_0, \beta_i, \beta_{ij}, \beta_{ii}, \beta_{\boldsymbol{\alpha}}$ | RSM and PCE coefficients |
| $\gamma$ | SVR threshold parameter |
| $\epsilon$ | Noise variable |
| $\overline{\varepsilon}^p$ | Equivalent plastic strain |
| $\lambda$ | Regularization term |
| $\mu$ | Friction coefficient |
| $\nu$ | Poisson coefficient |
| $\xi_i, \xi_i^*$ | SVR slack variables |
| $\sigma_{xx}, \sigma_{yy}, \sigma_{zz}, \tau_{xy}, \tau_{yz}, \tau_{xz}$ | Components of the Cauchy stress tensor |
| $\sigma_\epsilon^2$ | Noise variance |
| $\Psi_{\boldsymbol{\alpha}}$ | Orthogonal polynomial basis |
| $\varnothing$ | Activation Function in MLP |

## References


1. Wei, D.; Cui, Z.; Chen, J. Optimization and tolerance prediction of sheet metal forming process using response surface model. *Comput. Mater. Sci.* **2008**, *42*, 228–233. [CrossRef]
2. Naceur, H.; Ben-Elechi, S.; Batoz, J.L.; Knopf-Lenoir, C. Response surface methodology for the rapid design of aluminum sheet metal forming parameters. *Mater. Des.* **2008**, *29*, 781–790. [CrossRef]
3. Sun, G.; Li, G.; Li, Q. Variable fidelity design based surrogate and artificial bee colony algorithm for sheet metal forming process. *Finite Elem. Anal. Des.* **2012**, *59*, 76–90. [CrossRef]
4. Teimouri, R.; Baseri, H.; Rahmani, B.; Bakhshi-Jooybari, M. Modeling and optimization of spring-back in bending process using multiple regression analysis and neural computation. *Int. J. Mater. Form.* **2014**, *7*, 167–178. [CrossRef]
5. Wessing, S.; Rudolph, G.; Turck, S.; Klimmek, C.; Schäfer, S.C.; Schneider, M.; Lehmann, U. Replacing FEA for sheet metal forming by surrogate modeling. *Cogent Eng.* **2012**, *1*. [CrossRef]
6. Ambrogio, G.; Ciancio, C.; Filice, L.; Gagliardi, F. Innovative metamodelling-based process design for manufacturing: an application to Incremental Sheet Forming. *Int. J. Mater. Form.* **2017**, *10*, 279–286. [CrossRef]
7. Feng, Y.; Hong, Z.; Gao, Y.; Lu, R.; Wang, Y.; Tan, J. Optimization of variable blank holder force in deep drawing based on support vector regression model and trust region. *Int. J. Adv. Manuf.* **2019**, *105*, 4265–4278. [CrossRef]
8. Lin, J.D.; Huang, L.; Zhou, H.B. Forming defects prediction for sheet metal forming using Gaussian process regression. In Proceedings of the 29th Chinese Control and Decision Conference, Chongqing, China, 28–30 May 2017. [CrossRef]

9. Wiebenga, J.H.; Atzema, E.H.; An, Y.G.; Vegter, H.; van den Boogaard, A.H. Effects of material scatter on the plastic behavior and stretchability in sheet metal forming. *J. Mater. Process. Technol.* **2014**, *214*, 238–252. [CrossRef]
10. Blatman, G. Adaptive Sparse Polynomial Chaos Expansions for Uncertainty Propagation and Sensitivity Analysis. Ph.D. Thesis, Université Blaise Pascal, Clermont-Ferrand, France, 8 October 2009. [CrossRef]
11. Kaur, D.; Wilson, D.; Forrest, M.; Feng, L. Regression tree and neuro-fuzzy approach to system identification of laser lap welding. In Proceedings of the 2005 Annual Meeting of the North American Fuzzy Information Processing Society, Detroit, MI, USA, 26–28 June 2005. [CrossRef]
12. Segal, M.R. Machine Learning Benchmarks and Random Forest Regression. In *UCSF: Center for Bioinformatics and Molecular Biostatistics*; 2004; Available online: https://escholarship.org/uc/item/35x3v9t4 (accessed on 30 January 2020).
13. Cook, B.; Huber, M. Balanced k-Nearest Neighbors. In Proceedings of the Thirty-Second International Flairs Conference, Florida, FL, USA, 19–22 May 2019.
14. Smola, A.J.; Schölkopf, B. A tutorial on support vector regression. *Stat. Comput.* **2004**, *14*, 199–222. [CrossRef]
15. Welling, M. Kernel ridge regression. Max Welling's Classnotes in Machine Learning. 2013. Available online: https://www.ics.uci.edu/~{}welling/classnotes/papers_class/Kernel-Ridge.pdf (accessed on 30 January 2020).
16. Menezes, L.F.; Teodosiu, C. Three-dimensional numerical simulation of the deep-drawing process using solid finite elements. *J. Mater. Process. Technol.* **2000**, *97*, 100–106. [CrossRef]
17. Neto, D.M.; Oliveira, M.C.; Menezes, L.F. Surface Smoothing Procedures in Computational Contact Mechanics. *Arch. Comput. Meth. Eng.* **2017**, *24*, 37–87. [CrossRef]
18. Alves, J.L. Simulação Numérica do Processo de Estampagem de Chapas Metálicas. Ph.D. Thesis, University of Minho, Guimarães, Portugal, 7 November 2003.
19. GPy: A Gaussian Process Framework in Python. Available online: http://github.com/SheffieldML/GPy (accessed on 22 January 2020).
20. Pedregosa, F.; Varoquaux, G.; Gramfort, A.; Michel, V.; Thirion, B.; Grisel, O.; Blondel, M.; Prettenhofer, P.; Weiss, R.; Dubourg, V.; et al. Scikit-learn: machine learning in python. *J. Mach. Learn. Res.* **2011**, *12*, 2825–2830.

*metals*

MDPI

*Article*

# Machine Learning-Based Models for the Estimation of the Energy Consumption in Metal Forming Processes

**Irene Mirandola [1], Guido A. Berti [1], Roberto Caracciolo [1], Seungro Lee [2], Naksoo Kim [2] and Luca Quagliato [2,*]**

[1] Department of Management and Engineering, University of Padua, 36100 Vicenza, Italy; irene.mirandola@phd.unipd.it (I.M.); guido.berti@unipd.it (G.A.B.); roberto.caracciolo@unipd.it (R.C.)
[2] Department of Mechanical Engineering, Sogang University, Seoul 04107, Korea; lsr1310@sogang.ac.kr (S.L.); nskim@sogang.ac.kr (N.K.)
* Correspondence: lucaq@sogang.ac.kr

**Abstract:** This research provides an insight on the performances of machine learning (ML)-based algorithms for the estimation of the energy consumption in metal forming processes and is applied to the radial-axial ring rolling process. To define the mutual influence between ring geometry, process settings, and ring rolling mill geometries with the resulting energy consumption, measured in terms of the force integral over the processing time (FIOT), FEM simulations have been implemented in the commercial SW Simufact Forming 15. A total of 380 finite element simulations with rings ranging from 650 mm < DF < 2000 mm have been implemented and constitute the bulk of the training and validation datasets. Both finite element simulation settings (input), as well as the FI (output), have been utilized for the training of eight machine learning models, implemented with Python scripts. The results allow defining that the Gradient Boosting (GB) method is the most reliable for the FIOT prediction in forming processes, being its maximum and average errors equal to 9.03% and 3.18%, respectively. The trained ML models have been also applied to own and literature experimental cases, showing a maximum and average error equal to 8.00% and 5.70%, respectively, thus proving once again its reliability.

**Keywords:** ring rolling; process energy estimation; metal forming; thermo-mechanical FEM analysis; machine learning; artificial neural network

**Citation:** Mirandola, I.; Berti, G.A.; Caracciolo, R.; Lee, S.; Kim, N.; Quagliato, L. Machine Learning-Based Models for the Estimation of the Energy Consumption in Metal Forming Processes. *Metals* **2021**, *11*, 833. https://doi.org/10.3390/met11050833

Academic Editors: Pedro Prates and André Pereira

Received: 24 April 2021
Accepted: 18 May 2021
Published: 19 May 2021



## 1. Introduction

The radial axial-ring rolling (RARR) is a versatile forging process widely used in different industrial sectors such as automotive, agricultural, wind power, piping, and aerospace [1]. In recent years, several improvements have been introduced helping to obtain good surface quality, fine tolerances, and a considerable saving in material cost [2] with less production time compared to the machining process. Rings manufactured through RARR have high durability and structural strength, but the complexity of the process makes its settings and control hard to be handled without numerical simulations or prediction algorithms. For these reasons, several authors focused their attention on the development of algorithms and finite element models for a better understanding of the ring rolling process, as is hereafter summarized.

Lugora and Bramley [3] utilized Hill's general method for predicting the evolution of the ring during the process considering a rigid-perfectly plastic and incompressible material. Bruschi et al. [4] established a real-time control model, based on the artificial neural network (ANN) approach, to predict the geometrical accuracy of the ring, showing a good correlation between the ANN model and FEM results. Guo and Yang [5] defined the steady forming condition for the ring rolling process and built a mathematical model based on a constant velocity growth condition of the ring and considered the ring geometry in terms of average diameters. More recently, Quagliato and Berti [6] superseded this limit by proposing a more accurate mathematical approach for the determination of the ring geometry for a subsection of the ring geometry, defined as a slice.

As concerns the force prediction for the ring rolling process, Quagliato and Berti [7,8] proposed two mathematical models based on the slip line theory and estimated radial and axial forces with a deviation equal to ~5% and ~6%, respectively, in comparison to the relevant FEM simulations and experimental results. Furthermore, Ryoo et al. [9] defined the relationship between the parameters that affect the ring rolling process at high temperatures and investigated the influence of the main roll rotational speed and the mandrel feeding speed. Kalyani et al. [10] investigated radial and axial forces during the forming process of profiled rings in terms of time and temperature, calculated the forces with an analytical approach, and compared them with FEM simulations. Kim et al. [11] investigated the influence of process parameters in producing large rings, focusing on minimizing the load, but did not consider temperature and process parameters' reciprocal influence.

As concerns energy estimation and starvation algorithms for industrial process, due to the strong influence of the energy demand on production planning and control [12], several authors focused on this topic. Unver and Kara [13] introduced a decision support tool called HORUS 5.0 to determine the lowest energy-consuming route within the scope of sustainable energy efficiency. Meissner et al. [14] developed an indicator system considering the impact of the materials, energies, and economic attributes of energy efficiency, concluding that strategic decision-making concerning energy optimization is important to be competitive. Larkiola et al. [15] investigated the role of energy efficiency in the rolling processes employing an ANN-based approach and achieved an improvement estimated in 1.8% of the overall energy efficiency. Giorleo et al. [16] compared simulation analyses with an industrial case to evaluate the effect of utilizing different ring preform geometries to reduce the total energy required during the process but focused on a single material and single set of process parameters. Allegri et al. [17] defined a main roll speed law that allows maintaining a constant ring angular velocity and achieved a 35% fishtail defect reduction and a 9% energy consumption reduction.

As summarized so far, several authors developed models for the prediction of the kinematic expansion, the force, and torque but it seems that the impact of the process parameters on the energy consumption has not been thoroughly investigated in the literature. In the hot radial-axial ring rolling process, as in every metal forming process carried out at hot or warm forming conditions, a lower process force can be obtained by reducing the feeding, or the deformation, over time. On the other hand, a longer manufacturing time induces a higher temperature drop in the workpiece, which leads to an increase in the resistance to the deformation of the material. The energy integral over time can be estimated by employing numerical simulations [18–21] but, considering the complex tools-workpiece interaction in the RARR process, the computational time required for one single simulation might range between several hours and a few days.

In the literature, machine learning algorithms have been already applied to various manufacturing topics, such as for the prediction of joint strength of ultrasonic welding processes [22], to estimate the tool wear in milling operations [23], to diagnose the dimensional variation of additive manufactured parts [24], to classify the cutting phase of the natural fiber reinforced plastic composites [25] and to predict the tool life in the micro-milling process [26]. More recently, Wang et al. [27] developed a deep learning-based algorithm for the recognition of the defects in the strip rolling process, Marques et al. [28] investigated the performances of parametric and non-parametric models for the correlation of process and material variables to springback and wall thinning, Palmieri et al. [29] defined a metamodel to correlate the process parameters and key-quality indicators for the optimization of the blank-holding forces in the stamping process, and Winiczenko [30] utilized a hybrid response surface methodology combined with a genetic algorithm to simulate and optimize the friction welding parameters in AISI 1020-ASTM A536 joints.

Although ML algorithms have been applied to various manufacturing processes, they have not yet been utilized for the investigation of the influence of process, material, and geometrical parameters in metal forming processes and have not been applied yet to the RARR process. Accordingly, the research presented in this paper aims to fill this gap in the literature by investigating the influence of (i) process parameters, (ii) material properties, (iii) initial/final ring geometries, and (iv) processing conditions on energy consumption. Based on the implemented numerical simulation database, eight machine learning (ML) models have been trained and utilized for the prediction of the energy consumption during the process based on the above-mentioned (i, ii, iii, iv) parameters clusters. The mandrel forming force integral over time, (FIOT), has been utilized as the output variable in the analysis, and as response value for the training and validation of the ML algorithms. Based on the most recent applications of machine learning model, eight different models have been adopted in the research presented in this paper, namely: linear methods [31,32], the kernel methods [33,34], the ensemble methods [35–37], and the artificial neural network (ANN) methodology [38–40], respectively.

To create the dataset for the training and the validation of machine learning models, radial-axial ring rolling finite element simulations models have been implemented in the commercial software Simufact Forming 15: six ring final outer diameters, equal to 650, 800, 1100, 1400, 1700, and 2000 mm have been considered along with three different materials, largely utilized in the ring rolling process, namely the 42CrMo4 steel [4], the Inconel 718 superalloy [41], and the AA6082 (AlMgSi) aluminum alloy [42]. The material properties have been accounted for by its temperature-dependent elastic modulus and yield strength. Since the training and validation datasets have been all acquired through FE simulations, the implemented FEM model has been validated by comparing its results with a previously published once [8], showing a maximum deviation equal to 2.15% and 0.95% in the prediction of the radial forming force outer diameter of the ring.

A total of 380 numerical simulation models have been implemented and 80% of the results have been utilized for the training of the ML models, whereas the remaining 20% were for their validation. An additional validation phase has been carried out considering the previous literature experimental results published in [5,8,11]. Based on both validation phases, the Gradient Boosting method, belonging to the ensembles methods, has been shown to be able to accurately predict the force integral over time (FIOT) and is therefore considered to be the most reliable for the case of a complex thermo-mechanical forming process, such as the radial-axial ring rolling process.

## 2. Materials and Methods

### 2.1. Finite Element Simulation Model Definition

To create the database for the training of the machine learning-based force integral over time (FIOT) prediction models, presented in Section 3 of the paper, thermo-mechanical FEM simulations have been implemented in the commercial software Simufact Forming 15 following the general implementation scheme shown in Figure 1. In the numerical simulation models, the dies are considered as rigid with conductive, convective, and radiation heat transfer with the ring and the surrounding environment. The reason for introducing this approximation is justified by the fact that, although the elastic deformation in the rolls can slightly affect the final shape of the ring, its influence is negligible in comparison to the size of the rings considered in this paper. The dimensions for the tools of the ring rolling mill utilized in all the implemented FEM models are summarized in Table 1 along with the additional common process conditions. Friction has been modeled considering a shear friction law, Equation (1), and the utilized friction factor [6–8,18] is also reported in Table 1. In Equation (1), $k$ is defined as the ratio between the yield strength of the material and the square root of 3, according to the von Mises criterion.

$$\tau = m \cdot k \quad (1)$$

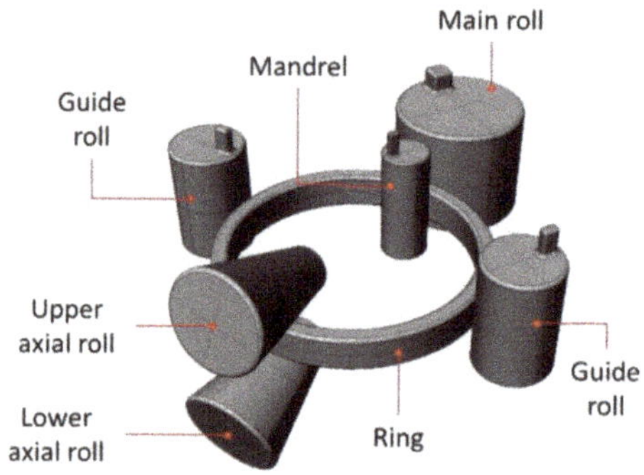


**Figure 1.** Configuration of the ring rolling process.

**Table 1.** Ring rolling mill characteristics and general process settings.

| Parameters | Values |
|---|---|
| Radius of the main roll [mm] | 325 |
| Radius of the mandrel [mm] | 125 |
| Length of the axial rolls [mm] | 595.9 |
| Half of axial rolls vertex angle [°] | 17.5 |
| Temperature of the environment [°C] | 50 |
| Tool initial temperature [°C] | 150 |
| Friction factor mandrel and main roll [−] | 0.85 |
| Friction factor axial and guide rolls [−] | 0.6 |

A higher friction factor has been considered for the contact conditions between the ring, mandrel, and the main roll due to higher thickness draft along the radial direction in comparison to the vertical deformation, carried out by the axial rolls. As concerns the centering rolls, their role is mainly to avoid excessive shifting of the ring during the process, thus their contact with the ring is limited and discontinuous over the processing time.

Friction influences the force calculation but, as will be shown in the results section, even though the training of the machine learning models has been carried out with a single set of friction constants (Table 1) when the model is applied to literature experimental cases, where different friction conditions are considered, an accurate FIOT prediction can still be achieved. Considering the training and validation (phase 1) datasets altogether, it is composed of 380 thermo-mechanical radial-axial ring rolling numerical simulations where the final outer diameter of the ring ($D_F$) ranges from 650 mm to 2000 mm.

As concerns the initial annular blanks, they have been defined in terms of initial outer diameter $D_0$, initial blank height $h_0$, and initial inner diameter $d_0$ according to the relevant final shape, by means of the procedure defined in Berti et al. [18]. A total of 16 different preform sizes have been utilized for the considered six final outer diameter geometries, as summarized in Table 2.

The ring preforms have been optimized considering four different mesh detail levels, and the best compromise between accuracy has been identified in (i) 1 element every 0.5° for the circumferential direction, 1 element every 2.5 mm for the radial direction, and 1 element every 5 mm for the vertical direction. These 16 geometries have been combined with different materials, Section 2.2, and process settings, Section 2.3, allowing obtaining the final database of 380 FEM simulations. Due to the impracticality of reporting the whole 380 settings in table form, a summary is added in Appendix A, whereas the whole dataset is made available as Supplementary Material.

**Table 2.** Initial and final ring geometries.

| Ring Initial Geometries [mm] | | | Ring Final Geometries | | |
|---|---|---|---|---|---|
| $D_0$ | $d_0$ | $h_0$ | $D_F$ | $d_F$ | $h_F$ |
| 490.5 | 325 | 188.8 | 650 | 530 | 180 |
| 510.9 | 325 | 156.2 | 650 | 505 | 145 |
| 549.3 | 325 | 112.2 | 650 | 450 | 100 |
| 518.9 | 325 | 195.4 | 800 | 680 | 180 |
| 550.9 | 325 | 175.4 | 800 | 645 | 155 |
| 575.3 | 325 | 218.7 | 1100 | 975 | 190 |
| 621.6 | 325 | 209.0 | 1100 | 930 | 170 |
| 621.6 | 325 | 209.0 | 1100 | 930 | 170 |
| 668.7 | 325 | 171.8 | 1100 | 855 | 122.5 |
| 876.2 | 600 | 208.2 | 1400 | 1220 | 180 |
| 909.3 | 600 | 221.3 | 1400 | 1190 | 190 |
| 897.7 | 600 | 209.4 | 1700 | 1530 | 170 |
| 875.7 | 600 | 226.3 | 2000 | 1875 | 190 |
| 944.5 | 600 | 232.4 | 2000 | 1820 | 180 |
| 944.7 | 600 | 232.4 | 2000 | 1820 | 180 |
| 1049.3 | 600 | 224.7 | 2000 | 1700 | 150 |

### 2.2. Materials

In the FEM models, presented in the previous section, three materials largely utilized in the hot ring rolling process [4,41,42] have been considered: (i) 42CrMo4 steel, (ii) Inconel 718 super alloy and (iii) AA6082 (AlMgSi) aluminum alloy. Due to the high diversity of mechanical behaviors, the consideration of these three materials allows widening the range of validity of the proposed investigation. For the definition of the plastic material behavior, the Hansel-Spittel flow stress model [43] has been utilized, as reported in Equation (2) whereas the relevant model constant ($C_1$, $C_2$, $n_1$, $n_2$, $L_1$, $L_2$, $m_1$, $m_2$), for the three considered materials, are reported in Table 3. In Equation (2), $\varepsilon$, $\dot{\varepsilon}$, and $T$ represent the considered strain, the strain rate, and temperature. The combined consideration of these three parameters during the FEM simulations allows estimating the flow stress of the material for each element of the mesh, thus accurately estimating the relevant forming force.

$$\sigma_F = C_1 \cdot e^{(C^2 \cdot T)} \cdot \varepsilon^{(n_1 \cdot T + n_2)} \cdot e^{\left(\frac{L_1 \cdot T + L_2}{\varepsilon}\right)} \cdot \dot{\varepsilon}^{(m_1 \cdot T + m_2)} \quad (2)$$

**Table 3.** Validity range and Hansel-Spittel flow stress model constants for the (i) 42CrMo4 steel, (ii) Inconel 718 super alloy and (iii) AA6082 (AlMgSi) aluminum alloy.

| Parameters | 42CrMo4 | IN 718 | AA6082 |
|---|---|---|---|
| Temperature range for the model [°C] | 800–1250 | 950–1100 | 200–530 |
| Strain range for the model [[–] | 0.05–2 | 0.05–2 | 0.05–0.9 |
| Strain rate range for the model [1/s] | 0.01–150 | 0.01–150 | 0.01–63 |
| $C_1$ | 5290.5 | 10501.1 | 378.5 |
| $C_2$ | −0.00370 | −0.00307 | −0.00492 |
| $n_1$ | −0.00033 | −0.00018 | −0.00011 |
| $n_2$ | 0.20612 | 0.54398 | −0.02573 |
| $L_1$ | $-8.26584 \times 10^{-5}$ | $-2.17606 \times 10^{-5}$ | $6.03612 \times 10^{-5}$ |
| $L_2$ | 0.02891 | 0.02376 | −0.02548 |
| $m_1$ | 0.000301 | $-2.67316 \times 10^{-6}$ | 0.000345 |
| $m_2$ | −0.15618 | 0.09746 | −0.031501 |

To be able to consider the influence of the material in the FIOT prediction models, the initial temperature of the ring, set as initial boundary conditions in the FEM models, as well as Young's modulus and yield strength at that temperature, have been considered as features in the analysis. The three considered temperatures, for each one of the three materials, are reported in Table 4 along with the two above-mentioned mechanical properties. All the elastic, plastic, and thermal-mechanical properties for the three considered materials have been acquired from the MATILDA® (Material Information Link and Database Service) database available in Simufact Forming 15.

**Table 4.** Ring rolling mill characteristics and general process settings.

| Parameters | 42CrMo4 | | | IN 718 | | | AA6082 | | |
|---|---|---|---|---|---|---|---|---|---|
| Initial temperature [°C] | 900 | 1050 | 1200 | 980 | 1025 | 1070 | 300 | 375 | 450 |
| Density [kg/m³] | | 7847 | | | 8190 | | | 2695 | |
| Young modulus [GPa] | 129 | 108 | 84 | 126 | 120 | 100 | 58 | 54 | 51 |
| Yield strength [MPa] | 126 | 50 | 40 | 216 | 187 | 161 | 101 | 82 | 67 |
| Thermal conductivity [W/(m·K)] | 28 | 29 | 30 | 29 | 30 | 31 | 200 | 208 | 214 |
| Specific heat capacity [J/(kg·K)] | 645 | 635 | 642 | 647 | 687 | 704 | 1032 | 1069 | 1120 |

The material features reported in Table 4 have been combined with the geometrical features, presented in previous Section 2.1, and with the process setting features, reported in the following Section 2.3, allowing the creation of the dataset utilized for the training and tests of the considered machine learning algorithms.

### 2.3. Radial-Axial Ring Rolling FEM Simulation Settings

The process parameters for the numerical simulations have been set considering the models proposed and validated in Berti et al. [18]. The three main parameters utilized in the analysis are reported in Equation (3), for the main roll rotational speed $\omega_R$, in Equations (4) and (5) for the mandrel initial $[v_M]_0$ and final $[v_M]_F$ feeding speeds, and in Equations (6) and (7) for the upper axial roll initial $[v_A]_0$ and final $[v_A]_F$ feeding speeds.

$$\frac{400}{R_R} < \omega_R < \frac{1600}{R_R} \tag{3}$$

$$\frac{\omega_R \cdot R_R \cdot 6.55 \cdot 10^{-3} \cdot (R_0 - r_0)^2 \cdot (\frac{1}{R_R} + \frac{1}{R_M} + \frac{1}{R_0} + \frac{1}{r_0})}{2\pi \cdot R_0} < [v_M]_0 < \frac{\omega_R \cdot R_R \cdot \frac{2 \cdot {\beta_R}^2}{(\frac{1}{R_R} + \frac{1}{R_M})^2} \cdot (\frac{1}{R_R} + \frac{1}{R_M} + \frac{1}{R_0} + \frac{1}{r_0})}{2\pi \cdot R_0} \tag{4}$$

$$\frac{\omega_R \cdot R_R \cdot 6.55 \cdot 10^{-3} \cdot (R_F - r_F)^2 \cdot (\frac{1}{R_R} + \frac{1}{R_M} + \frac{1}{R_F} + \frac{1}{r_F})}{2 \cdot \pi \cdot R_F} < [v_M]_F < \frac{\omega_R \cdot R_R \cdot \frac{2 \cdot {\beta_R}^2}{(\frac{1}{R_R} + \frac{1}{R_M})^2} \cdot (\frac{1}{R_R} + \frac{1}{R_M} + \frac{1}{R_F} + \frac{1}{r_F})}{2 \cdot \pi \cdot R_F} \tag{5}$$

$$\frac{\omega_R \cdot R_R \cdot \frac{0.0131 \cdot {h_0}^2}{\left(L_0 - \frac{s_0}{2}\right) \tan\left(\frac{\theta}{2}\right)}}{2\pi \cdot R_0} < [v_A]_0 < \frac{4\omega_R \cdot R_R \cdot {\beta_A}^2 \left(L_0 - \frac{s_0}{2}\right) \tan\left(\frac{\theta}{2}\right)}{2\pi \cdot R_0} \tag{6}$$

$$\frac{\omega_R \cdot R_R \cdot \frac{0.0131 \cdot {h_F}^2}{\left(L_F - \frac{s_F}{2}\right) \tan\left(\frac{\theta}{2}\right)}}{2\pi \cdot R_F} < [v_A]_F < \frac{4\omega_R \cdot R_R \cdot {\beta_A}^2 \left(L_F - \frac{s_F}{2}\right) \tan\left(\frac{\theta}{2}\right)}{2\pi \cdot R_F} \tag{7}$$

In Equations (4) and (7) $R_R$ is the radius of the main roll, $R_M$ the radius of the mandrel, $R_0$, $r_0$, and $h_0$ the outer radius, inner radius and height of the initial ring blank, $R_F$, $r_F$, and $h_F$ the outer radius, inner radius, and height of the final ring, $\theta$ half of the axial rolls vertex angle whereas $\beta_R$ and $\beta_A$ the friction angle in the contact between main roll and mandrel and axial rolls, with the ring, respectively. The friction angle is calculated based on the friction factors, Table 1, as $\beta_R = arctg(m)$.

For each one of the implemented numerical simulations, the above-mentioned process parameters have been set according to the range proposed in [18] and have been considered as input for the force integral over time (FIOT) estimation models, presented in Section 4.3 of the paper. Since the process parameter setting is based on a kinematic approach, different temperatures or materials result in the same set of speeds. The summary of the implemented study cases is reported in Appendix A and is fully disclosed in the Supplementary Material.

## 3. Machine Learning Models Definition, Preprocessing, and Training

Due to the complex interaction between the considered process, materials, and geometry parameters, eight machine learning (ML) algorithms, one of which is based on the artificial neural network (ANN), with different levels of complexity, have been considered in this paper. The target is to implement a methodology for the estimation of the energy consumption in the radial-axial ring rolling process based on a set of input variables composed of geometry, process conditions, and materials. The architecture of the implemented ANN model is shown in Figure 2 where four hidden layers have been considered. As concerns the remaining ML models, input and output layers are the same as shown in Figure 2 but are connected through the weights vectors, defined during the optimization process. All the considered models have been applied to the above-mentioned dataset, considering 80% of the set for the model training whereas the remaining 20% has been employed for the assessment of the model accuracy. Both sets are not predetermined but are randomly selected before the training. The employed algorithms belong to the (i) linear, (ii) kernel, (iii) ensemble and (iv) artificial neural network approaches.

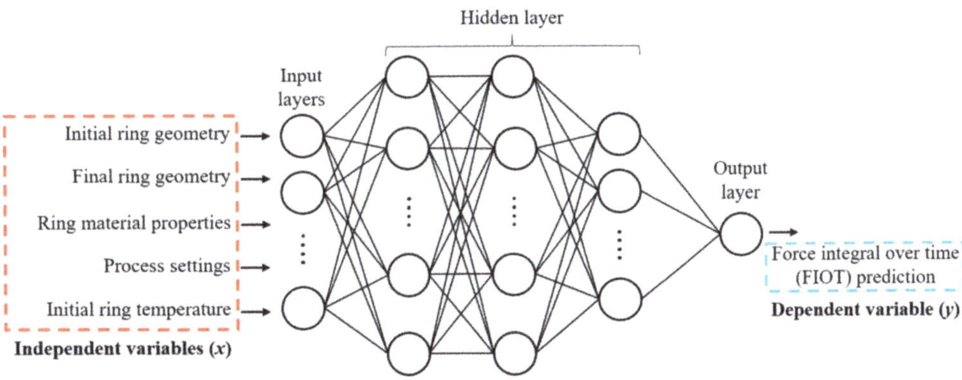


**Figure 2.** Artificial Neural Network model architecture schematic explaining the connection between input layers (input parameters considered in this research) and output layer, considered as the force integral of mandrel acting time.

### 3.1. Linear Methods

Linear regression methods [31,32] are utilized to model linear correlations between the independent variable $x$ and dependent variable $y$ as in Equation (8). The prediction calculated by the model is defined as $\hat{\mathbf{y}}$ and the aim is to minimize the Residual Sum of Squares (RSS) of the objective function, as shown in Equation (9). The subscript $D$ represents the number of considered features, whereas $N$ represents the size of the dataset.

Linear methods can be expanded to model the non-linear relationships by replacing $\mathbf{X}$ with non-linear functions. In this paper, to avoid the over-fitting problem, the regularized linear method has been utilized where constraints have been imposed on the weights vector ($\boldsymbol{w}$) of Equation (8).

$$\hat{\mathbf{y}} = w_0 + w_1 x_1 + \cdots + w_D x_D = w_0 + \boldsymbol{w}^{\mathrm{T}}\mathbf{X} \tag{8}$$

$$\mathrm{RSS}(\boldsymbol{w}) = \sum_{i=1}^{N} (y_i - \boldsymbol{w}^{\mathrm{T}} X_{i,D})^2 \quad (9)$$

Based on the general form of Equation (8), the Ridge model is defined to minimize the squared sum of weights, thus resulting in the objective function ($\Upsilon$), Equation (10). If the hyperparameter $\lambda$ of Equation (10) is equal to 0, we return to the original linear model of Equation (9). The hyperparameter, present in the Ridge model, as well as in other of the models subsequently presented, is a tuning parameter utilized to increase the accuracy of the prediction and is calculated, during the training, to maximize the correlation factor between independent and dependent variables [44].

$$\Upsilon(\boldsymbol{w}) = \mathrm{RSS}(\boldsymbol{w}) + \lambda \sum_{j=1}^{D} \|w_j\|^2 \quad (10)$$

Another variation of Equation (8) is defined as the Least Absolute Shrinkage and Selection Operator (LASSO) model where the absolute values of the weights are optimized to minimize the derivative of the target function $\Upsilon(\boldsymbol{w})$, defined as in Equation (11).

$$\Upsilon(\boldsymbol{w}) = \frac{1}{2N}\mathrm{RSS}(\boldsymbol{w}) + \lambda \sum_{j=1}^{D} \|w_j\| \quad (11)$$

Considering together the square of the weights, as in the Ridge model of Equation (10), and the norm of the weights, as in the LASSO algorithm of Equation (11), the third considered linear model is shown as in Equation (12) and is defined as the Elastic Net model.

$$\Upsilon(\boldsymbol{w}) = \frac{1}{2N}\mathrm{RSS}(\boldsymbol{w}) + \frac{1}{2}\lambda_2 \left\{ \frac{1}{2}(1-\lambda_1)\sum_{j=1}^{D} \|w_j\|^2 + \lambda_1 \sum_{j=1}^{D} \|w_j\| \right\} \quad (12)$$

In Equation (12), if the hyperparameter is set as $\lambda_1 = 1$ then the LASSO Equation (11) is obtained, whereas if $\lambda_1 = 0$ it results in the Ridge model, respectively. The $\lambda_1$ and $\lambda_2$ parameters represent the constants related to the first and second-order norms, respectively, and are calculated based on the random search method [44].

### 3.2. Kernel Methods

Linear methods can be expanded to model non-linear relationships between the independent and dependent variables by replacing $\mathbf{X}$, Equation (8), with the feature function $\phi(x)$. The feature function can be written with the Gram matrix ($\mathbf{K}$), as shown in Equations (13) and (14) where $\kappa(x_i, x_j)$ is the kernel function [33,34], defined to model the considered relationship. The Kernel Ridge (KR) model combines the kernel method with the Ridge model (10) and, in this paper, the polynomial kernel of Equation (15) is utilized for the Kernel Ridge model. The $c$ and $d$ constants in Equation (15) influence the feature functions and are determined through the random search method [44] during the training process.

$$\phi \cdot \phi^{T} = \mathbf{K} \quad (13)$$

$$\mathbf{K} = \begin{bmatrix} \kappa(x_1, x_1) & \cdots & \kappa(x_1, x_n) \\ \vdots & \ddots & \vdots \\ \kappa(x_n, x_1) & \cdots & \kappa(x_n, x_n) \end{bmatrix} \quad (14)$$

$$\kappa(x_i, x_j) = (\gamma {x_i}^{T} x_j + c)^{d} \quad (15)$$

Another Kernel method based on the squared norm of the weight factors is defined as the Support Vector Machine (SVM) model [45], Equation (16), where the $\mathrm{RSS}(\boldsymbol{w})$ function of Equation (10) is changed into the epsilon intensive loss function, Equation (17). In this paper, as for the case of the SVM model, the polynomial kernel function of Equation (15) has been utilized.

$$\Upsilon(\boldsymbol{w}) = C\sum_{i=1}^{N} L_\varepsilon(y_i, \hat{y}_i) + \frac{1}{2}\sum_{j=1}^{D}\|w_j\|^2 \quad (16)$$

$$L_\varepsilon(y, \hat{y}) = \begin{cases} 0 & \text{if } |y - \hat{y}| < \varepsilon \\ |y - \hat{y}| - \varepsilon & \text{otherwise} \end{cases} \quad (17)$$

### 3.3. Ensemble Methods

The ensemble methods [35–37] combine different approaches and apply them to randomly selected data sub-sets to improve the prediction performances. Among the ensemble approaches, the Random Forest (RF) model, utilized in this paper, trains $M$ decision trees and calculated the response for each one of them. For each tree, the response is defined considering different intervals for the estimator, allowing the subdivision of the problem into subclasses, which are classified by their accuracy by comparing their values with the true value. The final prediction is given by the average of the $M$ ones, calculated as the average of each one relevant for each one of the subsets of every tree as shown in Equation (18) where $\hat{\mathbf{y}}$ is the average prediction over $M$-trees and $\hat{\mathbf{y}}_m$ is the prediction of each tree.

$$\hat{\mathbf{y}} = \sum_{m=1}^{M} \frac{1}{M}\hat{\mathbf{y}}_m \quad (18)$$

Another ensemble method is defined as Gradient Boosting (GB) and, differently from the RF, in the beginning, only one tree ($f$) is created and it is progressively updated to minimize the objective function $\Upsilon(\boldsymbol{w})$, Equation (19). Therefore, the $m+1$ tree is based on the results of the $m$ tree compensated by the gradient residual of the previous tree by considering the learning rate $\eta$, as shown in Equation (20).

$$\Upsilon(\boldsymbol{w}) = L_\delta(\mathbf{y}, f) = \begin{cases} \frac{1}{2}(y - f)^2 & \text{if } |y - f| < \delta \\ \delta|y - f| - \frac{1}{2}\delta^2 & \text{otherwise} \end{cases} \quad (19)$$

The learning rate $\eta$ is defined as the speed by which the algorithm minimizes the loss function, $L_\delta$ for the ensemble methods and is present only for the case of ensemble and ANN methods. For the case of the linear method, like those of Sections 3.1 and 3.2, the learning rate is not considered since the optimized value is defined as the minimum of the loss function.

$$\begin{aligned} f_{m+1} &= f_m + \eta \cdot r_{m+1} \\ \text{where } r_{i,m+1} &= -\left|\frac{\partial L_\delta(y_i, f_i)}{\partial f_i}\right|_{f_m} \end{aligned} \quad (20)$$

### 3.4. Artificial Neural Network Methods

Artificial Neural Network (ANN) models [38–40] consist of: (i) input layers, (ii) hidden layers, and (iii) output layers, Figure 2. Input layers are connected to the hidden layers by the weight functions ($w_{ij}$) which are calculated during the training of the ANN algorithm. For each one of the nodes, the input coming from the previous layer is defined as $x_{ij}$ and are multiplied by the weight functions ($w_{ij}$) and summed out to the bias values ($w_{i0}$), and the output of the layer is derived through the activation function ($\Psi$), Equation (21).

$$\hat{y}_i = \Psi\left(\sum_{j=1}^{D} w_{ij} \cdot x_j + w_{i0}\right) \quad (21)$$

In the research presented in this paper, the weight matrix is updated considering the RMSprop algorithm [46], reported in Equation (22). The learning rate $\eta$ and the hyperparameter $\rho$ are optimized considering the random search method [44]. Finally, the activation function for the ANN algorithm, $\Psi$ of Equation (23), is defined as the threshold for the activation of a considered node in the hidden layers.

Only if $\Psi$ exceeds the threshold, the considered node in the i-layer is connected to the nodes of the i + 1 layer. The considered ANN algorithm is composed of four hidden layers made up of 200, 100, 50, and 25 nodes, respectively. The number of neurons for each layer has been optimized to minimize the loss function. The $L_\delta$ target function of Equation (19) has also been utilized for the case of the ANN model.

$$
\begin{aligned}
w_{i+1} &= w_i - \eta \frac{1}{\sqrt{h}} \cdot \frac{\partial L_\delta(\mathbf{y}_i, \hat{\mathbf{y}}_i)}{\partial w_i} \\
\text{where } h_i &= \rho \cdot h_{i-1} + (1-\rho)\left(\frac{\partial L_\delta(\mathbf{y}_i, \hat{\mathbf{y}}_i)}{\partial w_i}\right)^2
\end{aligned}
\tag{22}
$$

$$
\Psi(x) = \begin{cases} 0 \; for \; x \leq 0 \\ x \; for \; x > 0 \end{cases}
\tag{23}
$$

*3.5. Data Preprocessing and Machine Learning Algorithm Training*

For the training of the selected machine learning algorithms, presented in Sections 3.1–3.4, the input data for the 380 FEM simulations as well as the result, in terms of radial forming force integral over the mandrel time (FIOT), have been randomly arranged to avoid any bias. In both the training and test datasets, a single feature is defined as a row of the table composed of the following data: (i) main roll rotational speed, (ii) average mandrel feeding speed, (iii) initial ring geometry, (iv) final ring geometry, (v) initial ring temperature, (vi) material yield strength, (vii) material Young's modulus and (viii) force integral over the mandrel time. Since the parameters considered in this research have different intervals and measurement units, normalization has been applied to convert them to a 0 to 1 range. As concerns the FIOT, due to the skewness of the data distribution, the input data has been converted into log(1 + FIOT) before the normalization process. Similarly, the remaining parameters have also been converted by a box-cox transformation defined as $(x^{0.15} - 1)/0.15$, where x is the considered parameter.

This procedure allows reducing the computational burden during the training as well as increasing the accuracy. The hyperparameters of the prediction models described in chapter 3 have been obtained by applying the random search method aiming to maximize the correlation factor on both the training and the test datasets. The algorithms presented in the previous sections of chapter 3 have been implemented in a Windows OS environment utilizing the scikit-learn 0.22.2 and Keras 2.3.1 modules implemented in the Anaconda Spyder program with Python 3.7.4. As previously mentioned, 80% of the dataset, corresponding to 304 data, has been randomly selected from the whole database and the remaining 20%, 76 data, has been utilized as a test set.

For the evaluation of the accuracy of each model, three validation steps have been considered: (i) in the first step, the training dataset is fed once again to the model after the hyperparameters, if present, have been optimized; (ii) the test set is fed to the model and the accuracy, for the case of untrained data, is evaluated; (iii) finally, experimental values from reference papers and self-developed experiments are fed to the model and its accuracy is defined. The results concerning the accuracy of each of the considered methods, for the above-mentioned three validation steps, are reported in Section 4 of the paper along with the relevant optimized hyperparameters.

## 4. Results

To condense the vast amount of data relevant for the 380 numerical simulations composing the training and test datasets, the key results of three numerical simulations, in terms of equivalent plastic strain and mandrel force, have been summarized in Section 4.1. In addition to that, to prove the reliability of the numerical model implementation procedure, in Section 4.2 a validation has been carried out by comparing the outer diameter expansion and mandrel force over time. The results presented in Section 4.2 are from the authors' previous work [8] and have been briefly summarized.

Finally, in Section 4.3, the results of the optimization of the hyperparameters as well as the performances of the considered machine learning models, as presented in Section 3, are reported. To enhance the validation of the proposed FIOT estimation procedure, the four most accurate machine learning models, among the eight employed, have been utilized for the prediction of three experimental cases from literature papers. This second validation phase allowed confirming the reliability of the defined investigation procedure as well as the accuracy of the implemented solutions.

### *4.1. Thermo-Mechanical FEM Models Results*

To provide insight on the results of the numerical simulation implemented for all the 380 analyzed cases, in Figure 3 the equivalent plastic strain distribution at the end of the calibration phase, the outer diameter, and radial force evolution during the process are reported for the case of an 1100 mm final outer diameter ring made of 42CrMo4 steel with an initial temperature of 1200 °C. The radial (mandrel) forming forces relevant for all the 380 cases have been exported from the FEM simulations and utilized for the creation of the training and test database for the machine learning algorithms. Due to the large amount of data composing the database, they are not included in the manuscript but submitted along with the paper as Supplementary Material.

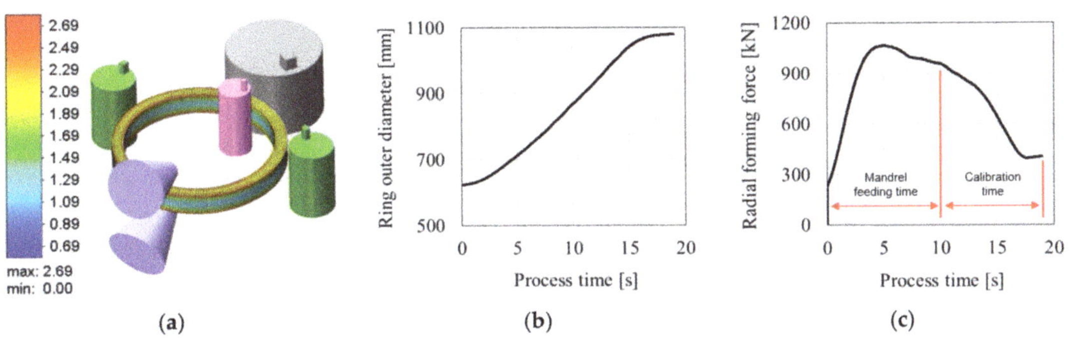


**Figure 3.** (**a**) Effective plastic strain distribution on the ring at the end of the calibration phase, (**b**) Ring outer diameter and (**c**) radial forming force evolution throughout the simulation process (DF = 1100 mm, 42CrMo4 steel, initial ring temperature of 1200 °C).

After the export of the results of the radial forming force from the Simufact Forming 15 numerical simulations, a script has been implemented in MS-Excel for the automatic calculation of the time integral of the force, allowing to calculate the FIOT, utilized as a sort of measure of the amount of energy required in the whole forming process.

For the calculation of the FIOT, only the mandrel time, Figure 3c, utilized as user input in the numerical simulation, has been considered. The overall simulation time is composed of mandrel time and calibration time but, since the latter one can be extended at will to increase the accuracy of the ring geometry, it has not been considered in the analysis. The mandrel time instead is the time during which the mandrel is actively translating towards the main roll, thus when most of the process energy is employed.

### 4.2. Thermo-Mechanical FEM Model Validation

To validate the developed numerical simulation model, the experimental results presented in the authors' previous research [8] have been utilized and are hereafter summarized. For the validation, a Pb75-Sn25 alloy has been utilized for the manufacturing of the ring preform with the initial dimensions $D_0$, $d_0$ and $h_0$ equal to 155 mm, 105 mm, and 42 mm, and final dimensions equal to $D_F$, $d_F$ and $h_F$ equal to 195 mm, 153 mm, and 37 mm, respectively (Figure 4).

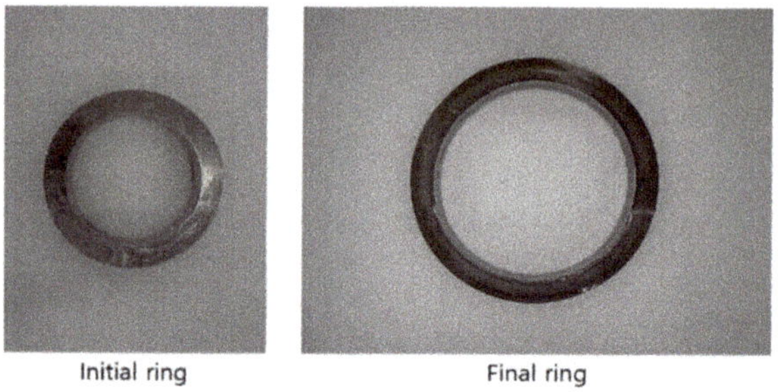


**Figure 4.** Initial and final Pb75-Sn25 ring.

Additional details concerning the material properties of the Pb75-Sn25 and the ring rolling machine utilized for the validation experiments are reported in Appendix B of the paper. The validation of the implemented thermo-mechanical finite element simulation has been carried out by comparing numerical and experimental results relevant for the expansion of the outer diameter of the ring during the process, Figure 5a, and of the mandrel forming force, Figure 5b.

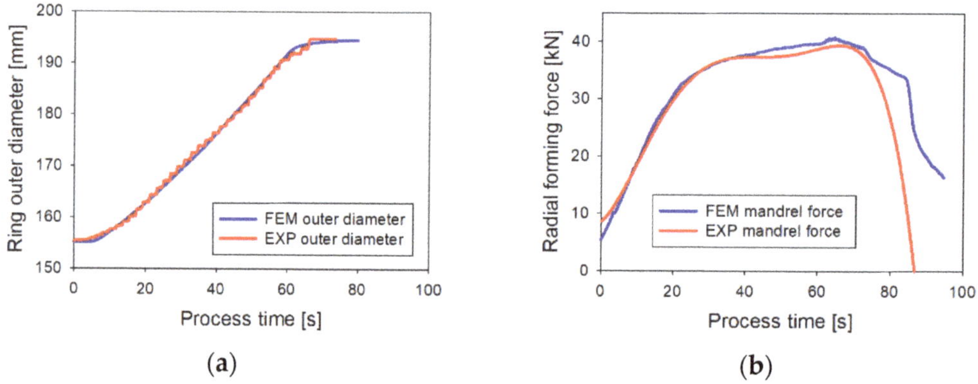


**Figure 5.** Comparison between the experimental and finite element (**a**) ring outer diameter and (**b**) radial forming force for the Pb75-Sn25 validation ring.

According to the results presented in Figure 5, the maximum deviation between experimental and finite element results is equal to 0.95% for the outer diameter 2.15% for the radial forming force, showing the reliability of the implemented numerical simulation model in replicating real process conditions. Since the FIOT estimation is based on the precise estimation of the forming force for the whole mandrel feeding time, the validation carried out against experimental results allows confirming the accuracy of the implemented finite element model solution, and thus the reliability of the input dataset for the training of the considered machine learning models.

### 4.3. Energy Prediction Models Results and Validation

By considering the setting parameters and FIOT results of the 380 implemented FEM simulations, as presented in Section 2, the hyperparameters relevant for the eight considered machine learning models have been calculated by means of the random search method and optimized during the training phase of the algorithms. The hyperparameters have been all set as random numbers at the beginning of the training process and optimized during the training to minimize the residual between prediction and true values. During the training phase, 80% of the whole 380 simulations have been utilized and this set is defined as the "train set". After the optimization of the hyperparameters, the trained machine learning models have been applied to the remaining 20% of the 380 simulations, not utilized during the training, and the accuracy in the estimation of the FIOT has been investigated. The accuracy of the training process has been verified by considering the correlation factor (R2). The optimized hyperparameters as well as the correlation factors for each model, relevant for the training dataset and the test dataset, calculated for the optimized hyperparameters, are reported in Table 5.

**Table 5.** Machine learning models' optimized hyperparameters and accuracy.

| | Model | Hyperparameters Values | Correlation Factor ($R^2$) | |
|---|---|---|---|---|
| | | | Train Set | Test Set |
| Linear method | Ridge | $\lambda = 0.6$ | 0.920 | 0.855 |
| | Lasso | $\lambda = 0.0005$ | 0.921 | 0.852 |
| | Elastic Net | $\lambda_1 = 0.0005$ , $\lambda_2 = 0.9$ | 0.953 | 0.921 |
| Kernel method | Kernel Ridge | $\lambda = 0.02$ , $\gamma = 1.0$ $d = 3.0$ , $c = 15.0$ | 0.985 | 0.970 |
| | Support Vector Machine | $C = 0.89$ , $\varepsilon = 0.03$ $\gamma = 1.0$ , $d = 3.0$ , $c = 15.0$ | 0.983 | 0.952 |
| Ensemble method | Random Foresting | $M = 5000$ | 0.995 | 0.971 |
| | Gradient Boosting | $M = 6900$ , $\delta = 0.9$ $\eta = 0.091$ | 0.998 | 0.996 |
| Artificial Neural Network method | Artificial Neural Network (ANN) | $\delta = 1.0$ $\eta = 0.001$ | 0.993 | 0.992 |

According to the results presented in Table 5, the Gradient Boosting method shows the best correlation factor ($R^2$) both in the training and test datasets. This high accuracy is related to the capability of the ensemble methods to subdivide the training dataset into sub-problems and thus, as concerns the research presented in this paper, to properly interpret the influence of different levels of the process, material, and geometrical parameters on the FIOT. Moreover, the higher accuracy of the Gradient Boosting method in comparison to the Random Forest method is related to the nature of the error minimization of the former. For the case of the Gradient Boosting method, only one tree is considered, and it is progressively optimized to minimize the residuals. The Random Forest method instead creates several trees and assigns a sub-problem to each one of them, optimizing the solutions for each one of them. However, the subdivision into sub-classes might lead to biases during the training process, a fact which is clear from the drop of the correlation factor between train test and test set for the case of the Random Forest method (Table 5).

To provide a more comprehensive evaluation of the performances of the four machine learning models that have shown the best results in terms of correlation factors (Table 5), the true values vs. prediction as well as the percentage residuals for the 76 cases of the test set are reported in Figure 6a,b, respectively. The true values in Figure 6a are the FEM result whereas the prediction values refer to the relevant ML models predictions.

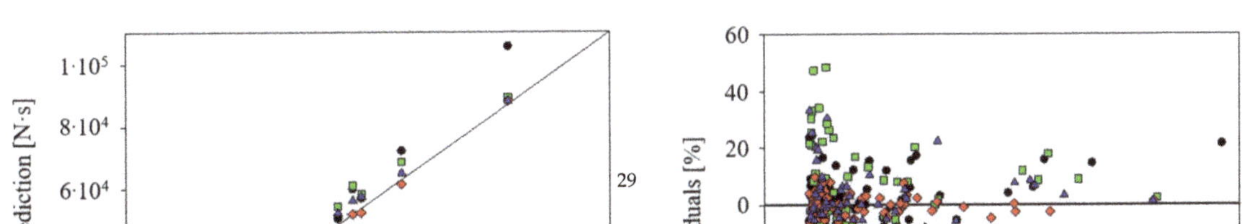

The analysis of the residuals, Figure 6b, shows that although the Kernel, Random Forest and ANN methods have a remarkably high correlation factor, their residuals are considerably high, especially for small prediction values. On the other hand, the Gradient Boosting method allows having low residuals for all FIOT levels. The maximum and average residuals, for the four methods summarized in Figure 6, are reported in Table 6.

**Table 6.** Machine learning models accuracy for the test data set.

| Model | Maximum Residual | Average Residual |
|---|---|---|
| Kernel Ridge | 24.28% | 6.93% |
| Random Forest | 48.44% | 13.08% |
| Gradient Boosting | 9.03% | 3.18% |
| Artificial Neural Network | 43.00% | 9.17% |

In addition to that, as previously mentioned, the accuracy in the prediction of the FIOT has also been evaluated for the case of three experimental ring rolling cases from the literature [5,8,11] by applying the four trained machine learning models that showed the highest correlation factors (Table 6). These experimental results are all relevant for experiments carried out on GH4169 nickel-based superalloy [5], the Pb-Sn alloy ring also utilized for the finite element model validation [8], and AISI-304 steel alloy [11]. All these three cases are completely different in terms of the geometry and material of the ring, process conditions, and size of the ring rolling mill and have been selected to provide additional insight into the accuracy of the predictions carried out by the proposed models. True prediction percentage residuals for these three cases are summarized in Table 7.

**Table 7.** Machine learning models accuracy for the literature experimental cases.

| Model | [5] FIOT Prediction Error | [8] FIOT Prediction Error | [11] FIOT Prediction Error |
|---|---|---|---|
| Kernel Ridge | 16.73% | 15.11% | 28.97% |
| Random Forest | 8.48% | 13.44% | 24.10% |
| Gradient Boosting | 7.01% | 2.09% | 8.00% |
| Artificial Neural Network | 7.23% | 6.27% | 9.11% |

Considering the results presented in Table 7, it is once again clear that the structure of the Gradient Boosting method can catch the complex nature of the interaction between geometrical, material, and process parameters in the radial-axial ring rolling process thanks to its ability to subdivide the given task into sub-problems but while keeping error minimization linked to a single residual function. Moreover, as mentioned in Section 2.1, although the process conditions relevant for the [5,8,11] are different from those utilized in the FEM simulations utilized for the training, where the same friction conditions have been considered in all the cases, the accuracy is still remarkably good and the computational time is almost real-time, allowing a considerable improvement in comparison to the computational time of the thermo-mechanical numerical simulations, which may range from 9~12 h, for the case of the 650 mm final ring outer diameter simulations, to 1.5 to 3 days for the case of 2000 mm final ring outer diameter simulations.

## 5. Discussion

Considering the results relevant for all the utilized machine learning models, as presented in Table 5, the relatively low correlation factor shown by the linear models is an indication of the fact that the relationship between the considered input and output parameters is not linear. For the same reason, the Kernel methods, which utilize a polynomial function, show better accuracy than the linear methods but still have high residuals, as shown in the detailed analysis of Figure 6 and Table 6.

Both linear and Kernel methods calibrate the components of the weights vector ($w$), Equation (8) by minimizing the residual of the objective function, thus they tend to show low residuals for the case of the training dataset but relatively high ones for untrained data. On the other hand, both the Gradient Boosting as well as the Neural Network algorithms calibrate the components of the weights vector ($w$) considering the learning rate which allows a more robust consistency in both training and test datasets, as well as for additional predictions. Considering altogether the three validation steps carried out considering the (i) training dataset, (ii) the test dataset and (iii) the literature experimental cases, the complex interaction between process, material, and geometry parameters is therefore representable neither by a linear nor by a polynomial function.

As concerns the applicability of the proposed procedure outside the ranges considered for the construction of the training dataset, the validation case relevant for [5] gives a remarkably interesting insight. Although geometry, process parameters, and material are all different in comparison to those considered in this paper, the robustness of the trained Gradient Boosting model allows obtaining a reasonable residual in the estimation of force integral, as shown in Table 7. On the other hand, as previously mentioned, the linear and polynomial correlations considered by the linear and Kernel methods render their prediction to be affected by a high residual if the requested prediction is outside the trained ranges. Furthermore, the choice of normalizing all the parameters is also an important step for the application of the proposed procedure outside its training ranges.

Finally, an interesting feature relevant to the machine learning methods concerns the balance of the training dataset. In the considered research, the amount of data relevant for low FIOT is considerably higher than that of high force integral, and, for the case of multivariable regression methods, this fact would have resulted in good predictions for the former scenario and bad for the latter one. In principle, the need for a balanced training dataset is also valid for the machine learning models but, for the case of the ensemble methods, as well as the neural network, their sensitivity to the data clustering is almost negligible and is therefore suitable for the application in sparse and not balanced data environments. Considering altogether the investigation proposed in this paper, the applicability of the machine learning-based algorithm for the prediction of energy consumption, measured in terms of force integral over time, in forming processes has been largely explored, and both the results and analysis reported in this paper might be helpful for the extension of its application to additional industrially relevant processes.

## 6. Conclusions

The research presented in this paper highlighted the importance of considering the material, geometrical, and process parameters when estimating the forming force during the radial-axial ring rolling process. Moreover, eight different Machine Learning-based algorithms have been utilized for the prediction of the mandrel force integral over time (FIOT) and showed that the Gradient Boosting (GB) algorithm, belonging to the ensemble methods, grants the best accuracy in the prediction of the FIOT, being the maximum residual equal to 9.03%. Since the validation has been carried out on previously published results where ring geometries, process conditions, and materials were not included in the training dataset, the proposed approach has proven its robustness in predicting the FIOT also outside the range of the training data set. The trained GB algorithm can be directly applied to the radial-axial ring rolling (RARR) process through the algorithm provided as Supplementary Material and applies also to other forming and forging processes where the contact between workpiece and tools is defined by a curved line, as in the RARR process. The application of the proposed procedure allows a significant reduction in the time required for the estimation of the energy consumption during forming processes, its calculation being almost real-time, in comparison to the case of FEM simulations where the computational time ranges between ~10 h (for 650 mm final outer diameter rings) to 3 days (for 2000 m final outer diameter rings). The procedure presented in this paper can also be extended to different metal forming and forging processes by considering the same geometry, material, and process parameters influence on the energy consumption, but the creation of a new training dataset might be required. For these reasons, the research presented in this paper might be of interest to researchers and process engineers interested in energy consumption in metal forming processes.

**Supplementary Materials:** The following are available online https://www.mdpi.com/article/10.3390/met11050833/s1 as supplementary material: Python code for Gradient Boosting model, Simulation settings, and results database.

**Author Contributions:** Conceptualization, L.Q. and I.M.; methodology, I.M.; software, S.L.; validation, I.M. and L.Q.; formal analysis, I.M.; investigation, I.M.; resources, I.M.; data curation, I.M.; writing—original draft preparation, L.Q. and I.M.; writing—review and editing, L.Q.; visualization, L.Q.; supervision, R.C. and N.K.; project administration, G.A.B.; funding acquisition, L.Q. and N.K. All authors have read and agreed to the published version of the manuscript.

**Funding:** This work was supported by the Basic Science Research Program through the National Research Foundation of Korea (NRF) funded by the Ministry of Education (grant number: 2019R1I1A1A 01062323) and by the National Research Foundation of Korea (NRF) grant funded by Korea government (MSIT) (grant number: 2019R1F1A1060567).

**Institutional Review Board Statement:** Not applicable.

**Informed Consent Statement:** Not applicable.

**Data Availability Statement:** All the results and Python scripts are available on request to the corresponding author.

**Acknowledgments:** This research was carried out with the help of the "HPC Support" Project, supported by Ministry of Science, ICT and NIPA of Korea. This support is gratefully acknowledged.

**Conflicts of Interest:** The authors declare no conflict of interest.

## Appendix A. Summary of the Geometrical Settings for the ML Model Database

| | | | | 42CrMo4 | | | AlMgSi1 | | | IN-718 | |
|---|---|---|---|---|---|---|---|---|---|---|---|
| | | **T [°C]** | **900** | **1050** | **1200** | **300** | **375** | **450** | **980** | **1025** | **1070** |
| $\omega_R$[rad/s] | $\overline{v}_M$[mm/s] | $D_0$[mm] | | $D_F$[mm] | | | $D_F$[mm] | | | $D_F$[mm] | |
| | 2.05 | 944.7 | 2000 | 2000 | 2000 | - | - | - | - | - | - |
| | 2.17 | 944.7 | - | - | - | 2000 | 2000 | 2000 | 2000 | 2000 | 2000 |
| | 2.24 | 897.7 | 1700 | 1700 | 1700 | 1700 | 1700 | 1700 | 1700 | 1700 | 1700 |
| | 2.55 | 909.3 | 1400 | 1400 | 1400 | 1400 | 1400 | 1400 | 1400 | 1400 | 1400 |
| | 2.56 | 876.2 | 1400 | 1400 | 1400 | 1400 | 1400 | 1400 | 1400 | 1400 | 1400 |
| | 2.75 | 549.3 | 650 | 650 | 650 | 650 | 650 | 650 | 650 | 650 | 650 |
| | 3.11 | 621.6 | 1100 | 1100 | 1100 | 1100 | 1100 | 1100 | 1100 | 1100 | 1100 |
| 2 | 3.37 | 897.7 | 1700 | - | - | - | - | - | - | - | - |
| | 3.60 | 550.9 | 800 | 800 | 800 | 800 | 800 | 800 | 800 | 800 | 800 |
| | 4.09 | 575.3 | 1100 | 1100 | 1100 | 1100 | 1100 | 1100 | 1100 | 1100 | 1100 |
| | 4.88 | 518.9 | 800 | 800 | 800 | 800 | 800 | 800 | 800 | 800 | 800 |
| | 5.65 | 490.5 | 650 | 650 | 650 | 650 | 650 | 650 | 650 | 650 | 650 |
| | 5.67 | 897.7 | 1700 | - | - | - | - | - | - | - | - |
| | 5.70 | 490.5 | 650 | 650 | 650 | 650 | 650 | 650 | 650 | 650 | 650 |
| | 5.73 | 490.5 | 650 | 650 | 650 | 650 | 650 | 650 | 650 | 650 | 650 |
| | 2.92 | 944.7 | 2000 | 2000 | 2000 | - | - | - | - | - | - |
| | 3.27 | 944.7 | - | - | - | 2000 | 2000 | 2000 | 2000 | 2000 | 2000 |
| | 3.37 | 897.7 | - | 1700 | 1700 | 1700 | 1700 | 1700 | 1700 | 1700 | 1700 |
| | 3.70 | 1049.3 | - | - | 2000 | - | - | - | - | - | - |
| | 3.75 | 549.3 | 650 | 650 | 650 | 650 | 650 | 650 | 650 | 650 | 650 |
| | 3.82 | 909.3 | 1400 | 1400 | 1400 | 1400 | 1400 | 1400 | 1400 | 1400 | 1400 |
| | 3.83 | 876.2 | 1400 | 1400 | 1400 | 1400 | 1400 | 1400 | 1400 | 1400 | 1400 |
| | 3.85 | 944.5 | - | - | 2000 | - | - | - | - | - | - |
| | 4.10 | 875.7 | - | - | 2000 | - | - | - | - | - | - |
| | 4.50 | 549.3 | 650 | 650 | 650 | 650 | 650 | 650 | 650 | 650 | 650 |
| | 4.60 | 621.6 | 1100 | 1100 | 1100 | - | - | 1100 | 1100 | 1100 | 1100 |
| 3 | 5.15 | 668.7 | 1100 | 1100 | 1100 | 1100 | 1100 | 1100 | 1100 | 1100 | 1100 |
| | 5.25 | 549.3 | 650 | 650 | 650 | 650 | 650 | 650 | 650 | 650 | 650 |
| | 5.39 | 550.9 | 800 | 800 | 800 | 800 | 800 | 800 | 800 | 800 | 800 |
| | 5.50 | 549.3 | 650 | 650 | 650 | 650 | 650 | 650 | 650 | 650 | 650 |
| | | 621.6 | 1100 | 1100 | 1100 | 1100 | 1100 | 1100 | 1100 | 1100 | 1100 |
| | 5.75 | 549.3 | 650 | 650 | 650 | 650 | 650 | 650 | 650 | 650 | 650 |
| | 6.00 | 490.5 | 650 | 650 | 650 | 650 | 650 | 650 | 650 | 650 | 650 |
| | 6.10 | 575.3 | 1100 | 1100 | 1100 | 1100 | 1100 | 1100 | 1100 | 1100 | 1100 |
| | 6.85 | 518.9 | 800 | 800 | 800 | 800 | 800 | 800 | 800 | 800 | 800 |
| | 7.50 | 510.9 | 650 | 650 | 650 | 650 | 650 | 650 | 650 | 650 | 650 |
| | 7.60 | 490.5 | 650 | 650 | 650 | 650 | 650 | 650 | 650 | 650 | 650 |
| | 8.50 | 490.5 | 650 | 650 | 650 | 650 | 650 | 650 | 650 | 650 | 650 |

| | | | | 42CrMo4 | | | AlMgSi1 | | | IN-718 | |
|---|---|---|---|---|---|---|---|---|---|---|---|
| | | T [°C] | 900 | 1050 | 1200 | 300 | 375 | 450 | 980 | 1025 | 1070 |
| $\omega_R$[rad/s] | $\overline{v}_M$[mm/s] | $D_0$[mm] | | $D_F$[mm] | | | $D_F$[mm] | | | $D_F$[mm] | |
| 5 | 4.09 | 575.3 | | 1100 | | | - | | | - | |
| | 4.85 | 944.7 | | 2000 | | | 2000 | | | 2000 | |
| | 5.43 | 944.7 | | - | | | - | | | - | |
| | 5.67 | 897.7 | | - | | | 1700 | | | 1700 | |
| | 6.35 | 909.3 | | 1400 | | | 1400 | | | 1400 | |
| | 6.38 | 876.2 | | 1400 | | | 1400 | | | 1400 | |
| | 7.30 | 549.3 | | 650 | | | 650 | | | 650 | |
| | 7.69 | 621.6 | | 1100 | | | 1100 | | | 1100 | |
| | 8.97 | 550.9 | | 800 | | | 800 | | | 800 | |
| | 10.04 | 575.3 | | - | | | 1100 | | | 1100 | |
| | 12.12 | 518.9 | | 800 | | | 800 | | | 800 | |
| | 13.70 | 490.5 | | 650 | | | 650 | | | 650 | |
| | 14.00 | 490.5 | | 650 | | | 650 | | | 650 | |
| | 14.15 | 490.5 | | 650 | | | 650 | | | 650 | |

## Appendix B

The material properties of the Pb75-Sn25 alloy have been determined by carrying out compression tests at four different strain rates at room temperature and the relevant flow stress curves have been derived considering the model presented in Equation (A1). The model constants for the flow stress model of Equation (A1) are reported in Table A1. The laboratory-size ring rolling machine, utilized for the validation experiment, is reported in Figure A1a whereas the comparison between experimental and numerical flow stress curves is shown in Figure A1b.

$$\sigma = K_0(a_0 + \varepsilon)^{a_1}\left(b_0 + \dot{\varepsilon}\right)^{b_1} \tag{A1}$$

**Table A1.** Material model constants for the Pb75-Sn25 material.

| **Parameters** | $K_0$ | $a_0$ | $a_1$ | $b_0$ | $b_1$ |
|---|---|---|---|---|---|
| Value | 92 | 0.1 | 0.03 | 0.015 | 0.17 |

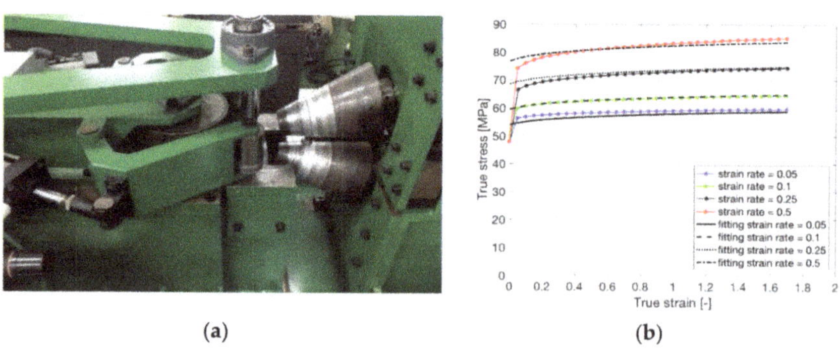


**Figure A1.** (**a**) Laboratory-size ring rolling machine utilized for the experiments and (**b**) Pb75-Sn25 alloy flow curves.

## References


1. Kim, N.; Machida, S.; Kobayashi, S. Ring rolling process simulation by the three dimensional finite element method. *Int. J. Mach. Tools Manuf.* **1990**, *30*, 569–577. [CrossRef]
2. Eruc, E.; Shivpuri, R. A summary of ring rolling technology-1. Recent trends in machines, process and production lines. *Int. J. Mach. Tools Manuf.* **1992**, *32*, 379–398. [CrossRef]
3. Lugora, C.F.; Bramley, A.N. Analysis of spread in ring rolling. *Int. J. Mech. Sci.* **1987**, *29*, 149–157. [CrossRef]
4. Bruschi, S.; Casptto, S.; Dal Negro, T.; Bariani, P.P. Real-time prediction of geometrical distortions of hot-rolled steel rings during cooling. *CIRP Ann. Manuf. Technol.* **2005**, *54*, 229–232. [CrossRef]
5. Guo, L.; Yang, H. Towards a steady forming condition for radial-axial ring rolling. *Int. J. Mech. Sci.* **2011**, *53*, 286–299. [CrossRef]
6. Quagliato, L.; Berti, G.A. Mathematical definition of the 3D strain field of the ring in the radial-axial ring rolling process. *Int. J. Mech. Sci.* **2016**, *115–116*, 746–759. [CrossRef]
7. Quagliato, L.; Berti, G.A. Temperature estimation and slip-line force analytical models for the estimation of the radial forming force in the RARR process of flat rings. *Int. J. Mech. Sci.* **2017**, *123*, 311–323. [CrossRef]
8. Quagliato, L.; Berti, G.A.; Kim, D.; Kim, N. Slip line model for forces estimation in the radial-axial ring rolling process. *Int. J. Mech. Sci.* **2018**, *138–139*, 17–33. [CrossRef]
9. Ryoo, J.S.; Yang, D.Y.; Johnson, W. The influence of process parameters on torque and load in ring rolling. *J. Mech. Work. Technol.* **1986**, *12*, 307–321. [CrossRef]
10. Kalyani, A.; Anand, M.; Amol, D. The Effect of Force Parameter on Profile Ring Rolling Process. *Int. J. Eng. Res.* **2015**, *V4*, 840–844.
11. Kim, N.; Kim, H.; Jin, K. Optimal design to reduce the maximum load in ring rolling process. *Int. J. Precis. Eng. Manuf.* **2012**, *13*, 1821–1828. [CrossRef]
12. Roesch, M.; Lukas, M.; Schultz, C.; Braunreuther, S.; Reinhart, G. An approach towards a cost-based production control for energy flexibility. *Procedia CIRP* **2019**, *79*, 227–232. [CrossRef]
13. Unver, U.; Kara, O. Energy efficiency by determining the production process with the lowest energy consumption in a steel forging facility. *J. Clean. Prod.* **2019**, *215*, 1362–1370. [CrossRef]
14. Meißner, M.; Massalski, L.; Wirtz, A.; Wiederkehr, P.; Myrzik, J. Representation of energy efficiency of energy converting production processes by process status indicators. *Procedia CIRP* **2019**, *79*, 221–226. [CrossRef]
15. Larkiola, J.; Myllykoski, P.; Korhonen, A.S.; Cser, L. The role of neural networks in the optimisation of rolling processes. *J. Mater. Process. Technol.* **1998**, *80–81*, 16–23. [CrossRef]
16. Giorleo, L.; Ceretti, E.; Giardini, C. Energy consumption reduction in Ring Rolling processes: A FEM analysis. *Int. J. Mech. Sci.* **2013**, *74*, 55–64. [CrossRef]
17. Allegri, G.; Giorleo, L.; Ceretti, E.; Giardini, C. Driver roll speed influence in Ring Rolling process. *Procedia Eng.* **2017**, *207*, 1230–1235. [CrossRef]
18. Berti, G.A.; Quagliato, L.; Monti, M. Set-up of radial–axial ring-rolling process: Process worksheet and ring geometry expansion prediction. *Int. J. Mech. Sci.* **2015**, *99*, 58–71. [CrossRef]
19. Davey, K.; Ward, M.J. A practical method for finite element ring rolling simulation using the ALE flow formulation. *Int. J. Mech. Sci.* **2002**, *44*, 165–190. [CrossRef]
20. Lim, T.; Pillinger, I.; Hartley, P. A finite-element simulation of profile ring rolling using a hybrid mesh model. *J. Mater. Process. Technol.* **1998**, *80–81*, 199–205. [CrossRef]
21. Kim, B.; Moon, H.; Kim, E.; Choi, M.; Joun, M. A dual-mesh approach to ring-rolling simulations with emphasis on remeshing. *J. Manuf. Process.* **2013**, *15*, 635–643. [CrossRef]
22. Zhao, D.; Zhao, K.; Ren, D.; Guo, X. Ultrasonic Welding of Magnesium-Titanium Dissimilar Metals: A Study on Influences of Welding Parameters on Mechanical Property by Experimentation and Artificial Neural Network. *J. Manuf. Sci. Eng. Trans. ASME* **2017**, *139*, 031019. [CrossRef]
23. Wu, D.; Jennings, C.; Terpenny, J.; Gao, R.X.; Kumara, S. A Comparative Study on Machine Learning Algorithms for Smart Manufacturing: Tool Wear Prediction Using Random Forests. *J. Manuf. Sci. Eng. Trans. ASME* **2017**, *139*, 071018. [CrossRef]
24. Tootooni, M.S.; Dsouza, A.; Donovan, R.; Rao, P.K.; Kong, Z.J.; Borgesen, P. Classifying the Dimensional Variation in Additive Manufactured Parts from Laser-Scanned Three-Dimensional Point Cloud Data Using Machine Learning Approaches. *J. Manuf. Sci. Eng. Trans. ASME* **2017**, *139*, 091005. [CrossRef]
25. Wang, Z.; Chegdani, F.; Yalamarti, N.; Takabi, B.; Tai, B.; El Mansori, M.; Bukkapatnam, S. Acoustic emission characterization of natural fiber reinforced plastic composite machining using a Random Forest machine learning model. *J. Manuf. Sci. Eng. Trans. ASME* **2020**, *142*, 1–13. [CrossRef]
26. Varghese, A.; Kulkarni, V.; Joshi, S.S. Tool life stage prediction in micro-milling from force signal analysis using machine learning methods. *J. Manuf. Sci. Eng. Trans. ASME* **2021**, *143*, 1–7. [CrossRef]
27. Wang, D.; Xu, Y.; Duan, B.; Wang, Y.; Song, M.; Yu, H.; Liu, H. Intelligent recognition model of hot rolling strip edge defects based on deep learning. *Metals* **2021**, *11*, 223. [CrossRef]
28. Marques, A.E.; Prates, P.A.; Pereira, A.F.G.; Oliveira, M.C.; Fernandes, J.V.; Ribeiro, B.M. Performance comparison of parametric and non-parametric regression models for uncertainty analysis of sheet metal forming processes. *Metals* **2020**, *10*, 457. [CrossRef]
29. Palmieri, M.E.; Lorusso, V.D.; Tricarico, L. Robust optimization and kriging metamodeling of deep-drawing process to obtain a regulation curve of blank holder force. *Metals* **2021**, *11*, 319. [CrossRef]

30. Winiczenko, R. Effect of friction welding parameters on the tensile strength and microstructural properties of dissimilar AISI 1020-ASTM A536 joints. *Int. J. Adv. Manuf. Technol.* **2015**, *84*, 941–955. [CrossRef]
31. Friedman, J.; Hastie, T.; Tibshirani, R. Regularization Paths for Generalized Linear Models via Coordinate Descent. *J. Stat. Softw.* **2010**, *33*, 1–22. [CrossRef]
32. Kim, S.J.; Koh, K.; Lustig, M.; Boyd, S.; Gorinevsky, D. An interior-point method for large-scale $\ell 1$-regularized least squares. *IEEE J. Sel. Top. Signal Process.* **2007**, *1*, 606–617. [CrossRef]
33. Murphy, K.P. *Machine Learning: A Probabilistic Perspective*; Chapter 14.4.3; The MIT Press: Cambridge, MA, USA, 2012; pp. 492–493.
34. Chang, C.-C.; Lin, C.J. A Library for Support Vector Machines. *ACM Trans. Intell. Syst. Technol.* **2019**, article 27.
35. Liu, Z.; Guo, Y. A hybrid approach to integrate machine learning and process mechanics for the prediction of specific cutting energy. *CIRP Ann.* **2018**, *67*, 57–60. [CrossRef]
36. Torres-Barrán, A.; Alonso, Á.; Dorronsoro, J.R. Regression tree ensembles for wind energy and solar radiation prediction. *Neurocomputing* **2019**, *326–327*, 151–160. [CrossRef]
37. Zhang, W.; Wu, C.; Zhong, H.; Li, Y.; Wang, L. Prediction of undrained shear strength using extreme gradient boosting and random forest based on Bayesian optimization. *Geosci. Front.* **2021**, *12*, 469–477. [CrossRef]
38. Stanke, J.; Feuerhack, A.; Trauth, D.; Mattfeld, P.; Klocke, F. A predictive model for die roll height in fine blanking using machine learning methods. *Procedia Manuf.* **2018**, *15*, 570–577. [CrossRef]
39. Olanrewaju, O.A.; Jimoh, A.A.; Kholopane, P.A. Integrated IDA-ANN-DEA for assessment and optimization of energy consumption in industrial sectors. *Energy* **2012**, *46*, 629–635. [CrossRef]
40. Ruiz, L.G.B.; Rueda, R.; Cuéllar, M.P.; Pegalajar, M.C. Energy consumption forecasting based on Elman neural networks with evolutive optimization. *Expert Syst. Appl.* **2018**, *92*, 380–389. [CrossRef]
41. Zhu, X.; Liu, D.; Yang, Y.; Hu, Y.; Zheng, Y. Optimization on cooperative feed strategy for radial-axial ring rolling process of Inco718 alloy by RSM and FEM. *Chin. J. Aeronaut.* **2016**, *29*, 831–842. [CrossRef]
42. Hawkyard, J.B.; Johnson, W.; Kirkland, J.; Appleton, E. Analyses for roll force and torque in ring rolling, with some supporting experiments. *Int. J. Mech. Sci.* **1973**, *15*, 873–893. [CrossRef]
43. Hensel, A.; Spittel, T. *Kraft und Arbeitsbedarf Bildsamer Formgebungsverfahren*, 1st ed.; Deutscher Verlag für Grundstoffindustrie: Leipzig, Germany, 1978. (In German)
44. Bergstra, J.; Bengio, Y. Random search for hyper-parameter optimization. *J. Mach. Learn. Res.* **2012**, *13*, 281–305.
45. Winiczenko, R.; Salat, R.; Awtoniuk, M. Estimation of tensile strength of ductile iron friction welded joints using hybrid intelligent methods. *Trans. Nonferrous Met. Soc. China* **2013**, *23*, 385–391. [CrossRef]
46. Tieleman, T.; Hinton, G. Lecture 6.5-rmsprop: Divide the gradient by a running average of its recent magnitude. *COURSERA Neural Netw. Mach. Learn.* **2012**, *4*, 26–31.

*metals*

MDPI

*Article*

# The Use of Machine-Learning Techniques in Material Constitutive Modelling for Metal Forming Processes

**Rúben Lourenço [1], António Andrade-Campos [1,*] and Pétia Georgieva [2]**

[1] Centre for Mechanical Technology and Automation (TEMA), Department of Mechanical Engineering, University of Aveiro, 3810-193 Aveiro, Portugal; rubenl@ua.pt

[2] Institute of Electronics and Informatics Engineering of Aveiro (IEETA), Department of Electronics, Telecomunications and Electronics, University of Aveiro, 3810-193 Aveiro, Portugal; petia@ua.pt

* Correspondence: gilac@ua.pt

**Abstract:** Accurate numerical simulations require constitutive models capable of providing precise material data. Several calibration methodologies have been developed to improve the accuracy of constitutive models. Nevertheless, a model's performance is always constrained by its mathematical formulation. Machine learning (ML) techniques, such as artificial neural networks (ANNs), have the potential to overcome these limitations. Nevertheless, the use of ML for material constitutive modelling is very recent and not fully explored. Difficulties related to data requirements and training are still open problems. This work explores and discusses the use of ML techniques regarding the accuracy of material constitutive models in metal plasticity, particularly contributing (i) a parameter identification inverse methodology, (ii) a constitutive model corrector, (iii) a data-driven constitutive model using empirical known concepts and (iv) a general implicit constitutive model using a data-driven learning approach. These approaches are discussed, and examples are given in the framework of non-linear elastoplasticity. To conveniently train these ML approaches, a large amount of data concerning material behaviour must be used. Therefore, non-homogeneous strain field and complex strain path tests measured with digital image correlation (DIC) techniques must be used for that purpose.

**Keywords:** mechanical constitutive model; machine learning; artificial neural network; finite element analysis; plasticity; parameter identification; full-field measurements

**Citation:** Lourenço, R.; Andrade-Campos, A.; Georgieva, P. The Use of Machine-Learning Techniques in Material Constitutive Modelling for Metal Forming Processes. *Metals* **2022**, *12*, 427. https://doi.org/10.3390/met12030427

Academic Editor: Ulrich Prahl

Received: 30 December 2021
Accepted: 23 February 2022
Published: 28 February 2022



## 1. Introduction

Considerable developments in the field of computational mechanics and the growth in computational power have made it possible for numerical simulation software to emerge and completely revolutionize the engineering industry. In fact, industry's high-demand-fuelled and fast-paced development cycles have accelerated the adoption of computational analysis software [1,2]. With a wide array of software solutions in the market, capable of virtualizing entire design workflows, numerical simulation has become a vital component of the product development process. Indeed, this virtualization has allowed companies to massively reduce the number of time-consuming experiments between design iterations, thus reducing delays and costs and allowing companies to reach higher levels of competitiveness [3]. In this context, finite element analysis (FEA) has been widely used as a powerful numerical simulation tool in engineering analyses, such as structural analysis, heat transfer and fluid flow. In FEA, the material behavior is defined by specific constitutive models for which the ability to adequately describe a given material directly impacts the reliability of the numerical results. As such, material characterization has received increasing attention given the need for computational analysis software for precise material data [3].

Conventionally, constitutive laws are established according to first-principle assumptions in order to generalize experimental observations made on simple loading regimes. In recent years, the fast development of complex materials exhibiting non-linear behavior has pushed the formulation of advanced and more complex constitutive models to accurately describe phenomena such as hardening and anisotropy. The challenge lies in reducing the error between the numerical predictions and the experimental value. To improve this differential, the models are enhanced with additional empirical parameters, albeit with the cost of adding complexity to the empirical expressions and making them more difficult to calibrate [4,5]. Although the advances in imaging techniques, such as DIC, have allowed for much richer information to be extracted from experimental observations, conventional models are sometimes incompatible with the abundance of data [6,7]. However, independently of a model's flexibility to adapt to the experimental results, it is always limited by its mathematical formulation, as it is written explicitly. Additionally, even if a model could be perfectly calibrated for a given set of experiments, there is no guarantee that it would provide a perfect result for a different set of experiments [8].

In recent years, the advent of Big Data combined with improved algorithms and the exponential increase in computing power led to an unprecedented growth in the field of artificial intelligence (AI) [9]. AI techniques, such as ML, allow computer systems to implicitly learn and recognize patterns purely based on data. For this reason, ML has become an important technology in various engineering fields [10]. Out of all ML algorithms, ANNs are extremely popular given their superior modelling performance in a wide range of applications, including as universal function approximators [10,11]. As such, neural networks have the potential to provide a radically different approach to material modelling, with the main advantage that it is not necessary to postulate a mathematical formulation or identify empirical parameters [12]. In general, data-driven approaches can better exploit the large volumes of data from modern experimental measurements, while avoiding the bias induced by constitutive models [6]. These ML techniques rely on different methods to determine the stress tensor corresponding to a given strain state (and other conditions such as temperature, strain rate, etc.). For instance, in [7,8], nearest neighbour interpolation is employed, while in [13–16], the authors employ a low-dimensional representation of the constitutive equation based uniquely on data. Bessa and co-workers [17] employ clustering techniques, while others employ spline approximations. ANNs were also used for modelling viscoplasticity behaviours [18,19]. Alternative solutions to pure data-driven models were given, namely hybrid models [15], that enhance/correct existing well-known models with information coming from data, thus performing a sort of data-driven correction. However, there is no consensus about the best data-driven strategy. Even regarding ML methods, different models were employed. In [18], a multi-layer feedforward neural network (FFNN) is presented. In turn, in [20], a nested adaptive neural network (NANN) is proposed as a variation of a standard FFNN. On [19,21], both authors use a FFNN with error backpropagation, but in [19], the algorithm is modified and based only on the error of the strain energy and the external work without the need for stress data. The major problem of data-driven and ML models is that their accuracy depends on the data used for training, including its quality and quantity. These models require massive amounts of necessarily dissimilar data [22]. However, most data obtained from mechanical tests, even heterogeneous tests, are similar in strain states [23]. Therefore, new mechanical tests are needed to extract the large quantity/quality of material behaviour data needed to accurately train these models.

The objective of this paper is to present, analyse and discuss the potential of ML models, particularly of ANNs, in material constitutive modelling for mechanical numerical simulations. The contribution of ANNs in material constitutive behaviour discussed in this paper includes the following approaches: (i) parameter identification inverse methodology, (ii) constitutive model corrector, (iii) data-driven constitutive model using empirical known concepts and (iv) general implicit constitutive model using data-driven learning approach. In this paper, special emphasis is given to implicit data-driven material modelling, and a dedicated state-of-the-art review is included. Nevertheless, examples of all previous approaches are given in the framework of non-linear elastoplasticity and full-field analysis.

The paper is structured as follows: in Section 2, the fundamentals of the conventional approach to material modelling are briefly explained and the mathematical formulations applied to elasticity and plasticity are introduced. In Section 3, the application potential of ML/ANNs to different sub-fields of material constitutive modelling is discussed, while also introducing the concept of implicit material modelling. In Section 4, a literature review on the latter topic is presented, as well as a discussion concerning its implementation into FEA codes; finally, in Section 5, concrete application examples of the different approaches discussed in Section 3 are presented.

## 2. Classical Material Modelling

### 2.1. Phenomenological Approach

Material modelling refers to the development of constitutive laws representing the material behavior at the continuum level. Conventionally, constitutive laws are developed based on simplified assumptions and empirical expressions. Such expressions rely on empirical parameters, computed or calibrated via experimental methods. The calibration process ensures the compatibility between the observed and numerically simulated mechanical responses [5,24]. As models become more advanced in their physical basis, empirical expressions get more complex, often requiring extensive experimental campaigns [2].

The highly complex nature of material behavior and the research of new materials in recent years has led to the development of a great number of constitutive models targeting different phenomena. In the framework of plasticity, one can highlight the classic models by Tresca, von Mises [25] or Hill [26]. Such models are formulated as flow rules based on the yield function of the material.

### 2.2. Elastoplasticity

A material under load suffers plastic deformation starting at the uniaxial yield stress $\sigma_{\mathrm{y}}$. Beyond this point, hardening phenomena take place. Reversing the load at a given strain $\varepsilon$, plastic deformation ceases and stress starts to linearly decrease with strain, with a gradient equal to the Young's modulus, $E$ [27]. Once the stress reaches zero, the strain remaining in the material is the plastic strain $\varepsilon^{\mathrm{p}}$, whereas the recovered strain $\varepsilon^{\mathrm{e}}$ is the elastic strain. The following relationships are easily derived [25]:

$$\varepsilon = \varepsilon^{\mathrm{e}} + \varepsilon^{\mathrm{p}} \,, \tag{1}$$

$$\sigma = E\varepsilon^{\mathrm{e}} = E(\varepsilon - \varepsilon^{\mathrm{p}}) \,. \tag{2}$$

There are plenty of developed yield criteria, with the most commonly used in engineering practice being that of von Mises. The criterion relies on the knowledge of an equivalent stress $\bar{\sigma}$, the von Mises stress, defined in tensor notation as [27]:

$$\bar{\sigma} = \left(\frac{3}{2}\boldsymbol{\sigma}' : \boldsymbol{\sigma}'\right)^{1/2} , \tag{3}$$

where $\boldsymbol{\sigma}'$ is the deviatoric stress tensor. Similarly, an effective plastic strain rate $\dot{p}$ can be established, such that:

$$\dot{p} = \left(\frac{2}{3}\dot{\boldsymbol{\varepsilon}}^{\mathrm{p}} : \dot{\boldsymbol{\varepsilon}}^{\mathrm{p}}\right)^{1/2} . \tag{4}$$

The yield function, thus, assumes the following representation [27]:

$$f(\boldsymbol{\sigma}, p) = \bar{\sigma} - \sigma_{\mathrm{y}} = \bar{\sigma}(\boldsymbol{\sigma}) - \sigma_{\mathrm{y}}(p) \,, \tag{5}$$

written here as a function of the stress $\boldsymbol{\sigma}$ and accumulated plastic strain $p$, such that:

$$p = \int \mathrm{d}p = \int \dot{p}\,\mathrm{d}t \,. \tag{6}$$

The performance of the material thus depends on the previous states of stress and strain, that is, plasticity is path-dependent [18]. In this case, the plastic range can be characterized by two internal variables: the back stress tensor $\chi$, representing kinematic hardening and the drag stress $R$ representing isotropic hardening. The yield function is, therefore, modified as [27]:

$$f = J(\sigma - \chi) - R - k \leq 0\,, \tag{7}$$

where $k$ is a material constant and $J$ is a distance in the stress space. For large deformations, the elastoplaticity can be formulated based on the co-rotational configuration. An in-depth overview of this can be seen in [28].

## 3. Machine Learning Approaches for Constitutive Modelling in Metal Forming

Material constitutive models are mathematical formulations representing the main characteristics, behaviours and functions of the materials. Generally, continuum mechanical models, such as the ones presented in Section 2, are written in a differential form and require a calibration procedure. However, even with a very robust calibration, known in the field as parameter identification, the accuracy of the model is always constrained by its mathematical formulation. Additionally, the development of analytical constitutive models requires a large effort in both time and resources. Today's state-of-the-art constitutive models required decades to develop and validate. However, there is still a margin for multiple improvement opportunities, particularly using ML techniques, such as the one presented in Appendix A. The most straightforward fields of application are the following:

### *3.1. Parameter Identification Inverse Modelling*

One category of inverse problems in metal forming is called parameter identification. The aim is to estimate material parameters for constitutive models, i.e., estimate parameters to calibrate the models. The development of new materials and the effort to characterize the existent ones led to the formulation of new complex constitutive laws. Nevertheless, many of these new laws demand the identification of a large number of parameters adjusted to the material whose behaviour is to be simulated. In cases where the number of parameters is high, it might be necessary to solve the problem as an inverse non-linear optimization problem.

The parameters' determination should always be performed confronting mathematical/numerical and experimental results. Therefore, the comparison between the mathematical model and the experimental results (observable data) is a function that must be evaluated and minimized. This is generally the objective function of these problems [29–31].

When using the finite element method (FEM) to reproduce the material behaviour, the most commonly used method is the finite element model updating (FEMU) technique, which consists of creating a FEM model of a mechanical test, collecting displacement or strain components at some points/nodes and calculating a cost function with the difference between numerical and experimental displacements at these nodes. Minimizing this cost function concerning the unknown constitutive parameters provides the solution to the problem. This method is very general and flexible and can be used with local, average or full-field measurements. Other methods have been also proposed, such as the constitutive equation gap method (CEGM), the equilibrium gap method (EGM), the reciprocity gap method (RGM) and the virtual fields method (VFM) [32]. A review of these methods can be found in [2].

Considering the direct problem, which is to obtain the strain, displacement and stresses given the input variables (such as boundary conditions, geometry, and material parameters), an inverse problem can be devised in a given way to obtain the material parameters and stresses, when the inputs are the geometry, boundary conditions, displacements, and strains. In this context, ML can be used to solve the inverse problem efficiently, modelling the inverse process using the direct problem results as training data. The solution for the direct problem can be easily obtained using FEA software. This approach is illustrated in Figure 1.

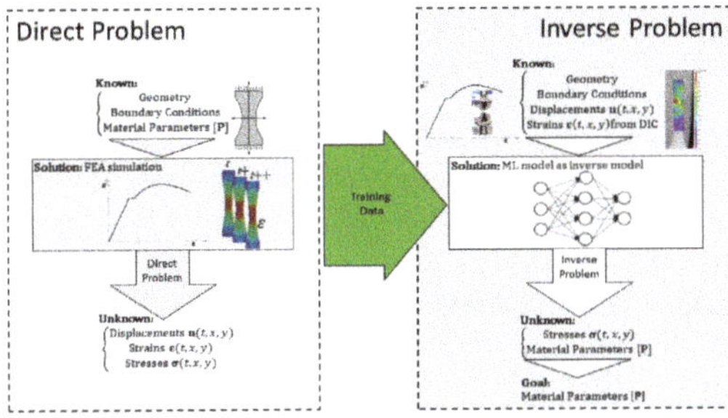


**Figure 1.** ML as an inverse model for parameter identification. The training data is obtained using the direct problem FEA results.

## 3.2. *Constitutive Model Corrector*

In the last decades, a large number of analytical constitutive models were developed. Although a large effort has been put in these successive developments, a model is always a simplification of reality and is always constrained by a mathematical formulation. Nevertheless, the effort and the knowledge gained in these last decades should not be neglected for the development of new constitutive models. Therefore, one approach to develop new constitutive models using ML techniques can take advantage of this knowledge, as a starting point, and can be used as what can be called a model corrector or model enhancement.

The approach of the model corrector starts after the calibration of the well-known analytical model. It consists of confronting the real observations, generally from experimental tests, with the numerical simulations. After the parameter identification process, the differences between the real and numerical values can be collected and evaluated in a way similar to a data decomposition process (see Figure 2). These differences do not come only from the errors of the measurement techniques but also from the limitations (non-mathematical flexibility) of the analytical model. Then, the ML model is used to model the differences as a different (refined) scale model.

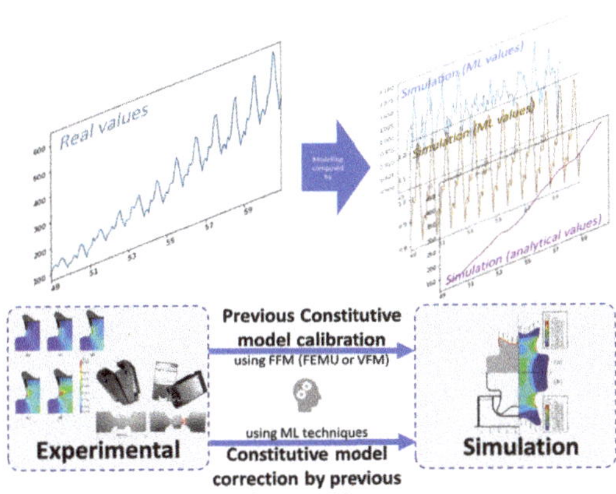


**Figure 2.** ML as a constitutive model corrector/improvement. The real behaviour (measured values) is decomposed into the analytical model (retrieved after a full-field calibration process) and the ML model. The training of the ML model uses the data obtained after the decomposition.

For accurate simulations, both analytical and ML corrector models must be introduced in the FEA software, resulting in a hybrid constitutive model.

### 3.3. Data-Driven Constitutive Model Using Empirical Known Concepts

An upper level of the previous approaches is using ML exclusively to model the behaviour of the material. However, the architecture and the development of the ML model considers some physical and empirical knowledge gained in the last decades in the development of analytical models, which are numerically implemented a posteriori. An example of this influence is the selection of the input and output features of the ML model. Generally, these ML material models that use empirical knowledge are developed in a differential point of view, such as the known analytical models, and all output features are time-derivatives of the input features. Additionally, all features, including internal variables, have specific physical or empirical representations or concepts, such as hardening, grain size, recovery, etc.

These ML material models can easily fully replace the existent constitutive models implemented in FEA software because both share the same type of structure and input/output features (see Figure 3). Although the majority of the implementations are done explicitly, these models can theoretically be implemented in implicit FEA codes, such as ABAQUS [22]. To train these models, a non-straightforward analysis and decomposition of the experimental data is required for each specific feature of the model. This procedure takes into account the meaning of each feature (i.e., of each internal variable).

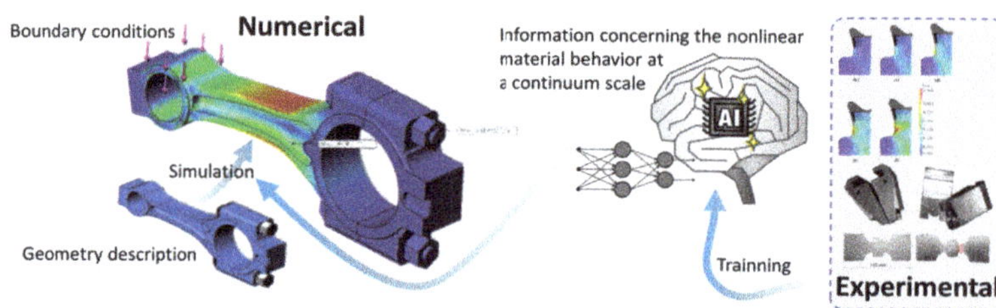


**Figure 3.** ML model as a full constitutive model, which is implemented into a FEA simulation software. The ML material model is trained indirectly using full-field measurements from experimental tests. This approach can optionally use empirical known concepts.

### 3.4. General Implicit Constitutive Model Using Data-Driven Learning Approach

The use of ML technology to fully model the behaviour of the material without any preconception, knowledge, analytical or mathematical formulation is the ultimate goal of this scientific thematic. Although not mature, this approach has a large potential because (i) the material behavior is caused by a multitude of phenomena too complex to be handled individually, as it is traditionally done in the development of analytical models; (ii) the amount of data now retrieved using full-field methods and heterogeneous tests is difficult to be analysed in the same way as the data previously acquired from classical homogeneous mechanical tests and (iii) no previous concepts or knowledge are required, and preconceptions can mean constraints for disruptive developments.

An illustration of the development of an ML constitutive model using full-field experimental data and its integration into FEA simulation software is depicted in Figure 3. The most difficult step in the development of this approach is training the model, which, in this case, does not use any physical knowledge as a training constraint. Additionally, output features of the model, such as components of the stress tensor, cannot be measured in real experimental mechanical tests.

## 4. Review on Implicit Data-Driven Material Modelling

The concept of implicit material modelling relies on the material behavior being captured within the connections of an ANN trained using experimental or even synthetic data. That is, the relationships between stress and strain are learned purely based on data without resorting to a mathematical model or any kind of assumptions. The neural network should not only be able to reproduce the data it was trained on, but also possess the ability to approximate the results of other experiments. In this regard, while explicit material models may often not be able to reproduce new sets of experiments [8], ANNs are highly flexible and adaptable in the sense that these can be improved by being retrained with new data [33,34].

The possibility of extending neural network concepts to constitutive modelling was first explored by Ghaboussi, Garret and Wu [35]. The authors trained feedforward neural networks with backpropogation using experimental data and tested them against several well-known analytical models describing the behavior of concrete in plane stress subjected to two load settings: monotonic biaxial loading and uniaxial compressive cyclic loading. For the first setting, both stress and strain-controlled models were developed in order to tackle the path dependency of the material behaviour. The models were trained to predict strain and stress increments, respectively, using the current states of stress and strain as inputs, with the addition of a stress or strain increment. For the cyclic load setting, the model was trained to predict the strain increment based on the current state and the previous two states of stress and strain. The authors reported promising results for both load settings with the ANN-based models showing, in some cases, better performance than some of the analytical formulations. With their research, Ghaboussi and coworkers effectively proved the powerful interpolation and generalization capabilities of ANNs applied to constitutive modelling tasks, thus laying the foundations for the further development of implicit material models. Their methodology was later applied to the constitutive modelling of concrete (Wu and Ghaboussi [36]), sand (Ghaboussi et al. [37] and Ellis et al. [38]) and composites (Ghaboussi et al. [39]).

An early application of the implicit modelling approach in the framework of metal plasticity is due to Furukawa and Yagawa [18]. The authors introduced the first mathematical description of the concept relying on a state-space representation and proposed an implicit viscoplastic model using neural networks based on the well-known Chaboche's model. The model was trained using synthetic data and was tested against experimental data. Due to the fact that the internal variables of the Chaboche's model are not experimentally obtainable, a simple technique was proposed in order to derive the states of kinematic and isotropic stresses from the experimental curves. Overall, the authors reported that the implicit model correlated well with the results provided by the conventional formulations when applied to testing data within the training range, certifying the great interpolative capacity of the neural network. However, considerable errors were observed in stress and stain curves when using testing data far off the limits of the training set, indicating the limited generalization capacity of the model.

Following a purely data-driven approach, several authors successfully applied very similar methodologies to train ANN-based constitutive models using experimental data to predict the flow behavior of different metals under hot deformation (i.e., compression, torsion). Many industrial and engineering applications require the hot deformation of materials which are associated with numerous metallurgic phenomena, affecting the structure of the material at the micro-scale. As such, several factors affecting the flow stress generate highly complicated and non-linear relationships, limiting the application range and accuracy of the conventional constitutive models. Li et al. [40] studied the high temperature flow characteristics of Ti–15–3 alloy for the determination of hot-forging processing parameters and used an ANN model to predict the flow stress, finding that the ANN was able to correctly reproduce the behavior in sampled and non-sampled data. Lin et al. [12] used an ANN to predict the flow stress of 42CrMo steel under hot compressive deformation tests, reporting that the predictions were in good agreement with the experimental results. Mandal et al. [41] stated that the ANN-based model was an efficient tool to evaluate and predict the deformation behavior of stainless steel type AISI 304L. Sun et al. [42] developed a constitutive relationship model for Ti40 alloy and indicated the predicted flow stress by the ANN-based model correlated well with experimental results. Li et al. [43] reported that an ANN-based model showed better performance and proved to be more accurate and effective in predicting the hot deformation behavior of modified 2.25Cr–1Mo steel than the conventional constitutive equations.

### 4.1. Neural Network Architectures and Generalization Capacity

The implicit approach to material modelling based on ANNs presents obvious advantages; however, important choices regarding network architecture and the training process have to be taken into account in order to avoid limited predictive capabilities. For example, non-linear material behavior is highly path-dependent, making it difficult for ANNs to learn and generalize well the full spectrum of material behavior. The issue is accounted by Ghaboussi and coworkers [35] in their original work. Aiming to improve the generalization capacity of the ANN-based material models, Ghaboussi and Sidarta [20] later proposed a new neural network architecture: nested adaptive neural networks (NANNs). Their proposal is based on the fact that most data have an inherent structure which, in the case of material behavior, is a nested structure that can be observable in terms of:

- *Dimensionality*—where constitutive funtions are taken as members of function spaces, each of which is contained in a space of subsequently higher dimensions;
- *Path-dependency*—where a state of the material behavior represented with a given number of history points is a subset of another state represented with a larger number of history points.

In line with this notion, the underlying concept of NANNs is to train a base module to represent the material behavior in the lowest function space and, subsequently, augment it with additional modules representing higher function spaces (Figure 4). All the nodes in each layer of the new added module are connected to all the nodes in the next layer of the lower levels. Each module is a standard FFNN trained using adaptive evolution; that is, neurons are subsequently added to the hidden layers as network approaches its capacity. With new nodes, new connections are generated, and the objective is for the new links to capture the knowledge that the old ones were not able to capture. As such, training is carried out for the new connections while the old ones are frozen. With this new technique, Ghaboussi and Sidarta [20] reported a significant error reduction when using high-level NANNs.

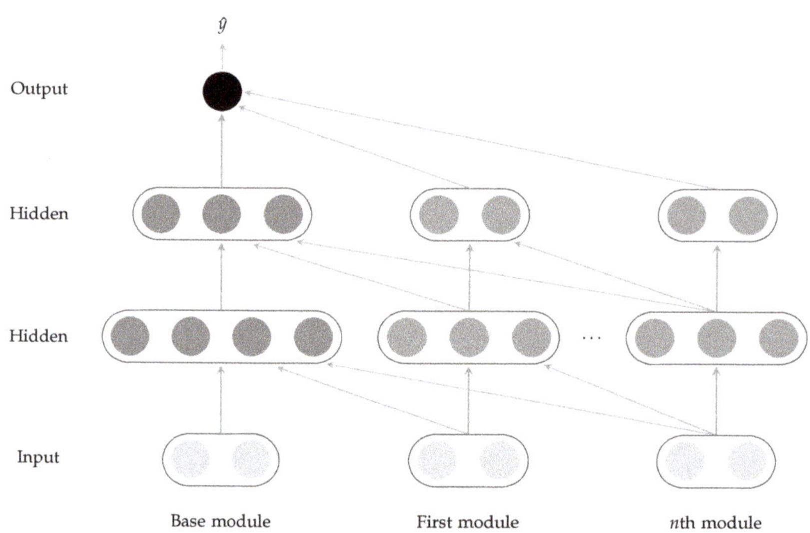


**Figure 4.** Nested adaptive neural network, as idealized by Ghaboussi and Sidarta [20].

In more recent works, regarding the scope of the path-dependency of material behavior, several authors (Abueidda et al. [44], Heider et al. [45], Ghavamian and Simone [46], Mozaffar et al. [47] and Gorji et al. [48]) demonstrated that RNNs can be particularly useful for modelling path-dependent plasticity, reporting superior performance compared to conventional ANNs. Contrarily to the latter, RNNs are designed to handle time sequences [47,49] and are equipped with history-dependent hidden states, enabling them to carry information from previous inputs onto future predictions. These internal variables have the potential to emulate the role of objects such as plastic strains and back stresses in physics-based plasticity. Nevertheless, standard RNNs suffer from vanishing/exploding gradients that hinder the error backpropagation process while dealing with long sequences. Long short-term memory (LSTM) and GRUs are more robust derivations of the standard RNNs (Figure 5) proposed to avoid those problems.

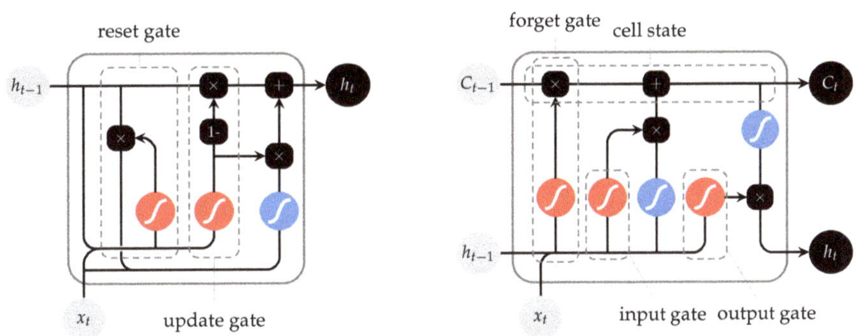


(**a**) Gated recurrent unit. (**b**) Long short-term memory unit.

**Figure 5.** Basic architectures of the most common topologies of RNNs: (**a**) the GRU and (**b**) the LSTM unit. The arrow lines indicate the flow of information through each unit; vector concatenation takes place in zones where two lines adjoin. The sum and multiplication operators refer to pointwise operations. Red and blue blocks represent sigmoid and hyperbolic tangent signal activations, respectively.

Apart from generalization issues, a much more fundamental problem is emphasized by several authors [20,35,50,51]: the "black-box" nature of ANNs does not guarantee the implicit material model to obey the conservation laws, symmetry and invariance requirements of conventional models. Nonetheless, if a neural network has been trained with a comprehensive dataset, it is reasonable to assume that it would be able to approximate the laws that the actual material obeys and use its generalization capability to appropriately predict stress-strain paths not included in the training data. However, building comprehensive datasets means acquiring large amounts of data. The cost of acquition is often prohibitive, and models end up being built in a small data regime.

To tackle these issues, Raissi et al. [51] proposed a new type of neural network: physics-informed neural networks (PINNs). The authors exploit the use of automatic differentiation to differentiate neural networks with respect to their input coordinates (i.e., space and time) and model parameters to obtain PINNs. The technique allows to constrain neural network models to respect symmetries, invariances, or conservation principles originating from the physical laws that govern the observed data. This type of information, encoded in such a way into the neural network, acts as a regularization agent that greatly reduces the space of admissible solutions. The methodology was proven by the authors as able to tackle a wide range or problems in computational science, allowing the use of relatively simple FFNN architecture trained with small amounts of data. The general methodology proposed by Raissi and coworkers [51] was applied to constitutive modeling in a recent work by Masi et al. [50] through the proposal of thermodynamics-based artificial neural networks (TANNs), a subtype of PINNs. In TANNs, basic laws of thermodynamics are encoded directly in the architecture of the neural network. In the framework of material modelling, these thermodynamic constraints concern the stresses and the internal state variables and their relationship with the free-energy and the dissipation rate. Masi and coworkers [50] report that TANN-based models were able to deliver thermodynamically consistent predictions both for seen and unseen data, achieving more accurate results than standard ANN material models. Moreover, the authors state that the inclusion of the laws of physics directly into the network architecture allows for the use of smaller training datasets given that the network already does not have to capture the underlying relationships from data.

Xu et al. [52], on the other hand, note that even though these physical constraints allow for improved accuracy in stress predictions, ANN-based constitutive models may still not yield a symmetric positive definite stiffness matrix. Therefore, nonphysical solutions and numerical instabilities arise when coupling ANN-based models with FEM solvers. The authors propose symmetric positive definite neural networks (SPD-NN) as a new architecture in order to deal with these issues. The neural network model is trained to predict the Cholesky factors of the tangent stiffness matrix, from which the stress is obtained. According to Xu et al. [52], the approach weakly imposes convexity on the strain energy function and satisfies the Hill's criterion and time consistency for path-dependent materials. The authors compared the SPD-NN with standard architectures in a series of problems involving elasto-plastic, hyper-elastic and multi-scale composite materials. The results showed that, overall, the SPD-NN was able to provide superior results, proving the effectiveness of the approach for learning path-dependent, path-independent and multi-scale material behaviour.

### *4.2. Integration in Finite Element Analysis*

An ANN-based constitutive model can be used in a way similar to conventional material models in FEA to solve boundary value problems. An important disadvantage regarding the numerical implementation of ANN-based material models in FEA was pointed out by Ghaboussi and Sidarta [20]. The issue is related to the fact that the ANN model already provides an updated state of stress, without the need for integration of the model's equations over the strain increment. Therefore, it is not possible to obtain an explicit form of the material's stiffness matrix. As a solution, the authors propose an explicit formulation of the material's stiffness matrix for the ANN-based material model, allowing for an efficient convergence of the finite element Newton–Raphson iterations.

Successful embedding of ANNs as material descriptions in finite element (FE) codes were recorded by several authors. Usually, the ANN architecture is implemented in the software's subroutines as a set of matrices containing the optimized weights obtained during training. Substantial gains in computational efficiency using ANNs have been reported due to the practically instantaneous mapping offered by the ANNs. Lefik and Schrefler [21] trained a simple ANN as an incremental non-linear constitutive model for a FE code and demonstrated that a well-trained, albeit simple, network model was able to reproduce the behavior from simulated material hyteresis loops.

Kessler et al. [53] developed an ANN-based constitutive model for aluminum 6061 based on experimental data for different temperatures, strains and strain rates. The authors implemented the model in Abaqus VUMAT and reported the ANN's superior ability to mirror the experimental results.

Jang et al. [10] proposed a constitutive model to predict elastoplastic behavior for J2-plasticity, where an ANN was implemented in Abaqus User MATerial (UMAT) to replace the conventional nonlinear stress-integration scheme under isotropic hardening. The ANN was used only for nonlinear plastic loading, while keeping linear elastic loading and unloading relegated to the physics-based model. The authors reported that the ANN-based model provided results in good agreement with those from the conventional model for single-element tensile test and circular cup drawing simulations.

Zhang and Mohr [54] rewrote a standard algorithm for von Mises plasticity such that the stress update was governed by a single non-linear function predicting the apparent modulus (i.e., the Young's modulus in case of elastic loading/unloading, and the elasto-plastic tangent matrix in case of plastic loading). This function was then replaced by an FFNN and implemented as a user material subroutine in Abaqus/explicit. The authors demonstrated that a neural network with 5 hidden layers and 15 neurons per layer was able to "learn" the yield condition, flow rule and hardening law from data, providing highly accurate responses to the predictions of a J2 plasticity model, including the large deformation response for arbitrary multi-axial loading paths and reverse loading.

### 4.3. Indirect/Inverse Training

The vast majority of the approaches documented in the literature for implicit constitutive modelling consists of feeding the ANN with paired data (usually, stress and strain) during the training process in order to assimilate the material behavior [55]. The process requires copious amounts of data, and obtaining comprehensive stress-strain relationships while relying on the standard simple mechanical tests poses a great challenge, especially when dealing with anisotropic materials. Moreover, in a complex experiment, certain variables such as stresses cannot be directly measured [56]. Therefore, the training process must be carried out in such a way that allows the ANN-based model to indirectly predict the stress using measurable data, e.g., displacements and global force, obtained from full-field measurements [56]. One of the first approaches to bypass these limitations was proposed by Ghaboussi et al. [39], consisting of an auto-progressive method to train an ANN to learn the behavior of concrete from experimental load–deflection curves. However, the method is inefficient due to the fact that FEM simulations are needed in order to compute the stress and strain.

The generalized SPD-NN architecture proposed by Xu et al. [52] was able to deal with both direct and indirect training procedures. For the latter, the authors used displacement and force data to indirectly train the SPD-NN-based model and predict the stress in an incremental form by coupling the ANN with a non-linear equation solver. Despite providing superior results compared to standard architectures, the authors point out the requirement of full-field data as the main limitation, especially for three-dimensional solid bodies.

Both Huang et al. [56] and Xu et al. [57] proposed a method to use an ANN to replace the constitutive model in a FEM solver in order to impose the physical constraints during training, enabling the ANN to learn material behaviour based on the experimental measurement. Nonetheless, the approach highly depends on a full-field data for force and displacements, while, in most experiments, only partially observed data are usually available, limiting its application to more complex three-dimensional bodies. A different method was reported by Liu et al. [55] who propose coupling an ANN with the FEM to form a coupled mechanical system, with the inputs and outputs being determined at each incremental step (Figure 6). With this, the ANN is able to learn the constitutive behavior based on partial force and displacement data. The authors applied the methodology to learn nonlinear in-plane shear and failure initiation in composite materials, achieving good models with results in excellent agreement with the analytical solutions. The main drawback is the need to write a FEM code based on an automatic differentiation package (e.g., Pytorch or TensorFlow), making the FEA step very time-consuming. In order to overcome this limitation, Tao et al. [58] propose a similar approach to the one proposed by Liu, with the ANN model being coupled with FEA software Abaqus. However, the authors propose a modified set of backward propagation equations that enable the communication between Abaqus and the ANN, keeping the whole training process internally in Abaqus.

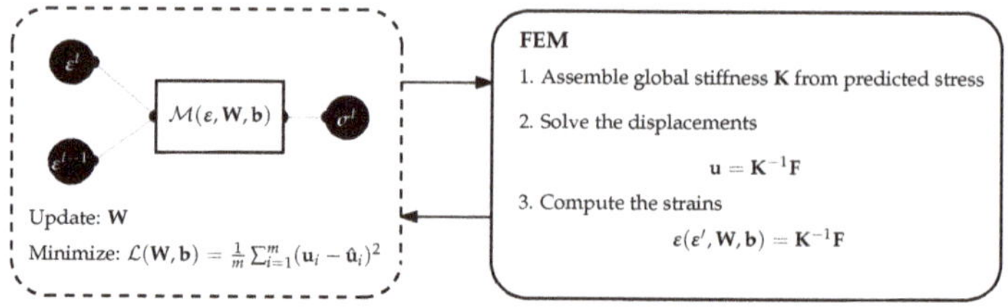


**Figure 6.** Coupled ANN-FEM model approach presented by Liu et al. [59].

A similar approach to Liu et al. [59] is here proposed; however, instead of coupling the ANN with a FEM solver to enforce the physical constraints, the virtual field method (VFM) is used to guarantee the global equilibrium (Figure 7). The VFM, first introduced by Grédiac [32], is known by its computational efficiency and does not require FEA in order to conduct any forward calculations [60]. The key elements behind the VFM are the principle of virtual work (PVW) and the choice of virtual fields. According to the PVW, the internal virtual work must be equal to the external virtual work performed by the external forces and is written by [61]:

$$-\int_V \sigma : \varepsilon^* \mathrm{d}V + \int_{\partial V} \mathbf{T} \cdot \mathbf{u}^* \mathrm{d}S = 0, \tag{8}$$

where $\varepsilon^*$ is the virtual strain, $\mathbf{u}^*$ is the virtual displacement and $\mathbf{T}$ is the global traction vector. The virtual fields are mathematical test functions which work as weights and can be defined independently of the measured displacements/strains. An infinite number of virtual fields can be used; nonetheless the following two conditions should be met [4,61]: the chosen virtual fields should be kinematically admissible; that is, the displacement boundary conditions must be satisfied and the virtual fields should be zero or constant along the boundary where the force is applied. By coupling the VFM with the ANN model, one can use force and displacement data, numerically generated or coming from full-field measurements, to indirectly train the ANN. The strain components are easily obtained from the displacements and are fed as inputs to the neural network, which will provide the stress components. The virtual strains are obtained from the virtual displacements, such that:

$$\varepsilon^* = \nabla \mathbf{u}^* + \nabla^{\mathrm{T}} \mathbf{u}^* . \tag{9}$$

Then, the stress equilibrium is evaluated globally by means of the PVW, and the parameters $(\mathbf{W}, \mathbf{b})$ are optimized until the equilibrium is respected, that is, by minimizing the loss:

$$\mathcal{L} = \frac{1}{n_V} \sum_{i=1}^{n_V} \left( -\int_V \hat{\sigma} : \varepsilon^* \mathrm{d}V + \int_{\partial V} \mathbf{T} \cdot \mathbf{u}^* \mathrm{d}S \right)^2, \tag{10}$$

where $n_V$ represents the number of virtual fields. This methodology is illustrated in Section 5.3.

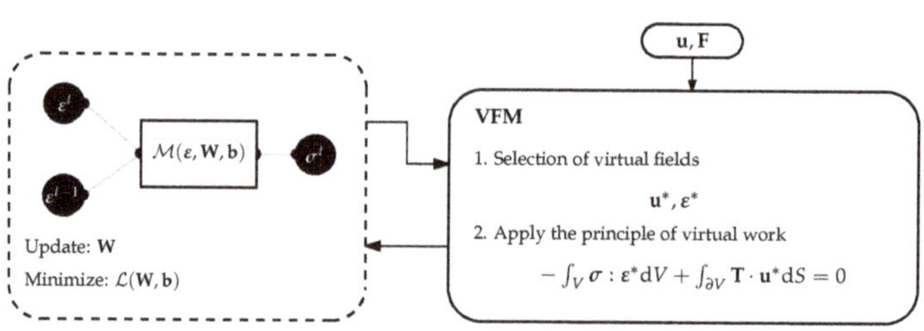


**Figure 7.** Coupled ANN-VFM method for implicit constitutive modelling.

## 5. Application Examples

This section presents an example of each approach, showing the large potential of ML in material constitutive modelling.

### 5.1. Parameter Identification

The parameter identification problem can be reduced to a curve-fitting problem if physical constraints would not be taken into account. However, most material constitutive models have physical constraints, such as material parameter boundary values and mathematical relations between them, in order to guarantee that the parameters have a physical meaning [62]. The formulation for the solution of the constitutive model parameters identification is as follows. First, a physical system with a behaviour that can be described by a numerical model and for which experimental results are available should be considered. A set of measurable variables that can be experimentally determined should also be considered. This set of variables can be defined as $\mathbf{Z}^{\mathrm{T}} = [z_1, z_2, \ldots, z_m]$. In the case where simple mechanical tests are considered, such as tensile and shear tests, these measurable variables would be the stresses or strains. Considering this information, it is then possible to formulate the solution of the identification problem as the minimization of a function that measures the difference between theoretical predictions $\mathbf{Z}^{\mathrm{num}}$ (obtained by the numerical model) and experimental data. This function is known as the objective function $\mathcal{L}(\mathbf{A})$, and can be formulated as [62]:

$$\mathcal{L}(\mathbf{A}) = \sum_{q=1}^{N} \mathcal{L}_q(\mathbf{A}), \tag{11}$$

with $\mathcal{L}_q(\mathbf{A})$ computed as:

$$\mathcal{L}_q(\mathbf{A}) = \frac{1}{(t_1 - t_0)} \int_{t_0}^{t_1} \left[\mathbf{Z}^{\mathrm{num}}(\mathbf{A}) - \mathbf{Z}^{\mathrm{exp}}\right]^{\mathrm{T}} \mathbf{D}_q \left[\mathbf{Z}^{\mathrm{num}}(\mathbf{A}) - \mathbf{Z}^{\mathrm{exp}}\right] \mathrm{d}t, \tag{12}$$

where $\mathbf{A} = [A_1, A_2, \ldots, A_r]^{\mathrm{T}}$ is the set of the $r \in \mathbb{N}$ constitutive model parameters, and $\mathbf{Z}^{\mathrm{exp}}$ is a known experimental value of $\mathbf{Z}$ in $N$ experimental tests. The tuple of time points $(t_0, t_1)$ is the time period of the generic test $q$. Additionally, $\mathbf{D}_q$ is a given weight matrix associated to the test $q$. The optimization problem consists on finding the minima of $\mathcal{L}(\mathbf{A})$, i.e.,

$$\begin{array}{llll} \underset{\mathbf{A}}{\text{minimize}} & \mathcal{L}(\mathbf{A}) & & \\ \text{subject to:} & g_m(\mathbf{A}) & \leq 0, & m = 1, \ldots, M \\ & h_l(\mathbf{A}) & = 0, & l = 1, \ldots, L \\ & A_i^{\min} & < A_i \; < A_i^{\max}, & i = 1, \ldots, r\,, \end{array} \tag{13}$$

where the $M$ inequalities $g_m(\mathbf{A})$ and the $L$ equalities $h_l(\mathbf{A})$ define the model constraints. The search-space is limited and, as a consequence, the optimization parameters should fit in it.

The previous classical methodology is indeed computationally expensive because an optimization problem must be solved. Therefore, ML techniques directly applied to solve the inverse problem without the need for iterative procedures can be beneficial. Additionally, the ML methodology is easier to implement and faster at obtaining results in comparison with the current state-of-the-art procedures, such as the VFM and FEMU methods. Considering that the parameters to be identified will be used in numerical FEA, the training phase can directly and accurately use a numerical FEA database. Even if the creation of this database is computationally time-consuming, it is less costly than an experimental database and it is done only once for each mechanical test. Then, the parameters are identified (i.e., predicted within the ML model) instantaneously and using one experimental test.

The example hereby presented is based on the work done in [63] and uses the biaxial cruciform test in order to generate data to train and identify the parameters for both the Swift's hardening law (parameters $\sigma_0$, $k$ and $n$) and the Hill48 anisotropic criterion (parameters $r_0$, $r_{45}$ and $r_{90}$). A Young's modulus $E = 210$ GPa and a Poisson's ratio $\nu = 0.3$ are used to characterize the elastic response of the material. As in [63], synthetic images are used for comparison and validation purposes. The reference parameters for the plastic anisotropy coefficients $r_\alpha$ at 0°, 45°and 90° from the rolling direction are listed in Table 1.

**Table 1.** Plastic anisotropy coefficients $r_\alpha$ used for obtaining the reference cruciform biaxial test [63].

| Hill'48 Parameters | Reference Values |
|---|---|
| $r_0$ | 1.680 |
| $r_{45}$ | 1.890 |
| $r_{90}$ | 2.253 |

5.1.1. Training the ML Using a Biaxial Cruciform Test

The full geometry of the cruciform test specimen is depicted in Figure 8a; however, only one-fourth is used. The solid geometry is discretized as shown in Figure 8b using a mesh composed by a total of 114 combined CPS4 and CPS3R elements. Symmetry boundary conditions are applied at $x = 0$ and $y = 0$ boundaries. A displacement condition $u_x = u_y = 2$ mm is applied along the $x$ and $y$ directions.

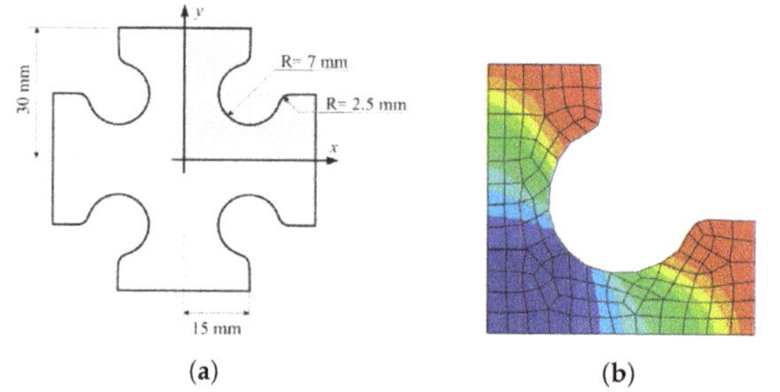


**Figure 8.** (**a**) Cruciform geometry [63] and (**b**) mesh for the parameter identification problem.

The range of the six constitutive parameters used to generate the database is shown in Table 2. Due to the large number of possible combinations, the Latin hypercube sampling method was used, resulting in a total of 6000 samples.

**Table 2.** Input space of constitutive parameters for the biaxial test.

| Constitutive Parameters | Input Space |
|---|---|
| $\sigma_0$ [MPa] | 120–200 |
| $n$ | 0.175–0.275 |
| $K$ [MPa] | 480–570 |
| $r_0$ | 1.0–2.5 |
| $r_{45}$ | 1.0–2.5 |
| $r_{90}$ | 1.0–2.5 |

The optimal ANN hyperparameters for the inverse model of the cruciform test were searched. The ideal architecture is listed in Table 3.

The input layer has 7203 input features, since 21 times steps were considered for each test sample and, for each time step, the global force and the strain tensor in all elements are used. Therefore, each sample has the following structure:

$$\left(\text{Force}, (\varepsilon_{xx}, \varepsilon_{yy}, \varepsilon_{xy})_{i=1,\ldots,\text{elements}}\right)_{j=1,\ldots,\text{times steps}}. \tag{14}$$

The suggested learning rate $\alpha = 1.512 \times 10^{-5}$ was also found using the *Hyperband* approach.

**Table 3.** ANN architecture for the biaxial test.

| Layer | Neurons | Activation | Kernel_Initializer |
|---|---|---|---|
| Input | 7203 | - | - |
| Dense 1 | 7000 | tanh | Orthogonal |
| Dense 2 | 3000 | tanh | Truncated normal |
| Dense 3 | 5000 | tanh | Truncated normal |
| Dense 4 | 5000 | selu | Variance scaling |
| Dense 5 | 6000 | selu | Variance scaling |
| Output | 6 | - | - |

In Figure 9, the learning curves are plotted, registering errors lower than 0.02 during training. The best result was achieved after 275 epochs with a MSE of 0.00083 and a MAE of 0.0072.

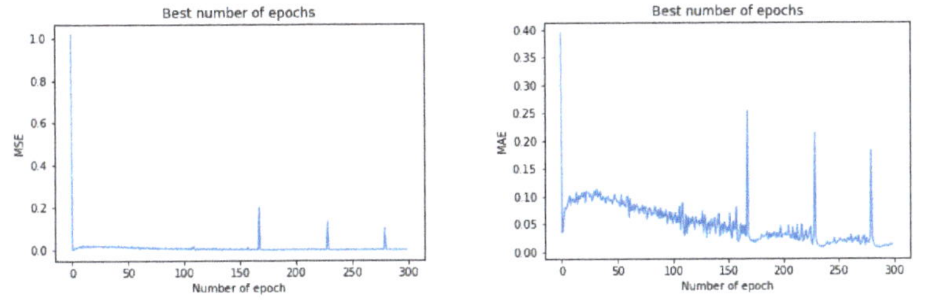


(**a**) (**b**)

**Figure 9.** (**a**) MSE and (**b**) MAE evolution during the training process.

Figure 10 presents the validation curves for the inverse cruciform test, where MSE curves show an exponential decay in the beginning and a final plateau with a very low error, while the MAE slowly decays with large oscillations.

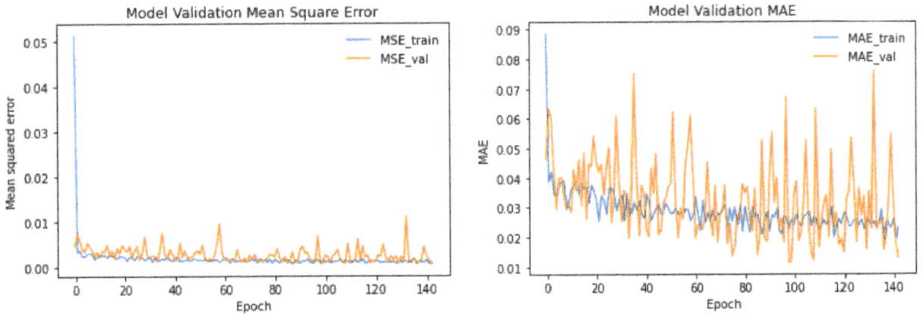


(**a**) (**b**)

**Figure 10.** Validation curves for cruciform inverse model: (**a**) MSE and (**b**) MAE.

Figure 11 shows the influence of the training set size in the MSE. Considering that a plateau was not achieved, it can be concluded that further information should be given to the model to avoid underfitting. The lack of convergence can also be due to unrepresentative data (lack of quality).

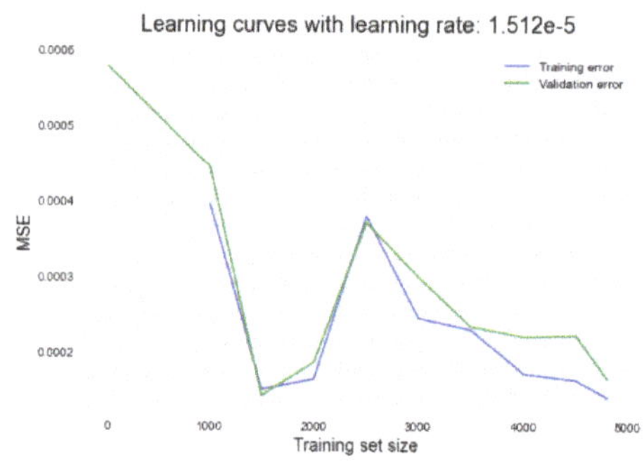


**Figure 11.** Influence of the training set size in the training and validation phases for the inverse biaxial test model.

5.1.2. Prediction and Comparison

In order to evaluate the performance of the ANN inverse model, three reference sets of material parameters were chosen (see reference values in Table 4) to create synthetic experimental data and used as test cases. Test case 2 uses the same parameters as in [63]. The predictions and corresponding errors are also depicted in Table 4. The smallest error was in test case 2, while the largest error was for test case 3. This accuracy difference between test case 2 and 3 is expected, with the largest error seen for test case 3, where the reference parameters are outside the range of the training space. Furthermore, the error in test cases 1 and 2 are very low, which contrasts with the discussion in the training and validation phase.

**Table 4.** Predictions and corresponding errors for inverse cruciform test model.

| | | $\sigma_0$ [MPa] | $n$ | $K$ [MPa] | $r_0$ | $r_{45}$ | $r_{90}$ |
|---|---|---|---|---|---|---|---|
| **Test 1** | Reference | 120.00 | 0.18 | 530.00 | 1.01 | 1.24 | 1.73 |
| | Predicted | 120.98 | 0.18 | 536.69 | 1.06 | 1.25 | 1.71 |
| | Error [%] | 0.82 | 1.67 | 1.26 | 4.95 | 0.81 | 1.16 |
| **Test 2** | Reference | 160.00 | 0.26 | 565.00 | 1.68 | 1.89 | 2.253 |
| | Predicted | 160.34 | 0.26 | 565.60 | 1.69 | 1.88 | 2.27 |
| | Error [%] | 0.21 | 1.15 | 0.11 | 0.60 | 0.53 | 0.75 |
| **Test 3** | Reference | 220.00 | 0.30 | 620.00 | 2.52 | 2.67 | 2.95 |
| | Predicted | 202.61 | 0.24 | 558.54 | 2.36 | 2.30 | 2.74 |
| | Error [%] | 7.90 | 21.67 | 9.91 | 6.35 | 13.86 | 7.12 |

The force-displacement and strain-stress curves resulting from the obtained parameters can be seen in Figure 12a,b, where a quite perfect fit can be observed.

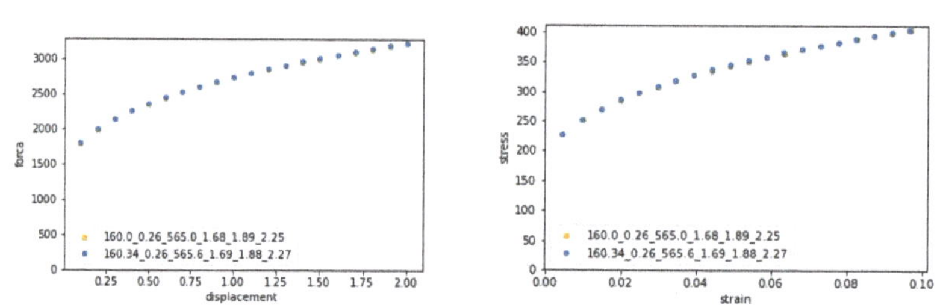


(a) (b)

**Figure 12.** (**a**) Force-displacement and (**b**) stress-strain curves obtained with the predicted parameters for the cruciform test case 2.

The results obtained using an ANN approach were compared with the results using other approaches in [63]. The comparison, presented in Table 5, show a minimal difference between the errors of the different approaches, meaning that the ANN is effective and can predict the parameters with the same precision as the other more established approaches. However, it should be noted that, although the identification took less than a second, the ANN approach required around 96 h to complete the 6000 sample database in a CPU Intel i7-8700 computer. The VFM solution took less than 5 min for the required 15 optimization iterations [63].

**Table 5.** Comparison of the ANN approach with other approaches presented in [63] for the parameter identification using the biaxial cruciform test for the test case 2.

| | | $\sigma_0$ [MPa] | $n$ | $K$ [MPa] | $r_0$ | $r_{45}$ | $r_{90}$ |
|---|---|---|---|---|---|---|---|
| | Reference | 160.00 | 0.26 | 565.00 | 1.68 | 1.89 | 2.25 |
| **ANN approach** | Predicted | 160.34 | 0.26 | 565.60 | 1.69 | 1.88 | 2.27 |
| (96 h) | Error [%] | 0.21 | 1.15 | 0.11 | 0.60 | 0.53 | 0.75 |
| **VFM** | Predicted | 159.76 | 0.26 | 566.16 | 1.67 | 1.91 | 2.24 |
| (<5 min) | Error [%] | 0.15 | 0.44 | 0.25 | 0.16 | 1.11 | 0.26 |

Figure 13 shows the force-displacement and stress-strain curves obtained with the learned material parameters for the test cases 1 and 3. As seen for test cases, even if the errors are larger, the curves show an excellent fit. However, for the test case 3, a large error amplitudes are observed. This can be explained by the fact that the reference data was outside the training space, meaning the network needed to extrapolate the information to predict a result.

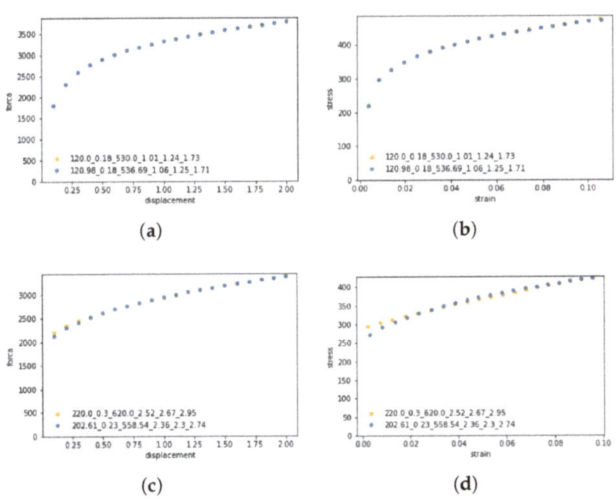


**Figure 13.** (**a**) Force displacement and (**b**) stress-strain curves for the cruciform test case 1. (**c**) Force displacement and (**d**) stress-strain curves for cruciform test case 3. The values depicted in the legends are the reference and predicted model parameters.

To test the robustness of the ANN parameter identification approach and to take into account possible errors coming from experimental measurements, the training dataset was polluted with random normal distribution noise (with a standard deviation of 1). Table 6 lists the predictions of the ANN after the training with the noise-polluted data. The predictions attained presented a larger error, demonstrating that this approach is influenced by the quality of the training data.

**Table 6.** Predictions with the ANN inverse cruciform model using the noise-polluted dataset.

| | | $\sigma_0$ [MPa] | $n$ | $K$ [MPa] | $r_0$ | $r_{45}$ | $r_{90}$ |
|---|---|---|---|---|---|---|---|
| **Test 1** | Reference | 120.00 | 0.18 | 530.00 | 1.01 | 1.24 | 1.73 |
| | Predicted | 146.82 | 0.197 | 568.02 | 1.66 | 1.72 | 1.98 |
| | Error | 22.35 | 9.44 | 7.17 | 64.36 | 38.71 | 14.45 |
| **Test 2** | Reference | 160.00 | 0.26 | 565.00 | 1.68 | 1.89 | 2.25 |
| | Predicted | 152.79 | 0.23 | 558.82 | 1.66 | 1.72 | 1.98 |
| | Error | 4.51 | 10.00 | 1.09 | 8.93 | 8.47 | 18.77 |
| **Test 3** | Reference | 220.00 | 0.30 | 620.00 | 2.52 | 2.67 | 2.95 |
| | Predicted | 166.61 | 0.23 | 554.62 | 1.77 | 1.73 | 1.93 |
| | Error | 24.27 | 24.27 | 10.55 | 29.76 | 35.21 | 34.58 |

#### 5.1.3. Remarks

This example shows the capabilities of a ML methodology applied to the calibration of numerical models of materials. Here, an easy to implement FFNN was trained as an inverse model with the aid of a large dataset generated from FEA numerical simulations. The ANN-based model was able to achieve its objectives and the results show that the network can predict material parameters, even though with some errors. It was also seen that the reliability of the ANN inverse model is highly dependent on the quantity and quality of the training data.

### *5.2. ML Constitutive Model Using Empirical Known Concepts*

The selection of the features of the ML model can be done using the knowledge already gained in the last decades. Therefore, concepts such as isotropic and kinematic hardening can be used as output/input of the ML. These concepts and their values can be retrieved in real material mechanical curves obtained with classical homogeneous tests using the result's decomposition, as done in [18], and used for training purposes. Additionally, the differential formulation of the majority of the known material model can also be maintained in the development of an ML plasticity model. In classical plasticity or viscoplasticity models with mixed isotropic-kinematic hardening, the inputs are the stress tensor $\sigma$, the plastic (or viscoplastic) strain tensor $\varepsilon^{vp}$, the backstress tensor $\chi$, the isotropic hardening $R$ and the equivalent plastic strain $p$, which is a scalar value representing the overall plasticity level. This results in 11 inputs, which can be used in a FFNN architecture with one single hidden layer. The eight outputs are the differential values of the inputs, i.e., the set $\left\{\dot{\varepsilon}^{vp}, \dot{\chi}, \dot{p}, \dot{R}\right\}$. The stress tensor can be updated by a hypoelastic relation given as

$$\sigma = C(\varepsilon - \varepsilon^{vp}). \tag{15}$$

where $C$ is the material stiffness matrix and $\varepsilon$ is the total strain tensor. Generally, the formulation in Equation (15) is used in its incremental form due to the nonlinearity of the plastic behaviour.

In this section, the material constitutive behaviour of a metal is modelled using ML/ANN techniques and additional empirical concepts to the ones previously introduced. Then, this model is used in a finite element analysis. In this example, the material ML model has an artificial neural network (ANN) architecture, and it is trained (fitted) using a virtual simulated material (as synthetic data). This virtual material has a a priori known behaviour, following the classical Chaboche elastoviscoplastic model. The advantage of knowing beforehand the behaviour is that this allows the ML capabilities and competences in material constitutive modelling to be evaluated at least to replace classical models [22].

#### 5.2.1. Creating the Dataset for Trainning

To train the ML/ANN model, a synthetic database representing the behaviour of a virtual material was artificially created. Although a large set of analytical elastoplastic constitutive models could be chosen, in this work, the elasto-viscoplastic Chaboche model was selected due to the challenge of adding rate-dependent phenomena in the behaviour of the material. Plane stress conditions were applied. The viscoplastic strain rate tensor and the plastic strain rate can be seen in Equation (A11). Additionally, the kinematic and isotropic hardening evolution are, respectively, defined by Equations (A13) and (A14). The stress-strain states of the differential equations governing the behaviour depend also on the elasto-plastic material properties (e.g., $E = 5000$ MPa, $\nu = 0.3$, $R_1 = 500$ MPa, $k = 0$, $K = 50$ MPa s$^{-1}$, $a = 7500$ MPa, $b = 0.6$, $c = 100$, $n = 3$).

The system of differential equations for two tensile tests ($T_{xx}$, $T_{yy}$) and one simple shear ($S_{xy}$) for three cyclic strain ranges of 0.05, 0.12 and 0.16 is solved using the Runge–Kutta method [64], and their results are added to the virtual material database. The 9 virtual experimental results are created and represented by Figure 14. Each test is simulated with 1000 data points, as done in [22].

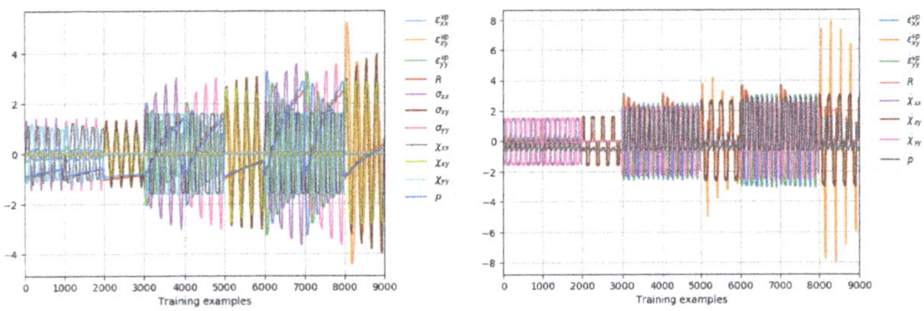


(a) (b)

**Figure 14.** Normalized/scaled and adimensional values of the (generated) dataset used for training. (**a**) the scaled input data ($\sigma$, $\chi$, $\varepsilon^{vp}$, $R$ and $p$) is transformed into the (**b**) scaled differential output data ($\dot{\chi}$, $\dot{\varepsilon}^{vp}$, $\dot{R}$ and $\dot{p}$) by the virtual material model.

5.2.2. Training the ANN Model

As supervised learning, the training finds the regression parameters to minimize the cost function. In this work, the best topology, activation functions and the optimization algorithm are found using a grid search function [65] and a 5-fold cross-validation scheme. For that purpose, (i) four topologies (four, ten, sixteen and twenty-two hidden neurons) for one hidden layer, (ii) three activation functions (logistic, tanh, ReLU), and (iii) three optimization algorithms (LBFGS, SGD, ADAM) were candidates. An automatic batch size of min (200, n° training samples) and a learning rate of 0.001 were used. For the output layer, a linear activation function is used.

Figure 15 presents the grid search results, where (i) the LBFGS optimization algorithm is responsible for the best result for all activation functions and size of the hidden layer, and (ii) the ReLU together with LBFGS seems to be the best combination for the training set.

The validation curve [66] for the ANN size can be seen in Figure 16a, where it can be seen that the optimal value ($R^2 = 0.965$ for training and $R^2 = 0.964$ for cross-validation) is obtained for 16 neurons. The influence of the training database size was also analysed [66] using the 16 hidden node configuration. However, Figure 16b shows that the increase in the training samples database does not improve the model because both the training and cross-validation $R^2$ seemed to converge after 4000 training samples. The regularization term $\lambda$ was also tuned using the same procedure [66]. It can be seen in Figure 16c that a regularization term lower than $10^{-2}$ should be used in both training and cross-validation.

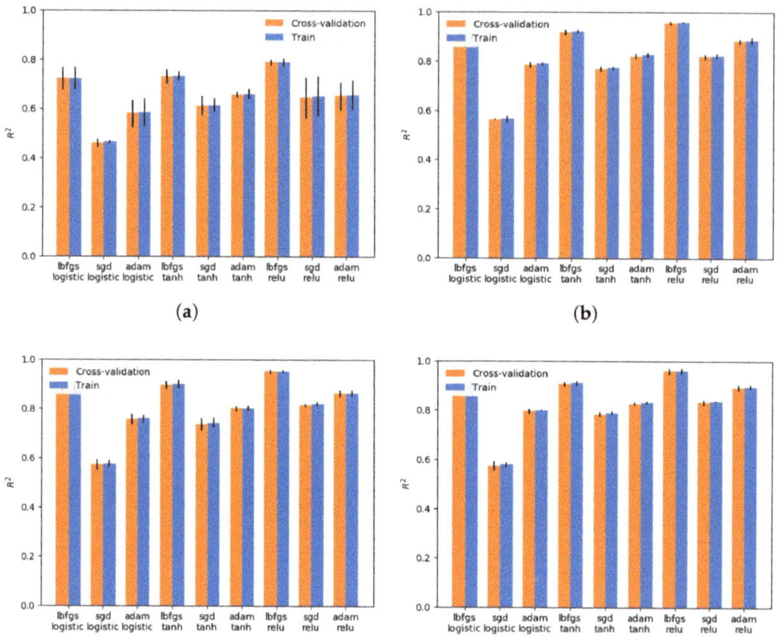


**Figure 15.** Combination of different optimization algorithms, activation functions and ANN hidden layer size used in a grid search approach. Topologies of (**a**) 4, (**b**) 10, (**c**) 16 and (**d**) 22 hidden neurons [22].

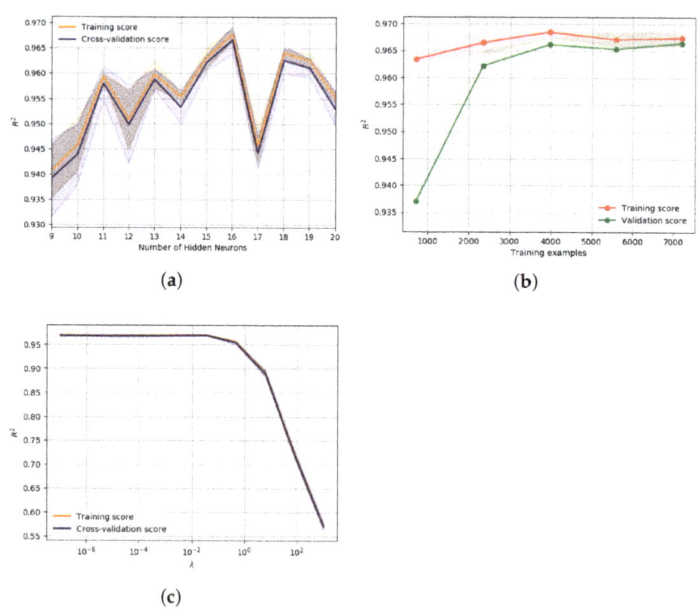


**Figure 16.** Analysis of (**a**) the ANN-model hidden layer size, (**b**) training sample database size and (**c**) regularization term $\lambda$ [22].

### 5.2.3. Testing the Trained Model in an FEA Simulation

The ANN model was implemented in the Abaqus standard (implicit) FEA code using a UMAT user routine. Although the UMATs are written in Fortran, here, a call for a python script was used. The advantage of using a differential ANN model is that the FEA time integration is straightforward. Nevertheless, small time increments were used due to the non-use of a consistent stiffness matrix.

The goal of the ANN model is to be tested in a simulation environment. Therefore, for that purpose, a heterogeneous mechanical test named adapted butterfly specimen [67] and illustrated in Figure 17a is used. Although not represented in Figure 17b, a tool displacement of 9.6 mm is imposed on the top of the specimen. Symmetry conditions are considered in both the $x$ and $y$ axis, allowing only $1/4$ of the specimen to be represented. Then, 2D 4-node bilinear elements (CPS4) are used with an element size of 0.75 mm, as can be seen in Figure 17c. A fixed increment size of 0.2 was used for a total time of 5 s.

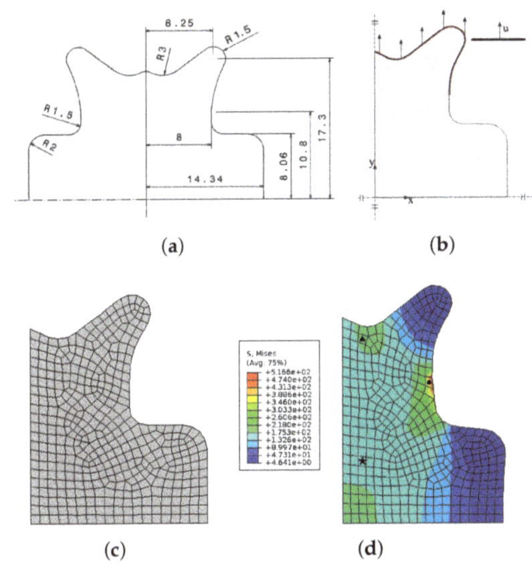


**Figure 17.** Adapted butterfly specimen [22,68]: (**a**) CAD geometry (in mm), (**b**) loading conditions, (**c**) numerical mesh and (**d**) the von Mises equivalent stress distribution obtained at the end of the test using the ANN model. Localization of the selected point for analysis.

This specimen has the advantage of producing heterogeneous stress fields, as seen in Figure 17d. Three points were selected to assess the accuracy of the ANN model and compared with the results obtained using the virtual material modelled by the Chaboche elasto-viscoplastic model. It must be noted that the ANN model was trained with the virtual material using only two tensile and one simple shear test.

Table 7 compares the values of the stress and backstress tensors and isotropic hardening/drag stress for the three selected points of Figure 17d at the end of the test. Here, the experiment values correspond to the virtual material experiments. In general, the ANN model can reproduce fairly well the virtual material. However, although lower errors can be seen for $\sigma_{xx}$ and $\sigma_{yy}$ (16% and 0.75%) and their correspondent backstress in the first two points, higher errors were obtained for the shear stress component. Nevertheless, these results can be explained that, for this point, $\varepsilon = \{-0.0914, 0.2008, 0.0442\}$, and the ANN model was never trained with such a complex strain field, only with a maximum strain of 0.16. For points 2 and 3, the performance of the ANN model is better because the strain values are lower. It must be highlighted that the larger error for point 2 is 8%.

Table 8 compares the equivalent plastic strain and the viscoplastic strain tensor at the end of the test for the selected three points. Again, the performance of the ANN model is fairly good; however, it is the shear strain that presents the worst results (errors larger than 20%).

**Table 7.** Comparison between the ANN model and the (virtual) experimental values at the end of the butterfly test for selected points: stress comparison [22].

| | | $\sigma_{xx}$ [MPa] | $\sigma_{yy}$ [MPa] | $\tau_{xy}$ [MPa] | $\chi_{xx}$ [MPa] | $\chi_{yy}$ [MPa] | $\chi_{xy}$ [MPa] | $R$ [MPa] |
|---|---|---|---|---|---|---|---|---|
| Point 1 ● | ANN | 283.88 | 533.31 | 69.08 | −100.18 | 101.41 | 3.49 | 81.97 |
| | Experiment | 243.42 | 537.35 | 37.83 | −95.94 | 95.94 | 23.94 | 82.81 |
| | Error [%] | 16.62 | 0.75 | 82.61 | 4.42 | 5.7 | 85.42 | 1.01 |
| Point 2 ▲ | ANN | −29.96 | 180.21 | 0.96 | −64.34 | 64.08 | 1.72 | 52.49 |
| | Experiment | −25.78 | 180.33 | 1.05 | −69.35 | 69.35 | 0.77 | 53.35 |
| | Error [%] | 0.16 | 0.07 | 8.57 | 7.22 | 7.60 | 0.01 | 1.61 |
| Point 3 ★ | ANN | 0.62 | 163.06 | 7.47 | −48.62 | 48.28 | 1.25 | 51.80 |
| | Experiment | −0.66 | 168.67 | 6.61 | −52.11 | 52.11 | 4.18 | 52.07 |
| | Error [%] | 193.94 | 3.33 | 13.01 | 6.7 | 7.35 | 70.1 | 0.52 |

**Table 8.** Comparison between the ANN model and the (virtual) experimental values at the end of the butterfly test for selected points: strain comparison [22].

| | | $\varepsilon_{xx}^{vp}$ | $\varepsilon_{yy}^{vp}$ | $\varepsilon_{xy}^{vp}$ | $p$ |
|---|---|---|---|---|---|
| Point 1 ● | ANN | $-1.162 \times 10^{-1}$ | $1.111 \times 10^{-1}$ | $8.266 \times 10^{-3}$ | $1.313 \times 10^{-1}$ |
| | Experiment | $-1.079 \times 10^{-1}$ | $1.079 \times 10^{-1}$ | $2.452 \times 10^{-2}$ | $1.262 \times 10^{-1}$ |
| | Error[%] | 7.69 | 2.97 | 66.29 | 4.04 |
| Point 2 ▲ | ANN | $-9.948 \times 10^{-3}$ | $1.055 \times 10^{-2}$ | $1.798 \times 10^{-4}$ | $1.162 \times 10^{-2}$ |
| | Experiment | $-1.077 \times 10^{-2}$ | $1.077 \times 10^{-2}$ | $1.400 \times 10^{-4}$ | $1.244 \times 10^{-2}$ |
| | Error[%] | 7.63 | 2.04 | 28.43 | 6.59 |
| Point 3 ★ | ANN | $-7.230 \times 10^{-3}$ | $7.836 \times 10^{-3}$ | $9.220 \times 10^{-5}$ | $8.538 \times 10^{-3}$ |
| | Experiment | $-6.630 \times 10^{-3}$ | $6.630 \times 10^{-3}$ | $5.400 \times 10^{-4}$ | $7.670 \times 10^{-3}$ |
| | Error[%] | 9.05 | 18.19 | 82.93 | 11.32 |

The evolution of $\sigma$ during the test for the selected points when using the ANN model can be seen in Figure 18. The experimental evolution of the virtual material is also added for comparison purposes. Although it is observed that both $\sigma_{xx}$ and $\sigma_{yy}$ evolution reproduces well the behaviour of the virtual material, ANN-$\tau_{xy}$ fails to foresee the experimental curve. This fact can be attributed to the absence of complex shear states in the training data.

The other internal state variables were also assessed and compared. Figure 19 presents the time evolution of the viscoplastic strain tensor, equivalent plastic strain, backstress tensor and the isotropic hardening/drag stress. In a general overview, it can be stated that the ANN model curves can fit the virtual experiments. However, larger computational times (>20×) were observed in comparison with the reference model.

5.2.4. Final Remarks

In this section, it can be seen that an ML/ANN model can be used to model the behaviour of an elasto-viscoplastic material. Here, an ANN model was effectively trained using virtually created (synthetic) data of cyclic tensile compression and shear tests. This was successfully reproduced for 2D (plane-stress) conditions, thus achieving the main objective.

Subsequently, the ANN model was applied in an FEA simulation of complex stress-strain states using a heterogeneous mechanical test denoted the butterfly test. Even knowing that the complexity of the stress-strain field used for analysis was not used for training, the ANN model was quite able to simulate the stress-strain evolution. In conclusion, the proposed ANN model, even with different combinations of strain never used in training, is capable of reproducing the material behaviour defined by the virtual material.

Moreover, the ANN implementation in the FEA code is straightforward and no integration problems were observed even in complex geometries This example shows the capability of the ML/ANN to model the material behaviour using some concepts already used in analytical models. Although this example used synthetic data, real experimental data could be used under the same conditions.

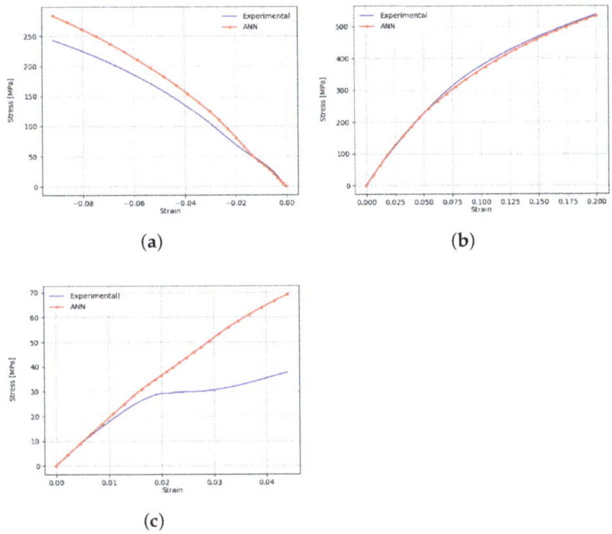


**Figure 18.** Stress-strain curves obtained for the selected point of the butterfly test: (**a**) $\sigma_{xx} - \varepsilon^{t}_{xx}$, (**b**) $\sigma_{yy} - \varepsilon^{t}_{yy}$ and (**c**) $\tau_{xy} - \varepsilon^{t}_{xy}$ curves. Comparison between the ANN model and (virtual) experimental values [22].

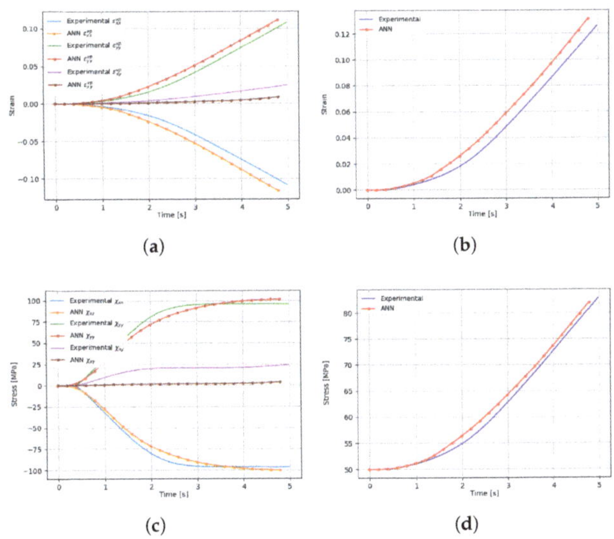


**Figure 19.** Evolution of the internal state variables for the selected points in the butterfly test: (**a**) viscoplastic strain tensor $\varepsilon^{vp}$, (**b**) back stress tensor $\chi$, (**c**) plastic strain $p$ and (**d**) isotropic (hardning/drag) stress $R$. Comparison between the ANN model and (virtual) experimental values [22].

### 5.3. Implicit Elastoplastic Modelling Using the VFM

The heterogeneous test developed by Martins et al. [2] was used to generate numerical experimental data to train the ANN model. The configuration consists of a solid $3 \times 3\ \text{mm}^2$ plate with a thickness $t = 0.1$ mm. The domain is discretized by 9 4-node bilinear plane stress elements. The initial mesh, geometry and boundary conditions are depicted in Figure 20. Symmetry boundary conditions are applied to the boundaries at $x = 0$ and $y = 0$, and a surface traction is defined at $x = 3$ mm. The traction follows a non-uniform distribution, composed by a single component along the $x$-direction, which varies linearly in the $y$-direction, according to $f_x(y) = my + b$, where $m$ and $b$ respectively control the distribution.

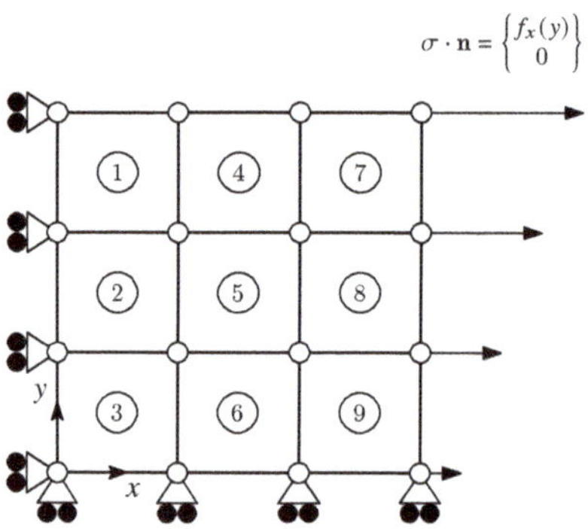


**Figure 20.** Heterogeneous test: initial geometry, mesh and boundary conditions.

The numerical simulations were conducted using the commercial finite element code Abaqus. The model was built with CPS4R elements (bilinear reduced integration plane stress). The material was simulated employing a non-linear isotropic elasto-plastic model, with the isotropic hardening response obeying Swift's law, given by:

$$\sigma_y = K(\varepsilon_0 + \bar{\varepsilon}^{\text{p}})^n, \tag{16}$$

where $\sigma_y$ is the flow stress, $K$ is a hardening coefficient, $n$ is the hardening exponent, $\sigma_0$ is the yield stress and $\varepsilon_0$ the deformation at the yielding point, computed as:

$$\varepsilon_0 = \left(\frac{\sigma_0}{K}\right)^{1/n}. \tag{17}$$

The elastic parameters were defined as $E = 210$ GPa and $\nu = 0.3$, while $\sigma_0 = 160$ MPa, $K = 565$ MPa and $n = 0.26$ were adopted for Swift's hardening law.

#### 5.3.1. Data Generation and ANN Training

To generate the training data, all simulations were performed using a small displacement formulation, with the time period set to 1 and using a fixed time increment $\Delta t = 0.005$. For each time increment, the deformation components at the centroid were extracted for all the elements, and the global force was determined by means of computing the equilibrium of the internal forces, such that:

$$FL = \sum_{i=1}^{n} \sigma_i A_i t\,, \tag{18}$$

where $F$ is the global force, $L$ is the length of the solid, $A$ is the element's area and $t$ is the thickness.

The training data were generated for different load distributions, keeping the slope fixed at $m = 10$ N/mm and varying the intercept parameter, such that: $b = \{50, 170, 270\}$ N.

Prior to training and for each mechanical trial, the dataset was organized into batches of 9 elements per time increment and shuffled before being split into training (80%) and test data (20%). The input features were transformed in order to follow a normal distribution with zero mean and unit variance. Two models were trained and aimed at predicting the linear elastic and elasto-plastic responses of the material. Once trained, the models were validated with different mechanical trials using the following load distributions: $m = 12$, $b = 100$ for the elastic model and $m = 12$, $b = 200$ for the elasto-plastic model.

The neural network model consisted of a simple FFNN with one hidden layer, with 4 hidden neurons for the elastic case and 8 hidden neurons for the elasto-plastic case. In both cases, the parametric rectified linear unit (PReLU) activation function [69] was used. The inputs given to the models were the deformation components in the current and previous time increments, $\varepsilon^t$ and $\varepsilon^{t-1}$, respectively, and the outputs were the stress tensor components at the current time increment, $\sigma^t$. The ADAM algorithm was used to optimize the network weights, with the initial learning rate set to 0.1, scheduled to be reduced using a multiplier of 0.2, if no improvement in the training loss was registered after 5 epochs. For the elastic model, the network was trained during 30 epochs, and for the elasto-plastic model, training was set to occur during a maximum of 150 epochs.

5.3.2. Results and Discussion

The learning curves for both models are depicted in Figure 21. The plots show the loss decreasing during the training epochs with convergence being achieved earlier for the elastic model. In this case, four virtual fields were used during training, while for the elasto-plastic response model, a total of six virtual fields were used. The complete set of virtual fields is presented in Table 9 and illustrated in the Appendix C.

**Table 9.** Virtual fields used to train both ANN models. Virtual fields 1 through 4 were used to train the elastic model, while the whole set was used to trained the elastoplastic model.

| Virtual Field | 1 | 2 | 3 | 4 | 5 | 6 |
|---|---|---|---|---|---|---|
| $u_1^{*(i)}$ | $\frac{x}{L}$ | $\sin\left(\frac{2\pi x}{3L}\right)$ | $\frac{\pi y^2}{L^2}\sin\left(\frac{2\pi x}{L}\right)$ | $\frac{y^2}{L^2}\sin\left(\frac{\pi x}{L}\right)$ | 0 | 0 |
| $u_2^{*(i)}$ | 0 | 0 | $\frac{-2y}{L}\sin^2\left(\frac{\pi x}{L}\right)$ | $\frac{x^2}{L^2}\sin\left(\frac{\pi y}{L}\right)$ | $\sin\left(\frac{2\pi y}{3L}\right)$ | $\frac{y}{L}\sin^2\left(\frac{\pi x}{L}\right)$ |
| $\varepsilon_{11}^{*(i)}$ | $\frac{1}{L}$ | $\frac{2\pi}{3L}\cos\left(\frac{2\pi x}{3L}\right)$ | $\frac{2\pi^2 y^2}{L^3}\cos\left(\frac{2\pi x}{L}\right)$ | $\frac{\pi y^2}{L^3}\cos\left(\frac{\pi x}{L}\right)$ | 0 | 0 |
| $\varepsilon_{22}^{*(i)}$ | 0 | 0 | $\frac{-2}{L}\sin^2\left(\frac{\pi x}{L}\right)$ | $\frac{\pi x^2}{L^3}\cos\left(\frac{\pi y}{L}\right)$ | $\frac{2\pi}{3L}\cos\left(\frac{2\pi y}{3L}\right)$ | $\frac{1}{L}\sin^2\left(\frac{\pi x}{L}\right)$ |
| $\varepsilon_{12}^{*(i)}$ | 0 | 0 | 0 | $\frac{2y}{L^2}\sin\left(\frac{\pi x}{L}\right)+\frac{2x}{L^2}\sin\left(\frac{\pi y}{L}\right)$ | 0 | $\frac{2\pi y}{L^2}\sin\left(\frac{\pi x}{L}\right)\cos\left(\frac{\pi x}{L}\right)$ |

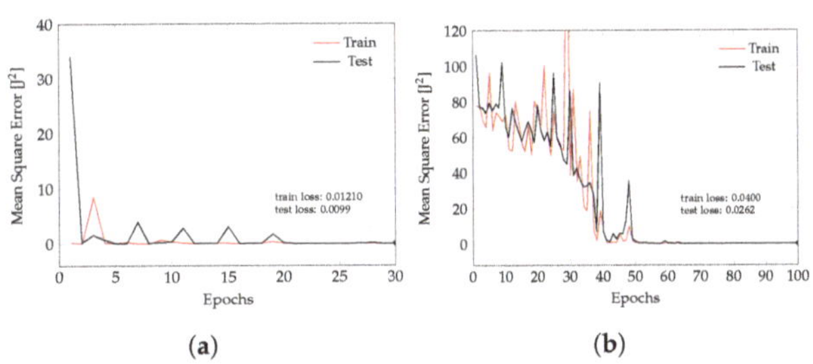


**Figure 21.** Learning curves for (**a**) the elastic and (**b**) the elasto-plastic response models.

The validation results (Figure 22) show that, in general, the ANN was able to learn both elastic and elasto-plastic behaviors. In the former case, the ANN provided near-excellent stress predictions, with, however, a higher error being registered for the stress component in the $y$-direction for the element 9 (Figure 22c). In the latter case, the ANN provided overall excellent predictions for the stress response along the $x$-direction. Although the use of additional virtual fields slightly improves the shear stress response for the plastic model (Figure 22b,d), it seems the ANN has a noticeable lower sensitivity to the stress responses along $y$ and $xy$. The issue may be due to fact that either the number of chosen virtual fields is not enough to capture the material behavior or the chosen set of virtual fields provides more weight to the stresses along $x$, to the detriment of the remaining components. Another factor at play is that the virtual fields were chosen manually. This strategy is often used for non-linear models and is the easiest to implement; nonetheless, it does not guarantee that the chosen virtual fields produce the best results and is tied to the user's expertise [2]. Additionally, it may be possible that the mechanical test considered in this example may be too simple, thus not providing enough dissimilar behaviour data. Albeit with a wide margin for further improvements in the future, this application example has effectively proven that the new methodology works.

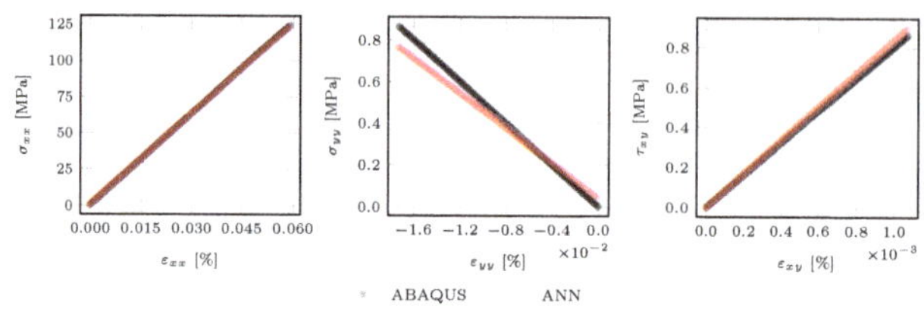


(**a**) Elastic model—element #1: $m = 12$, $b = 100$.

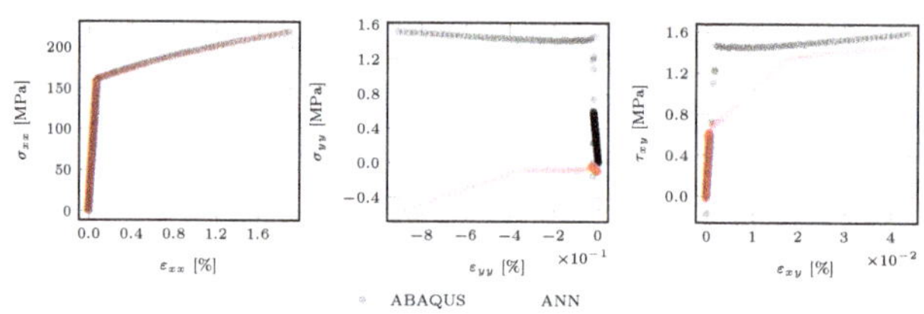


(**b**) Elastoplastic model—element #1: $m = 12$, $b = 200$.

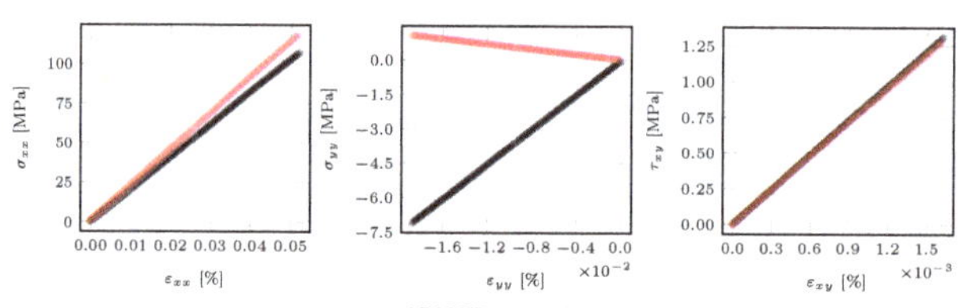


(**c**) Elastic model—element #9: $m = 12$, $b = 100$.

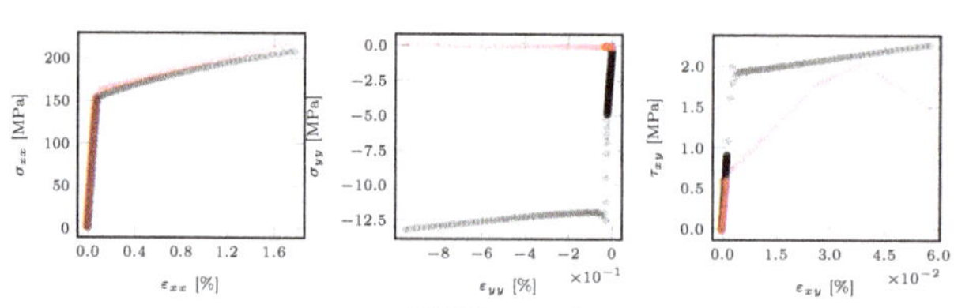


(**d**) Elastoplastic model—element #9: $m = 12$, $b = 200$.

**Figure 22.** Stress-strain curves corresponding to the elements 1 and 9 for the ANN-based elastic and elastoplastic response models.

## 6. Conclusions

This paper presents and discusses the recent advances and applications of ML, particularly ANNs, in metal forming processes, from which the following points can be highlighted:

- ANNs provide an opportunity for a complete change of paradigm in the field of constitutive modelling. Their powerful capabilities allow material models to be devised without the need to formulate complicated symbolic expressions that, conventionally, require extensive parameter calibration procedures;
- Nevertheless, even in calibration/parameter identification procedures, the power and potential of ANNs can still be used. This paper showed that an ANN can be trained as an inverse model for parameter identification;
- ANNs are paving the way for the development of implicit material models capable of reproducing complex material behavior purely from data. Additionally, the the network architectures themselves can be designed based on the structure of existing well-known constitutive models;
- A great number of methodologies employ ANNs as "black boxes" that are not guaranteed to respect the basic laws of physics governing the dynamics of the systems, thus requiring large amounts of training to do so. This is an important issue due to the cost of data acquisition being often prohibitive and time-consuming;
- Considerable efforts have been done to improve the generalization capacity of the ANN-based implicit constitutive models, either by the development of new architectures, more robust training procedures or through the augmentation of the datasets. The use of NANNs, for example, is an interesting approach to capture the full material response, but does not solve the underlying dependency on large datasets. On the other hand, PINNs, particularly TANNs, have the potential to leverage the knowledge obtained from training data through the direct encoding of basic thermodynamic laws in the neural network, thus resulting in models able to perform very well on low data regimes. The SPD-NN goes a step further by imposing time consistency, strain energy function convexity and the Hill's criterion, thus solving numerical instabilities derived from coupling standard ANN-based models with FEM solvers;
- Another powerful attribute of implicit constitutive models is the ease of implementation within FE subroutines. The architecture of ANNs can be stored in the code via matrices with the appropriate dimensions, containing the optimized weights obtained during training. The ANN operations are reduced to a sequence of matrix operations that provide an almost instantaneous mapping between input and output values, providing substantial gains in computational efficiency;
- A major part of the ANN-based methodologies to implicit constitutive modelling relies on labeled data pairs (strain and stress) to train the network. However, the stress tensor is not measurable in complex experimental tests, thus requiring the ANN-based model to be trained indirectly, using only measurable data, such as displacements and global force;
- Indirect training approaches for ANN-based constitutive modelling reported in the literature resort to coupling with FEA software in order for the numerical solver to impose the necessary physical and boundary contraints, while also allowing full-field data to be taken advantage of;
- A new disruptive methodology to train an ANN-based implicit constitutive model was proposed using the VFM. The new approach allows an ANN to learn both linear elastic and elastoplastic behaviors through the application of the PVW, using only measurable data, e.g., displacements and global force, to feed the training process. An interesting advantage of this new approach is that it does not require the FEM for further calculations.

**Author Contributions:** Conceptualization, R.L. and A.A.-C.; methodology, R.L. and A.A.-C.; software, R.L.; validation, R.L. and A.A.-C.; formal analysis, R.L. and A.A.-C.; investigation, R.L. and A.A.-C.; resources, R.L.; data curation, R.L.; writing—original draft preparation, R.L. and A.A.-C.; writing—review and editing, R.L., A.A.-C. and P.G.; visualization, R.L. and A.A.-C.; supervision, P.G. and A.A.-C.; project administration, A.A.-C.; funding acquisition, R.L. and A.A.-C. All authors have read and agreed to the published version of the manuscript.

**Funding:** This project has received funding from the Research Fund for Coal and Steel under grant agreement No 888153.

**Institutional Review Board Statement:** Not applicable.

**Informed Consent Statement:** Not applicable.

**Data Availability Statement:** The data presented in this study are available on request from the corresponding author.

**Acknowledgments:** Rúben Lourenço acknowledges the Portuguese Foundation for Science and Technology (FCT) for the financial support provided through the grant 2020.05279.BD, co-financed by the European Social Fund, through the Regional Operational Programme CENTRO 2020. The authors also gratefully acknowledge the financial support of the Portuguese Foundation for Science and Technology (FCT) under the projects PTDC/EME-APL/29713/2017 (CENTRO-01-0145-FEDER-029713), PTDC/EME-EME/31243/2017 (POCI-01-0145-FEDER-031243) and PTDC/EME-EME/30592/2017 (POCI-01-0145-FEDER-030592) by UE/FEDER through the programs CENTRO 2020 and COMPETE 2020, and UIDB/00481/2020 and UIDP/00481/2020-FCT under CENTRO-01-0145-FEDER-022083. This project has also received funding from the Research Fund for Coal and Steel under grant agreement No 888153.

**Conflicts of Interest:** The authors declare no conflict of interest.

## Abbreviations

The following abbreviations are used in this manuscript:

| | |
|---|---|
| ADAM | Adaptive moment estimation |
| ANN | Artificial neural network |
| CEGM | Constitutive equation gap method |
| CNN | Convolutional neural network |
| DIC | Digital image correlation |
| FEA | Finite element analysis |
| ECM | Equilibrium gap method |
| FEM | Finite element method |
| FEMU | Finite element model updating |
| FFNN | Feedforward neural network |
| GRU | Gated recurrent unit |
| LBFGS | Limited memory Broyden—Fletcher—Goldfarb—Shanno |
| LSTM | Long short-term memory unit |
| MAE | Mean absolute error |
| ML | Machine learning |
| MSE | Mean square error |
| NANN | Nested adaptive neural network |
| PINN | Physics-informed neural network |
| PReLU | Parametric rectified linear unit |
| ReLU | Rectified linear unit |
| RGM | Reciprocity gap method |
| RNN | Recurrent neural network |
| SELU | Scaled exponential linear unit |
| SGD | Stochastic gradient descent |
| SPD-NN | Symmetric positive definite neural network |
| TANN | Thermodynamics-based neural network |
| VFM | Virtual fields method |

## Appendix A. Artificial Neural Networks

### *Appendix A.1. Basic Principle and Components*

ANNs draw inspiration from the signal processing scheme used by the neural systems that constitute animal brains. In ANNs, the signal is transmitted between the neurons over links with an associated weight, providing strength to the connection. The greatest advantage of ANNs is their ability to learn from examples and describe complex non-linear, multi-dimensional relationships in a dataset without any prior assumptions [41]. ANNs not only possess the ability to make decisions based on incomplete and disordered data, but are also able to generalize rules and apply them to new, previously unseen, data. Both the network architecture and the learning method adopted to train the ANN model have a significant impact in the accuracy of the model's predictions [49,70].

### *Appendix A.2. Network Topology*

Multi-layer neural networks are, in essence, perceptrons with more than one computational layer (Figure A1b) [71]. The input layer receives the data and the output layer sends the information out, while the hidden layers provide the required complexity for nonlinear problems [33,70]. The signal flows uniquely in one direction, therefore multi-layer networks are also referred to as feedforward neural networks (FFNNs). FFNNs are commonly used for function approximation, defining a mapping of the type $y = f(\mathbf{x}; \mathbf{W})$ in which the network will learn the value of the parameters $w_{ij}$ that provide the best approximation [49]. FFNNs may be thought of as a composition of several different functions, connected in a chain structure of the form [33,49]:

$$f(\mathbf{x}) = f^{(k)}( \ \ldots \ f^{(2)}(f^{(1)}(\mathbf{x}))) \, , \tag{A1}$$

where $f^{(1)}$ corresponds to the first layer, $f^{(2)}$ to the second layer and so on, for an arbitrary network with $k$ layers [33,49]. The length of the chain provides the depth of the model, that is, the number of layers [33,49]. Each hidden layer is typically vector-valued and their dimensionality, given by the number of neurons, determines the width of the model. According to Cybenko [72], a neural network requires at most two hidden layers to approximate any function to an arbitrary order of accuracy and only one hidden layer to approximate a bounded continuous function to arbitrary accuracy. For this reason, neural networks are often referred to as universal function approximators [11]. In practice, however, the main issue is that the number of hidden units required to do so is rather large, which increases the number of parameters to be learned. This results in problems in training the network with a limited amount of data. Deeper networks with a lower number of hidden units in each layer are often preferred [33].

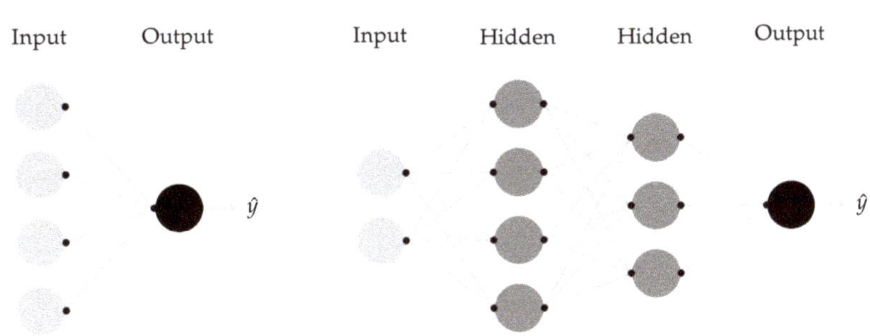


(**a**) Perceptron (**b**) Multi-layer feedforward network

**Figure A1.** Basic architectures of (**a**) the perceptron and (**b**) a multi-layer feedforward network.

Different network topologies were formulated as variants of FFNNs and became popular in the last two decades [73]: convolutional neural networks (CNNs) and long short-term memory neural networks (LSTMs). Typically, in CNNs, a convolutional layer moves across the data, similar to a filter in computer vision algorithms, requiring only a few parameters, as the convolving layer allows for effective weight replication [74]. LSTMs, a variation of recurrent neural networks (RNNs), offer specialized memory neurons with the purpose of dealing with the vanishing gradient problem. The latter often leads to sub-optimal local minima, especially as the number of neuronal layers increases [73].

*Appendix A.3. Mathematical Representation*

Consider a case where each training instance of a given dataset has the form $(\mathbf{x}, y)$, where $\mathbf{x} \in \mathbb{R}^n$ is an input vector containing the set of input features $[x_1, \ldots x_n]$ and $y$ is the observed value, obtained as a part of the training data. Given a FFNN of arbitrary architecture, as shown in Figure A2, the dimensionality of the feature vector $\mathbf{x}$ dictates the number of neurons in the input layer [33].

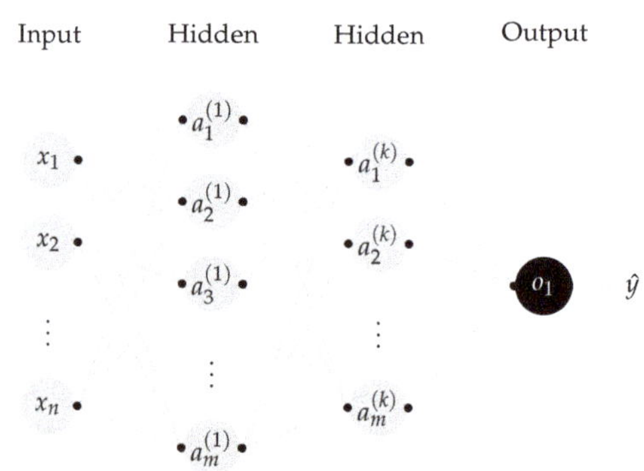


**Figure A2.** Multi-layer feedforward neural network with one output and $n$ inputs.

These inputs are mapped to the next layers with $m$ neurons, triggering a set of activations $\mathbf{a}^{(k)} = \{a_1^{(1)}, \ldots a_m^{(k)}\} \in \mathbb{R}^m$, with $a_m^{(k)}$ denoting the activation of the $m$th neuron in the network's $k$th layer. The activation potential from layer $(k-1)$ to layer $k$ is controlled by a matrix of parameters $\mathbf{W}^{(k)} \in \mathbb{R}^{m \times n}$ and computed as the weighted sum of the output values $\{x_1, \ldots x_n\}$ from the incoming connections. A function $g(\cdot)$ is then applied to this weighted sum, leading to the activations of the $k$th layer being computed as:

$$\mathbf{a}^{(k)} = g\left(\mathbf{W}^{(k)}\mathbf{x} + \mathbf{b}^{(k)}\right). \tag{A2}$$

A set of biases $\mathbf{b} \in \mathbb{R}^m$ is added as the weight of a link, which in practice means incorporating a bias neuron that conventionally transmits a value of 1 to the output node [33]. In essence, in a FFNN, the $n$-dimensional input vector $\boldsymbol{x}$ is transformed into the outputs using the following recursive relations [33,75]:

$$\begin{cases} \mathbf{a}^{(1)} = g\left(\mathbf{W}^{(1)}\mathbf{x} + \mathbf{b}^{(1)}\right) & \text{input to hidden layer} \\ \mathbf{a}^{(p+1)} = g\left(\mathbf{W}^{(p+1)}\mathbf{a}^{(p)} + \mathbf{b}^{(p+1)}\right) \ \forall p \in \{1, \ldots k-1\} & \text{hidden to hidden layer} \\ \mathbf{o} = g\left(\mathbf{W}^{(k+1)}\mathbf{a}^{(k)} + \mathbf{b}^{(k+1)}\right) & \text{hidden to output layer} \end{cases}. \tag{A3}$$

In the above expressions, $g(\cdot)$ is an activation function, a mathematical entity that receives the output of the weighted sum as an input and converts that into the final output of a node [33,49]. There are several different types of activation functions, and their selection is an important part of the design process of a neural network. The most common activation functions [33,75] are shown in Figure A3. Worthy of note is the fact that all the activation functions shown are monotonic. Moreover, other than the identity function, most of them saturate at large absolute values of the argument. The use of nonlinear activation functions greatly increases the modelling capabilities of a neural network [33,49].

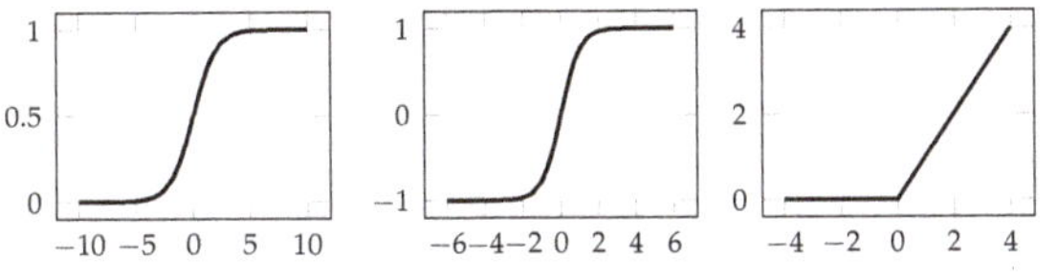


**Figure A3.** Most commonly used activation functions: (**a**) sigmoid, (**b**) hyperbolic tangent and (**c**) rectified linear unit (ReLU).

Historically, both the sigmoid and *tanh* functions have been the popular choices to incorporate non-linearity in the neural network. However, in recent years, the preference has been directed towards a family of piecewise linear functions, such as the rectified linear unit (ReLU) and its variations. Due to their simplicity, these functions are easier to differentiate, thus providing a faster training process [33].

### *Appendix A.4. Learning Paradigms and Training Procedures*

The training of an ANN is accomplished through a learning procedure during which sample data is fed to the network and a learning algorithm is in charge of modifying a set of parameters in order for the ANN to provide the best possible prediction [9]. There are essentially two learning paradigms: unsupervised learning and supervised learning [49]. Unsupervised learning is concerned with finding patterns in unlabeled datasets containing many features. Supervised learning is the most widespread method used to train ANNs and uses labeled datasets, where each training sample is linked to a label [49,75]. The process is similar to a standard fitting procedure, accompanied by the selection of a learning algorithm that is used to fit target value [9,49]. The main task relates to data generation and preparation in order to ensure the dataset is both accurate and consistent. The main dataset is split into various subsets, mainly a training and a test set. Additionally, a validation set, different from the latter ones, may be used to optimize the hyper-parameters [9]. The following step is concerned with translating the raw information into meaningful features that can be used as inputs for the ANN. Finally, the model is trained by optimizing its performance, that is, finding the set of parameters $(\mathbf{W}, \mathbf{b})$ that minimizes a given loss function [9]. For regression tasks, the mean squared error (MSE) is commonly employed, being written as:

$$\mathcal{L}(\mathbf{W}, \mathbf{b}) = \frac{1}{m}\sum_{i=0}^{m}\left(y^{(i)} - \hat{y}^{(i)}\right)^2 + \frac{\lambda}{2m}\sum_{i=1}^{k-1}\sum_{i=1}^{s_k}\sum_{j=1}^{s_{k+1}}(\mathbf{W}_{ji}^{(k)})^2\,, \tag{A4}$$

where $m$ is the total number of training instances, $k$ is the number of layers, $s_k$ is the number of neurons on the $k$th layer and $s_{k+1}$ is the number of neurons on the $(k+1)$th layer. There is a wide variety of optimization algorithms available to minimize $\mathcal{L}(\mathbf{W}, \mathbf{b})$. A gradient-based algorithm (e.g., gradient descent or adaptive moment estimation (ADAM)) is normally used to minimize the loss by subsequently moving in the direction of the negative gradient [33].

The parameters' update is driven by first taking the partial derivatives of $\mathcal{L}(\mathbf{W}, \mathbf{b})$, such that [49]:

$$\mathrm{d}\mathbf{W}^{(k)} = \frac{\partial \mathcal{L}}{\partial \mathbf{W}^{(k)}}, \tag{A5}$$

$$\mathrm{d}\mathbf{b}^{(k)} = \frac{\partial \mathcal{L}}{\partial \mathbf{b}^{(k)}} \tag{A6}$$

and repeatedly applying [49]:

$$\mathbf{W}^{(k)} := \mathbf{W}^{(k)} - \alpha \mathrm{d}\mathbf{W}^{(k)}, \tag{A7}$$

$$\mathbf{b}^{(k)} := \mathbf{b}^{(k)} - \alpha \mathrm{d}\mathbf{b}^{(k)}, \tag{A8}$$

until convergence is achieved, with $\alpha \in [0, 1]$ being the learning rate, a hyper-parameter controlling how quickly the model is adapted to the problem [49]. Small learning rates slow down training and can cause the optimizer to reach premature convergence. Larger learning rates speed up training and require fewer training epochs; however the model can overshoot the minimum, fail to converge or even diverge [33,49]. More complex algorithms have the ability to subsequently adapt the learning rate on each iteration.

An important challenge to overcome during neural network training is overfitting. The phenomenon happens when a particular model perfectly fits the available data but lacks the generalization power necessary to make predictions from unseen data [34]. A model's propensity to overfit can be controlled by changing its capacity, which is determined by the number of learnable parameters of the neural network (e.g., the number of layers and the number of neurons per layer) [33,34]. Generally speaking, increasing the number of training instances may improve the generalization power of the model, whereas increasing the capacity of the model often reduces its generalization power [34]. Weight regularization is commonly used to prevent overfitting and consists of adding a penalty of the form $\lambda \|\mathbf{W}^p\|$ to the loss function, with $\lambda$ being a regularization parameter. Depending on the value of $p$, there are two types of regularization [33]:

- *L1 regularization*: the penalty is proportional to the L1 norm of the weighting coefficients;
- *L2 regularization*: the penalty is proportional to the L2 norm of the weighting coefficients.

An example of L2 regularization can be seen in the last part of Equation (A4), where the penalty has the value of $p$ set to 2.

## Appendix B. Viscoplasticity

In viscoplasticity, the time-dependent effect of viscous forces is unified with the plastic deformation via a viscoplastic term, such that [18,27,76]:

$$\varepsilon = \varepsilon^{\mathrm{e}} + \varepsilon^{\mathrm{p}} + \varepsilon^{\mathrm{v}} = \varepsilon^{\mathrm{e}} + \varepsilon^{\mathrm{vp}}, \tag{A9}$$

where $\varepsilon^{\mathrm{v}}$ and $\varepsilon^{\mathrm{vp}}$ are, respectively, the viscous and viscoplastic strains. The Chaboche's model is a popular viscoplastic model in which the viscoplastic potential is written in terms of a power function of $f$ from Equation (5) [76]. The model takes into account both kinematic and isotropic hardening effects under stationary temperature conditions. The viscoplatic strain rate tensor $\dot{\boldsymbol{\varepsilon}}^{\mathrm{vp}}$ and the plastic strain rate $\dot{p}$ are defined as follows [27,76]:

$$\dot{\boldsymbol{\varepsilon}}^{\mathrm{vp}} = \frac{3}{2}\,\dot{p}\,\frac{\boldsymbol{\sigma}' - \boldsymbol{\chi}'}{J(\boldsymbol{\sigma}' - \boldsymbol{\chi}')}, \tag{A10}$$

$$\dot{p} = \left\langle \frac{J(\boldsymbol{\sigma} - \boldsymbol{\chi}) - R - k}{K} \right\rangle^{n}, \tag{A11}$$

where $k$, $R$, $K$ and $n$ are the initial yield stress, isotropic hardening and two material constants, respectively, whereas $\sigma'$ and $\chi'$ are the deviatoric parts of the stress and backstress tensors [27,76]. The $J(\sigma' - \chi')$ invariant is computed through the following formula:

$$J(\sigma' - \chi') = \sqrt{\frac{3}{2}(\sigma' - \chi') : (\sigma' - \chi')} \, . \tag{A12}$$

The kinematic hardening rate $\dot{\chi}$ is defined by means of a partial differential equation, given as:

$$\dot{\chi} = \frac{2}{3}\, a\, \dot{\varepsilon}^{\mathrm{vp}}\, c\, \chi\, \dot{p} \, , \tag{A13}$$

while the isotropic hardening rate is calculated from:

$$\dot{R} = b(R_1 - R)\dot{p}. \tag{A14}$$

The variables $a$, $b$, $c$ and $R_1$ are material parameters, which must be identified for each material.

## Appendix C. Virtual Fields Used to Train the Models

Figure A4 illustrates the virtual fields used to train both the elastic and the elastoplastic models.

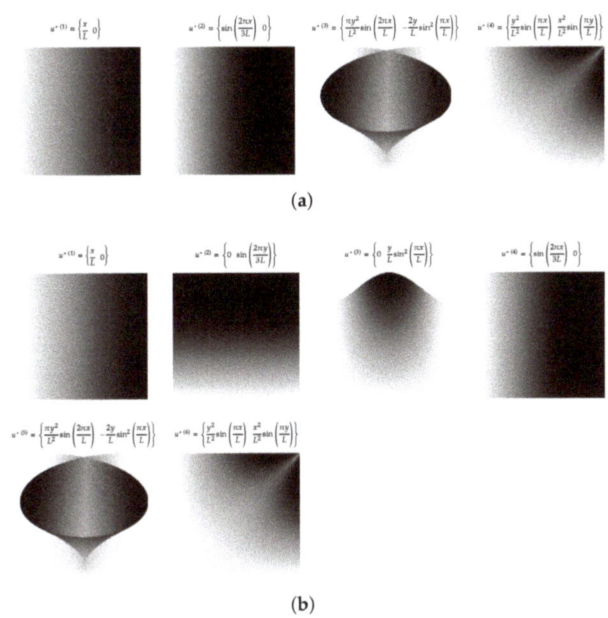


**Figure A4.** Virtual fields used to train the (**a**) elastic and (**b**) the plastic response models.

## References

1. Frank, M.; Drikakis, D.; Charissis, V. Machine-Learning Methods for Computational Science and Engineering. *Computation* **2020**, *8*, 15. [CrossRef]
2. Martins, J.; Andrade-Campos, A.; Thuillier, S. Comparison of inverse identification strategies for constitutive mechanical models using full-field measurements. *Int. J. Mech. Sci.* **2018**, *145*, 330–345. [CrossRef]
3. Andrade-Campos, A.; Thuillier, S.; Martins, J.; Carlone, P.; Tucci, F.; Valente, R.; Paulo, R.M.; de Sousa, R.J.A. Integrated Design in Welding and Incremental Forming: Material Model Calibration for Friction Stir Welded Blanks. *Procedia Manuf.* **2020**, *47*, 429–434. [CrossRef]

4. Grédiac, M.; Pierron, F.; Avril, S.; Toussaint, E. The Virtual Fields Method for Extracting Constitutive Parameters From Full-Field Measurements: A Review. *Strain* **2006**, *42*, 233–253. [CrossRef]
5. Versino, D.; Tonda, A.; Bronkhorst, C.A. Data driven modeling of plastic deformation. *Comput. Methods Appl. Mech. Eng.* **2017**, *318*, 981–1004. [CrossRef]
6. Avril, S.; Bonnet, M.; Bretelle, A.S.; Grédiac, M.; Hild, F.; Ienny, P.; Latourte, F.; Lemosse, D.; Pagano, S.; Pagnacco, E.; et al. Overview of Identification Methods of Mechanical Parameters Based on Full-field Measurements. *Exp. Mech.* **2008**, *48*, 381. [CrossRef]
7. Kirchdoerfer, T.; Ortiz, M. Data-driven computational mechanics. *Comput. Methods Appl. Mech. Eng.* **2016**, *304*, 81–101. [CrossRef]
8. Ibáñez, R.; Abisset-Chavanne, E.; González, D.; Duval, J.L.; Cueto, E.; Chinesta, F. Hybrid constitutive modeling: Data-driven learning of corrections to plasticity models. *Int. J. Mater. Form.* **2019**, *12*, 717–725. [CrossRef]
9. Schmidt, J.; Marques, M.R.G.; Botti, S.; Marques, M.A.L. Recent advances and applications of machine learning in solid-state materials science. *Npj Comput. Mater.* **2019**, *5*, 2057–3960. [CrossRef]
10. Jang, D.P.; Fazily, P.; Yoon, J.W. Machine learning-based constitutive model for J2- plasticity. *Int. J. Plast.* **2021**, *138*, 102919. [CrossRef]
11. Hornik, K.; Stinchcombe, M.; White, H. Multilayer feedforward networks are universal approximators. *Neural Netw.* **1989**, *2*, 359–366. [CrossRef]
12. Lin, Y.C.; Zhang, J.; Zhong, J. Application of neural networks to predict the elevated temperature flow behavior of a low alloy steel. *Comput. Mater. Sci.* **2008**, *43*, 752–758. [CrossRef]
13. González, D.; Aguado, J.V.; Cueto, E.; Abisset-Chavanne, E.; Chinesta, F. kPCA-Based Parametric Solutions Within the PGD Frameworks. *Arch. Comput. Methods Eng.* **2018**, *25*, 69–86. [CrossRef]
14. Lopez, E.; Gonzalez, D.; Aguado, J.V.; Abisset-Chavanne, E.; Cueto, E.; Binetruy, C.; Chinesta, F. A Manifold Learning Approach for Integrated Computational Materials Engineering. *Arch. Comput. Methods Eng.* **2018**, *25*, 59–68. [CrossRef]
15. Ibañez, R.; Abisset-Chavanne, E.; Aguado, J.V.; Gonzalez, D.; Cueto, E.; Chinesta, F. A Manifold Learning Approach to Data-Driven Computational Elasticity and Inelasticity. *Arch. Comput. Methods Eng.* **2018**, *25*, 47–57. [CrossRef]
16. González, D.; Chinesta, F.; Cueto, E. Thermodynamically consistent data-driven computational mechanics. *Contin. Mech. Thermodyn.* **2019**, *31*, 239–253. [CrossRef]
17. Bessa, M.; Bostanabad, R.; Liu, Z.; Hu, A.; Apley, D.W.; Brinson, C.; Chen, W.; Liu, W.K. A framework for data-driven analysis of materials under uncertainty: Countering the curse of dimensionality. *Comput. Methods Appl. Mech. Eng.* **2017**, *320*, 633–667. [CrossRef]
18. Furukawa, T.; Yagawa, G. Implicit constitutive modelling for viscoplasticity using neural networks. *Int. J. Numer. Methods Eng.* **1998**, *43*, 195–219. [CrossRef]
19. Man, H.; Furukawa, T. Neural network constitutive modelling for non-linear characterization of anisotropic materials. *Int. J. Numer. Methods Eng.* **2011**, *85*, 939–957. [CrossRef]
20. Ghaboussi, J.; Sidarta, D.E. New Nested Adaptive Neural Networks (NANN) for Constitutive Modeling. *Comput. Geotech.* **1998**, *22*, 29–52. [CrossRef]
21. Lefik, M.; Schrefler, B.A. Artificial neural network as an incremental non-linear constitutive model for a finite element code. *Comput. Methods Appl. Mech. Eng.* **2003**, *192*, 3265–3283. [CrossRef]
22. Gaspar, M.; Andrade-Campos, A. Implicit material modelling using artificial intelligence techniques. *AIP Conf. Proc.* **2019**, *2113*, 120004. [CrossRef]
23. Souto, N.; Andrade-Campos, A.; Thuillier, S. A numerical methodology to design heterogeneous mechanical tests. *Int. J. Mech. Sci.* **2016**, *107*, 264–276. [CrossRef]
24. Wang, K.; Sun, W.; Du, Q. A cooperative game for automated learning of elasto-plasticity knowledge graphs and models with AI-guided experimentation. *Comput. Mech.* **2019**, *64*, 467–499. [CrossRef]
25. Simo, J.; Hughes, T. *Computational Inelasticity*; Springer: Berlin/Heidelberg, Germany, 1998.
26. Hill, R.; Orowan, E. A theory of the yielding and plastic flow of anisotropic metals. *Proc. R. Soc. Lond. Ser. A Math. Phys. Sci.* **1948**, *193*, 281–297. [CrossRef]
27. Dunne, F.; Petrinic, N. *Introduction to Computational Plasticity*; Oxford University Press: Oxford, UK, 2006.
28. Crisfield, M. *Non-Linear Finite Element Analysis of Solids and Structures*; John Wiley & Sons: Hoboken, NJ, USA, 1996.
29. de Carvalho, R.; Valente, R.; Andrade-Campos, A. Optimization strategies for non-linear material parameters identification in metal forming problems. *Comput. Struct.* **2011**, *89*, 246–255. [CrossRef]
30. Ponthot, J.P.; Kleinermann, J.P. A cascade optimization methodology for automatic parameter identification and shape/process optimization in metal forming simulation. *Comput. Methods Appl. Mech. Eng.* **2006**, *195*, 5472–5508. [CrossRef]
31. Andrade-Campos, A.; Pilvin, P.; Simões, J.; Teixeira-Dias, F. Software development for inverse determination of constitutive model parameters. In *Software Engineering: New Research*; Nova Science Publishers, Inc.: Hauppauge, NY, USA, 2009; Chapter 6, pp. 93–124.
32. Grédiac, M. Principe des travaux virtuels et identification. *Comptes Rendus L'Académie Des. Sci.* **1989**, *309*, 1–5.
33. Aggarwal, C.C. *Neural Networks and Deep Learning*; Springer: Berlin/Heidelberg, Germany, 2018.
34. Chollet, F. *Deep Learning with Python*; Manning: Shelter Island, NY, USA, 2018.

35. Ghaboussi, J.; Garrett, J.H.; Wu, X. Knowledge-Based Modeling of Material Behavior with Neural Networks. *J. Eng. Mech.* **1991**, *117*, 132–153. [CrossRef]
36. Wu, X.; Ghaboussi, J. *Neural Network-Based Material Modeling*; Technical Report; University of Illinois at Urbana-Champaign: Champaign, IL, USA, 1995.
37. Ghaboussi, J.; Sidarta, D.; Lade, P. Neural network based modelling in geomechanics. In Proceedings of the 8th International Conference on Computer Methods and Advances in Geomechanics, Morgantown, WV, USA, 22–28 May 1994.
38. Ellis, G.W.; Yao, C.; Zhao, R.; Penumadu, D. Stress-Strain Modeling of Sands Using Artificial Neural Networks. *J. Geotech. Eng.* **1995**, *121*, 429–435. [CrossRef]
39. Ghaboussi, J.; Pecknold, D.A.; Zhang, M.; Haj-Ali, R.M. Autoprogressive training of neural network constitutive models. *Int. J. Numer. Methods Eng.* **1998**, *42*, 105–126. [CrossRef]
40. Ping, L.; Kemin, X.; Yan, L.; Jianrong, T. Neural network prediction of flow stress of Ti-15-3 alloy under hot compression. *J. Mater. Process. Technol.* **2004**, *148*, 235–238. [CrossRef]
41. Mandal, S.; Sivaprasad, P.V.; Venugopal, S.; Murthy, K.P. Artificial neural network modeling to evaluate and predict the deformation behavior of stainless steel type AISI 304L during hot torsion. *Appl. Soft Comput. J.* **2009**, *9*, 237–244. [CrossRef]
42. Sun, Y.; Zeng, W.D.; Zhao, Y.Q.; Zhang, X.M.; Shu, Y.; Zhou, Y.G. Modeling constitutive relationship of Ti40 alloy using artificial neural network. *Mater. Des.* **2011**, *32*, 1537–1541. [CrossRef]
43. Li, H.Y.; Hu, J.D.; Wei, D.D.; Wang, X.F.; Li, Y.H. Artificial neural network and constitutive equations to predict the hot deformation behavior of modified 2.25Cr-1Mo steel. *Mater. Des.* **2012**, *42*, 192–197. [CrossRef]
44. Abueidda, D.W.; Koric, S.; Sobh, N.A.; Sehitoglu, H. Deep learning for plasticity and thermo-viscoplasticity. *Int. J. Plast.* **2021**, *136*, 102852. [CrossRef]
45. Heider, Y.; Wang, K.; Sun, W.C. SO(3)-invariance of informed-graph-based deep neural network for anisotropic elastoplastic materials. *Comput. Methods Appl. Mech. Eng.* **2020**, *363*, 112875. [CrossRef]
46. Ghavamian, F.; Simone, A. Accelerating multiscale finite element simulations of history-dependent materials using a recurrent neural network. *Comput. Methods Appl. Mech. Eng.* **2019**, *357*, 112594. [CrossRef]
47. Mozaffar, M.; Bostanabad, R.; Chen, W.; Ehmann, K.; Cao, J.; Bessa, M.A. Deep learning predicts path-dependent plasticity. *Proc. Natl. Acad. Sci. USA* **2019**, *116*, 26414–26420. [CrossRef]
48. Gorji, M.B.; Mozaffar, M.; Heidenreich, J.N.; Cao, J.; Mohr, D. On the potential of recurrent neural networks for modeling path dependent plasticity. *J. Mech. Phys. Solids* **2020**, *143*, 103972. [CrossRef]
49. Goodfellow, I.; Bengio, Y.; Courville, A. *Deep Learning*; MIT Press: Cambridge, MA, USA, 2016.
50. Masi, F.; Stefanou, I.; Vannucci, P.; Maffi-Berthier, V. Thermodynamics-based Artificial Neural Networks for constitutive modeling. *J. Mech. Phys. Solids* **2021**, *147*, 104277. [CrossRef]
51. Raissi, M.; Perdikaris, P.; Karniadakis, G.E. Physics-informed neural networks: A deep learning framework for solving forward and inverse problems involving nonlinear partial differential equations. *J. Comput. Phys.* **2019**, *378*, 686–707. [CrossRef]
52. Xu, K.; Huang, D.Z.; Darve, E. Learning constitutive relations using symmetric positive definite neural networks. *J. Comput. Phys.* **2021**, *428*, 110072. [CrossRef]
53. Scott Kessler, B.; El-Gizawy, A.S.; Smith, D.E. Incorporating neural network material models within finite element analysis for rheological behavior prediction. *J. Press. Vessel Technol. Trans. ASME* **2007**, *129*, 58–65. [CrossRef]
54. Zhang, A.; Mohr, D. Using neural networks to represent von Mises plasticity with isotropic hardening. *Int. J. Plast.* **2020**, *132*, 102732. [CrossRef]
55. Liu, X.; Tao, F.; Yu, W. A neural network enhanced system for learning nonlinear constitutive law and failure initiation criterion of composites using indirectly measurable data. *Compos. Struct.* **2020**, *252*, 112658. [CrossRef]
56. Huang, D.Z.; Xu, K.; Farhat, C.; Darve, E. Learning constitutive relations from indirect observations using deep neural networks. *J. Comput. Phys.* **2020**, *416*, 109491. [CrossRef]
57. Xu, K.; Tartakovsky, A.M.; Burghardt, J.; Darve, E. Learning viscoelasticity models from indirect data using deep neural networks. *Comput. Methods Appl. Mech. Eng.* **2021**, *387*, 114124. [CrossRef]
58. Tao, F.; Liu, X.; Du, H.; Yu, W. Learning composite constitutive laws via coupling Abaqus and deep neural network. *Compos. Struct.* **2021**, *272*, 114137. [CrossRef]
59. Liu, X.; Tao, F.; Du, H.; Yu, W.; Xu, K. Learning nonlinear constitutive laws using neural network models based on indirectly measurable data. *J. Appl. Mech. Trans. ASME* **2020**, *87*, 1–8. [CrossRef]
60. Marek, A.; Davis, F.M.; Pierron, F. Sensitivity-based virtual fields for the non-linear virtual fields method. *Comput. Mech.* **2017**, *60*, 409–431. [CrossRef] [PubMed]
61. Pierron, F.; Grédiac, M. *The Virtual Fields Method: Extracting Constitutive Mechanical Parameters from Full-Field Deformation Measurements*; Springer Science & Business Media: Berlin/Heidelberg, Germany, 2012.
62. Andrade-Campos, A. Development of an Optimization Framework for Parameter Identification and Shape Optimization Problems in Engineering. *Int. J. Manuf. Mater. Mech. Eng. (IJMMME)* **2011**, *1*, 57–59. [CrossRef]
63. Martins, J.; Andrade-Campos, A.; Thuillier, S. Calibration of anisotropic plasticity models using a biaxial test and the virtual fields method. *Int. J. Solids Struct.* **2019**, *172–173*, 21–37. [CrossRef]
64. Scipy. scipy.integrate.odeint—SciPy v1.1.0 Reference Guide. Avaliable online: https://docs.scipy.org/doc/scipy-1.1.0/reference/ (accessed on 15 February 2022).

65. Scikit-Learn. 3.2. Tuning the Hyper-Parameters of an Estimator—Scikit-Learn 0.20.0 Documentation. Avaliable online: https://scikit-learn.org/stable/modules/grid_search.html (accessed on 15 February 2022).
66. Scikit-Learn. 3.5. Validation Curves: Plotting Scores to Evaluate Models—Scikit-Learn 0.20.1 Documentation. Avaliable online: https://scikit-learn.org/stable/modules/learning_curve.html (accessed on 15 February 2022).
67. Aquino, J.; Andrade-Campos, A.; Souto, N.; Thuillier, S. Design of heterogeneous mechanical tests—Numerical methodology and experimental validation. *AIP Conf. Proc.* **2018**, *1960*, 1–20. [CrossRef]
68. Aquino, J.; Andrade-Campos, A.G.; Martins, J.M.P.; Thuillier, S. Design of heterogeneous mechanical tests: Numerical methodology and experimental validation. *Strain* **2019**, *55*, e12313. [CrossRef]
69. He, K.; Zhang, X.; Ren, S.; Sun, J. Delving deep into rectifiers: Surpassing human-level performance on imagenet classification. In Proceedings of the IEEE International Conference on Computer Vision, Santiago, Chile, 11–18 December 2015; pp. 1026–1034. [CrossRef]
70. Ertel, W. *Introduction to Artificial Intelligence*; Springer: Berlin/Heidelberg, Germany, 2017.
71. Rosenblat, F. The perceptron: A probabilistic model for information storage and organization in the brain. *Psychol. Rev.* **1958**, *65*, 386–408. [CrossRef] [PubMed]
72. Cybenko, G. Approximation by Superpositions of a Sigmoidal Function. *Math. Control Signals Syst.* **1989**, *2*, 303–314. [CrossRef]
73. Bock, F.E.; Aydin, R.C.; Cyron, C.J.; Huber, N.; Kalidindi, S.R.; Klusemann, B. A Review of the Application of Machine Learning and Data Mining Approaches in Continuum Materials Mechanics. *Front. Mater.* **2019**, *6*, 110. [CrossRef]
74. LeCun, Y.; Haffner, P.; Bottou, L.; Bengio, Y. Object Recognition with Gradient-Based Learning. In *Shape, Contour and Grouping in Computer Vision*; Springer: Berlin/Heidelberg, Germany, 1999; pp. 319–345. [CrossRef]
75. Alaloul, W.S.; Qureshi, A.H. Data Processing Using Artificial Neural Networks. In *Dynamic Data Assimilation—Beating the Uncertainties*; Harkut, D.G., Ed.; IntechOpen: London, UK, 2020.
76. Ambroziak, A.; Kłosowski, P. The elasto-viscoplastic Chaboche model. *Task Q.* **2006**, *10*, 49–61.

metals

*Article*

# Application of Machine Learning to Bending Processes and Material Identification

**Daniel J. Cruz [1,*], Manuel R. Barbosa [2], Abel D. Santos [2], Sara S. Miranda [1] and Rui L. Amaral [1]**

1 INEGI, Institute of Science and Innovation in Mechanical and Industrial Engineering, R. Dr. Roberto Frias 400, 4200-465 Porto, Portugal; smiranda@inegi.up.pt (S.S.M.); ramaral@inegi.up.pt (R.L.A.)
2 FEUP, Faculty of Engineering, University of Porto, R. Dr. Roberto Frias, 4200-465 Porto, Portugal; mbarbosa@fe.up.pt (M.R.B.); abel@fe.up.pt (A.D.S.)
* Correspondence: dcruz@inegi.up.pt

**Abstract:** The increasing availability of data, which becomes a continually increasing trend in multiple fields of application, has given machine learning approaches a renewed interest in recent years. Accordingly, manufacturing processes and sheet metal forming follow such directions, having in mind the efficiency and control of the many parameters involved, in processing and material characterization. In this article, two applications are considered to explore the capability of machine learning modeling through shallow artificial neural networks (ANN). One consists of developing an ANN to identify the constitutive model parameters of a material using the force–displacement curves obtained with a standard bending test. The second one concentrates on the springback problem in sheet metal press-brake air bending, with the objective of predicting the punch displacement required to attain a desired bending angle, including additional information of the springback angle. The required data for designing the ANN solutions are collected from numerical simulation using finite element methodology (FEM), which in turn was validated by experiments.

**Keywords:** artificial neural networks (ANN); machine learning (ML); press-brake bending; air-bending; three-point bending test; sheet metal forming

**Citation:** Cruz, D.J.; Barbosa, M.R.; Santos, A.D.; Miranda, S.S.; Amaral, R.L. Application of Machine Learning to Bending Processes and Material Identification. *Metals* **2021**, *11*, 1418. https://doi.org/10.3390/met11091418

Academic Editor: Lijun Zhang

Received: 30 July 2021
Accepted: 3 September 2021
Published: 7 September 2021



## 1. Introduction

Sheet metal forming has been employed for centuries in diverse manufacturing industries to create a wide range of products that may be used in several applications. Among different forming techniques, sheet bending and stamping can be considered the most important variants in forming industry. These techniques have been continuously improved in recent decades to meet the growing need for lightweight metallic components in the automotive sector in order to address environmental concerns about energy efficiency and emissions [1,2]. To bend a sheet metal material, different methods can be used such as air bending, coining, and bottom bending. Air bending, Figure 1a, is a process in which the punch deforms the sheet by bending without the sheet being coined against the bottom die. Therefore, it is frequently the preferred bending method because it provides a high level of flexibility, as it is possible to obtain different bending angles using the same set of tools by only controlling the punch stroke. However, this process is characterized by strong nonlinear behavior, considering its parameters and their interrelationships [3].

In bending operations, one of the most important issues to consider is the springback effect. In fact, the removal of the tools causes the release of the installed residual stresses, leading to elastic recovery of the material and a change in the final bending angle. Consequently, estimating the springback effect becomes a vital requirement for achieving an accurate and regulated procedure. To address this issue, several authors tried to estimate the springback behavior in bending operations in order to develop compensation methods based on experimental, analytic and numerical approaches. The authors of [4–8] proposed analytic solutions to reproduce the evolution of the bending angle with the punch displacement for a certain combination of sheet material and tools. However, these analytical approaches are often established upon simplifications and assumptions based on the material properties and tool geometry, which sometimes lead to inconsistent results [9]. To overcome these limitations, finite element analysis (FEA) is widely used as a process modeling tool always supported by experimental validation [10,11]. In this context, [12] studied the springback effect of different types of high-strength steels using FEA models. The authors of [13] developed the smooth displacement adjustment (SDA) method and the surface controlled overbending (SCO) method to optimize the tool shape for a forming process, in order to increase the geometry accuracy of the product after springback. The authors of [14] proposed a springback compensation method for singly curved products which are established based on the numerical solution of springback behavior. Numerical methods generally present a reliable prediction of the resultant outcome of a given forming process, although their use requires the creation of an appropriate model, which easily becomes a complex process due to the need to adjust multiple parameters. Furthermore, the process of creating and running models to achieve results can be a time-consuming task and demands a high computational cost, especially for complex models and when modifications are required to evaluate various alternatives [15,16].

Recently, there has been an increasing use of machine learning (ML) algorithms in various applications related to sheet metal forming to improve decision making and achieve cost-effective, defect-free, and optimal manufacturing quality [17,18]. The ML algorithms can be divided mainly in three categories: supervised learning, unsupervised learning [19], and reinforcement learning [20]. Generally, supervised learning is preferred and is used in classification or regression problems, encompassing support vector machine (SVM) algorithms, naive Bayes classifier, decision tree, the *K*-nearest neighbor (*K*NN) algorithm and artificial neural networks (ANN). The authors of [21] used SVM to estimate the springback of a micro W-bending process with high prediction accuracy and generalization performance. The authors of [22] compared the performance of different machine learning algorithms (multilayer perceptron type ANN, random forest, decision tree, naive Bayes, SVM, *K*NN, and logistic regression) in predicting springback and maximum thinning in two different forming geometries, namely U-channel and square cup. The authors concluded that the multilayer perceptron algorithm was the best in identifying the springback, with a slightly higher score than SVM.

Artificial neural networks, among the various types of learning algorithms, are widely used in sheet metal forming processes due to their ability to overcome the limitations imposed by nonlinearities and the multiple parameters involved in forming problems. Several articles on air bending have been published, following the use of artificial neural networks [23–27]. The authors of [28] studied the use of ANN on modeling the air V-bending processes using both an analytical and experimental data set and demonstrated the capability of ANN to model the springback problem. The authors of [29] implemented a neural network in order to predict the stepped binder force trajectory for different punch displacements, in a plane strain channel forming process. The authors of [30] evaluated the applicability of ANN to the problem of choosing a tool geometry to bend a component with a defined shape. A finite element model created to simulate the bending process and a genetic algorithm (GA) were used to optimize the weights of an artificial neural network, thus reducing the deviation between the predicted tool and the experimental solution. The authors of [31] analyzed the performance of a multilinear regression model and an ANN in predicting the springback angle in air bending processes. The results show ANN outperforming the regression models approach for the evaluated cases. The authors of [32] also investigated the effect of bending and springback angles in bending processes. In this case, experimental data obtained from FEA models was used to design and train the developed ANN models. The results confirm the validity of the FEA analysis and consequently their capability to provide data for developing ANN. The authors of [33] developed a combination of error backpropagation neural network and spline function (BPNN-Spline) in order to estimate the springback angle in a V-die bending process. The results showed that the proposed BPNN-Spline model outperforms the traditional ANN in predicting the bending angles for different punch displacements. The authors of [3] developed a methodology based on ANN and FEA, capable of establishing the specific punch displacement for bending a sheet metal material according to the desired forming angle in press brake bending. The results showed that the developed methodology can successfully predict the required punch penetration to achieve a given bending angle by considering results both for geometry after springback and also geometry before springback. More recently, the authors of [34] proposed a novel theory-guided regularization method for deep neural network (TG-DNN) training, which uses the material Swift's law as the guidance to predict the deformed workpiece geometry after springback and the corresponding forming parameter of loading stroke. The authors conclude that the proposed TG-DNN outperforms the conventional pure data-driven DNN for its superior generalization accuracy, especially when only scarce and scattered experimental data are available for training.

In all these studies, the material is known and its mechanical characterization is described usually through work hardening laws (e.g., Swift law), determined in advance. Thus, the bending test is not usually the preferred method to determine experimentally the hardening behavior of a given material and normally tensile tests are used to determine such behavior. It is known that bending tests entail inhomogeneous stress and strain distributions in the material, making it impossible to infer the stress–strain relationship directly from the experiment [35]. Typically, the material parameters identification is performed by inverse methodologies where the objective is to find the parameters that provide an optimal fit for a wide range of experiments. The main drawback of this strategy is being time-consuming, particularly when several test conditions are used [36]. Recently, neural networks have been used to replace or improve the constitutive model obtained by inverse analysis [36–40]. For example, [41] developed a machine-learning based Johnson-Cook (JC) plasticity model to capture the non-monotonic effect of the temperature and strain rate on the hardening response for DP800 steel. The authors concluded that by combining the neural network and existent material model, all the experimental data were described with high accuracy. None of these studies, however, are directly connected to the use of the bending process to evaluate a material hardening behavior. This paper thus appears as an attempt to respond to this research opportunity.

The purpose of the current study is to evaluate the applicability of machine learning algorithms on bending procedures. Therefore, different ANN will be developed in order to explore the modeling capabilities and to resolve two different problems directly related to bending solicitations herein called Problem (a) and Problem (b). Problem (a) is related to using neural networks to determine the hardening behavior of a given material using only the three-point bending test results. On a different perspective, Problem (b) is related to the implementation parameters of the air bending process where it is intended to develop a methodology that estimates the punch displacement required to obtain a given bending angle. In both approaches it is proposed to combine the use of a learning tool with a simulation and data generation tool (FEA) in order to train the developed ANN.

This paper is organized as follows. In Section 2, we provide an overview of the corresponding problem statements and establish the adopted methodologies to solve both problems. In Section 3, we include the neural network formulation and the implementations that showed the best performance. In Section 4, the neural network results obtained for the two problems are analyzed and discussed.

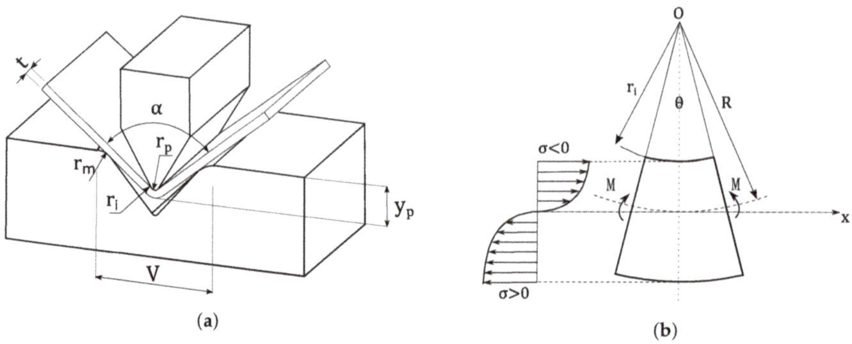


**Figure 1.** (**a**) Air bending process parameters: die opening ($V$), die radius ($r_m$), inside bending radius ($r_i$), punch nose radius ($r_p$), punch penetration ($y_p$), sheet bending angle ($\alpha$), sheet thickness ($t$); (**b**) stress distribution in plastic bending deformation.

## 2. Materials and Methods

### 2.1. Problem Statement

Air bending is one of the most frequent plastic deformation methods used on parts made from flat sheets. As represented in Figure 1a, the main objective of this process is to obtain a desired bending angle, $\alpha$, in a sheet metal material by applying a specific punch displacement, $y_p$. The main process parameters include: (a) the tools geometry, which comprises the die opening, $V$, die radius, $r_m$, and punch radius, $r_p$; (b) the blank material, which includes not only the sheet thickness, $t$, but also the material properties (e.g., type of material, hardening law, yield stress, ultimate tensile strength, elongation) [3].

When a sheet metal material is bent, it is subjected to different stress stages throughout its depth (Figure 1b). The applied moment, $M$, causes the sheet to take the form of an arc of a circle centered at point $O$ with a radius $r_i$. On the convex side, the longitudinal fibers expand, and on the concave side, they compress. These two types of fibers are separated by a third set of fibers that retain their initial length and constitute the so-called neutral axis. The neutral axis divides the straight section into two parts: one part in tension ($\sigma > 0$), and one in compression ($\sigma < 0$). On the neutral axis the longitudinal stress is zero ($\sigma = 0$) [4].

As mentioned earlier, the springback effect plays an important role in bending processes. As represented in Figure 2 the removal of the tools leads to elastic recovery of the material, which results in different bending angles before and after springback. The springback angle, $\Delta\alpha_{SB}$, corresponds to the difference between the angle after elastic recovery, $\alpha_f$, and the bending angle defined when the punch contacts the part, $\alpha_i$. The springback is entirely intercorrelated with the stress distribution on sheet metal as residual stresses [42]. Its behavior is also affected by material properties such as strain hardening, elastic property evolution, the presence of Bauschinger effects, elastic and plastic anisotropy, and tribology between contacting surfaces [43]. Although there are mathematical models for predicting springback in bending situations, most of them are simplistic and do not take into account all influential factors.

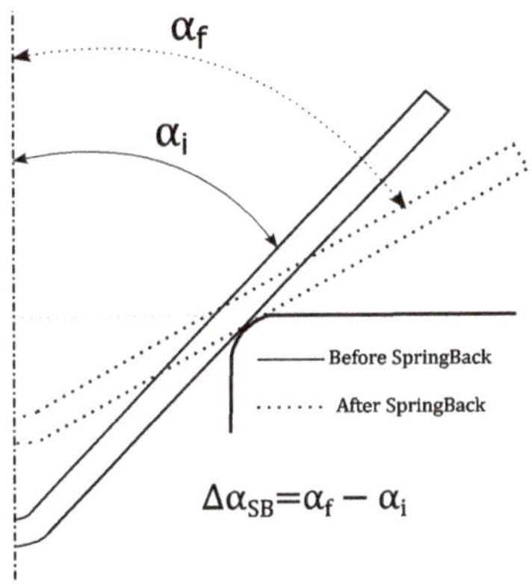


**Figure 2.** Definition of elastic recovery (springback) in bending process.

#### 2.1.1. Problem (a)—Material Characterization

The three-point bending test [44] is a classic experiment used to evaluate the behavior of a material when subjected to bending. This test is in every way similar to the air bending process, as represented in Figure 3, however the main objective within this test will be to evaluate the behavior of a certain material to pure bending loading. This test is quite simple, as it does not require any prior sample preparation (e.g., machining), and can easily be performed on a universal tensile testing machine. The challenges with this test resides in the involvement of axial and transverse forces in the bending deformation. Furthermore, friction and local deformation beneath the contact points can also affect the results [45]. To convert the measured output from these tests (punch displacement, $y_p$ and punch force, $F_p$) into stress–strain ($\sigma$-$\varepsilon$) response, inverse fitting models are usually used in literature [46–48]. These methods require accurate modeling of the test with a predetermined hardening model and costly optimization loops. The main drawback of this strategy is being time-consuming, particularly when several experimental tests are used. Another way is by using analytical approaches as for example the derivation proposed by [49].

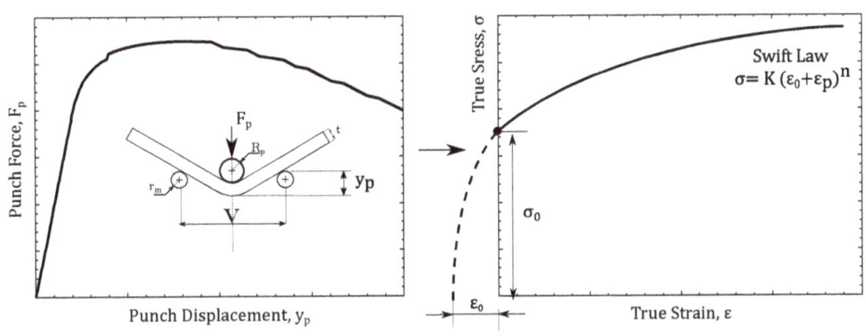


**Figure 3.** Schematic illustration of the three-point bending test and definition of the main objectives for problem (a).

The objective of this first problem is to develop a new methodology based on neural networks to characterize the hardening behavior of a material using the results obtained in a three-point bending test. Accordingly, it is proposed to implement a new procedure to replace the traditionally inverse and analytical fitting methods in order to easily characterize the hardening behavior of given material. The developed neural networks should consider as input the punch force displacement curve obtained in a three point bending test and provide the characteristic parameters of a Swift hardening law, as represented in

Equation (1). The methodology should characterize materials with a $K$ parameter between [400–1600] and $n$ parameter between [0.05–0.35]. These limits are consistent with the objectives of characterizing materials widely used in industrial applications such as sheet metal steel, ranging from mild steels to AHSS. Accordingly, these parameters for $K$ (strength coefficient) and $n$ (work hardening exponent) will include every behavior for materials of interest, both for strength and also for different hardening behavior. In this problem, the value of $\varepsilon_0$ was taken as fixed with a constant value of 0.01. Additionally, only 0.8 mm sheet thickness will be considered. The Swift law parameters and their limits considered in this bending problem (a) are summarized in Table 1. This table also includes the geometry of the chosen setup to perform the three-point bending test.

$$\sigma = K(\varepsilon_0 + \varepsilon_p)^n \quad (1)$$

**Table 1.** Swift parameters and bending test geometry used in problem (a).

| **Swift Parameters** | | | **3-Point Bending Test Geometry** | | |
|---|---|---|---|---|---|
| $K$ | $n$ | $\varepsilon_0$ | $R_m$ [mm] | $R_p$ [mm] | $V$ [mm] |
| 400–1600 | 0.05–0.35 | 0.01 | 4 | 4 | 50 |

2.1.2. Problem (b)—Air Bending, Forming and Springback Prediction

The second problem of this work will address the air bending process, considering the influence of springback on the process. As referred, the major advantage of this bending technique is the ability to use the same set of tools (punch and die) to achieve multiple bending angles in different materials. In this context, establishing and controlling the amount of punch penetration becomes the most important process parameter to establish and control. However, obtaining the required bending angle with adequate accuracy can become a major challenge due to the many parameters involved in the process. The main objective of Problem (b) is the development of a new methodology based on neural networks to estimate the required punch displacement, $y_p$, to produce a given bending angle, $\alpha$, using well-defined process conditions. These bending conditions include not only the tools geometry ($V$, $r_m$, $r_p$) but also the material thickness, $t$.

Due to the problem's geometric simplicity, analytical relationships involving variables in the bending process, such as punch displacement as a function of bending angle, die radius, die opening, and sheet thickness, can be defined. Previous research has shown that J. Bessa Pacheco's analytical model ($Y_{JBP}$) [8] is the one that best reproduces the behavior ($y_p = f(\alpha)$) seen in press brake air bending. This proposal, represented in Equation (2), incorporates the sheet metal thickness ($t$), bending angle ($\alpha$), die opening ($V$), inside bending radius ($r_i$), and die radius ($r_m$) and will be utilized as a supplement to the results analysis in the current work.

$$Y_{JBP} = \frac{V}{2 \cdot \tan(\frac{\alpha}{2})} + (r_i + t + r_m) \times \frac{1 - \sin(\frac{\alpha}{2})}{\sin(\frac{\alpha}{2})} \quad (2)$$

The die opening is a significant influencing factor in the bending process. The typical bender's practice is to choose the die opening based on the sheet thickness. The ideal combination of sheet thicknesses and die opening can be defined by practical guidelines derived from industrial process experience. Linear relationships between those variables (Equation (3)) characterized by a scalar $k_{vt}$ factor can be utilized to determine the boundaries of suitable combinations. Normally, $k_{vt}$ factors with values between 6 and 10 establish the limits of appropriate combinations for a correct bending operation.

$$V = k_{vt} \cdot t \quad (3)$$

Figure 4 represents industrial practices for press-brake bending, in which standard die dimensions ($V$ opening) are considered ($V$ = 11.5, 18.3, 23.1, 34.2, 43.7, 53.7 mm) and the intended sheet metal thickness ($t$ between 0.5 and 6.0 mm) for practical applications. It is seen also in Figure 4 the recommended industrial practice of $V/t$ ratios between 6 and 10, a region defined by such straight lines. Press-brake bending performed outside this region of recommended ratios will either give rise for high springback results (case A, $V/t > 10$) or punch indentations (case C, $V/t < 6$). A recommended ratio is represented by case B, for a $6 < V/t < 10$. Accordingly, when comparing cases A and B the same die is used ($V = 43.7$) for a thickness of 1 mm (non-recommended $V/t = 43.7$) and 5 mm (recommended $V/t = 8.7$); as a consequence it is seen an excessive curvature geometry for case A, thus resulting an higher springback, after bending. This also means that when comparing these two situations, case A (non-recommended $V/t$) will have a different relation between bent angle and punch penetration when compared to case B (recommended $V/t$). On the other hand, $V/t$ ratios less than 6 should be avoided since greater pressures are created at tool/blank interfaces, and localized deformations (i.e., indentations) emerge at such contact zones, increasing the probability of fracture. For example, comparing cases C and B, the same sheet thickness is used ($t = 5$ mm) to be processed by different die openings: case B uses $V = 43.7$ mm (recommended $V/t = 8.7$), while case C uses $V = 23.1$ mm (non-recommended $V/t = 4.6$). It is seen (Figure 4) that for case C the punch indents the material, causing a superficial defect and also contributing to a different relation between bent angle and punch, when compared to recommended case B.

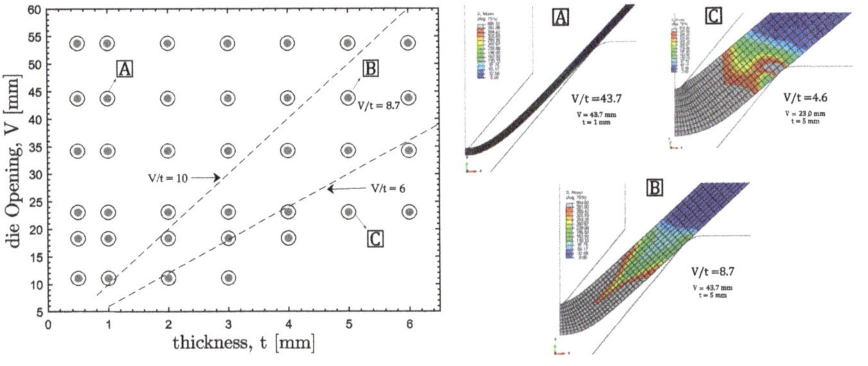


**Figure 4.** $V$-$t$ diagram: combinations of tested $V$ and $t$ values (points) and limits (straight lines) representing industrial practice reference rules.

The influence of springback in the $y_p$-$\alpha$ relation, for A, B, and C is represented in Figure 5. In each graph it is presented the punch displacement, $y_p$, needed to produce a desired bending angle, $\alpha$, between 90° and 180° and considering two distinct situations: before and after springback. This representation confirms that the springback effect is more evident for case A since the difference between the punch displacement before and after springback is higher. Thus, in this bending condition the springback need to be taken into account when predicting the punch displacement, in order to achieve a proper bending result. In addition, the punch displacement predicted by the $Y_{JBP}$ analytical model is also shown, for each case, in Figure 5. Generally speaking, the analytical model prediction is in closer agreement with the numerical reference value before springback for all cases. This is quite expected since the mathematical formulation of this model only includes the tool geometry and the material sheet thickness, ignoring the springback effect. Thus, the results obtained using the analytical approach are suitable for regions with reduced springback effect ($V/t < 10$) while for higher ratios the accuracy of results is reduced. Nevertheless, even for this zone ($V/t < 10$), the analytical approaches can be synonymous of errors.

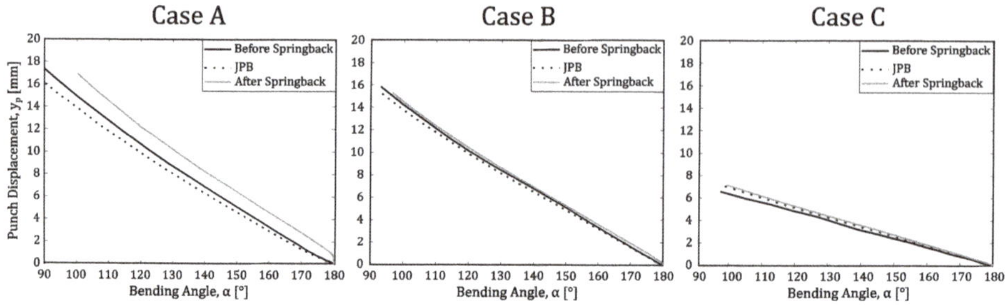


**Figure 5.** Punch displacement ($y_p$) graphs as a function of required bending angle ($\alpha$) for cases A–C.

The springback angle, $\Delta\alpha_{SB}$ is represented in Figure 6 for different $V/t$ ratios and assuming different desired bending angles, $\alpha$. The values for the cases A and B are also represented. This representation reinforces that higher $V/t$ ratios promotes the material elastic recovery, which is translated in a higher difference between the angle before and after tooling removal. Additionally, it is visible that springback angle increases during the bending operation. Therefore, for a final bending angle of 90°, the springback angle for cases A and B are 14° and 3° respectively.

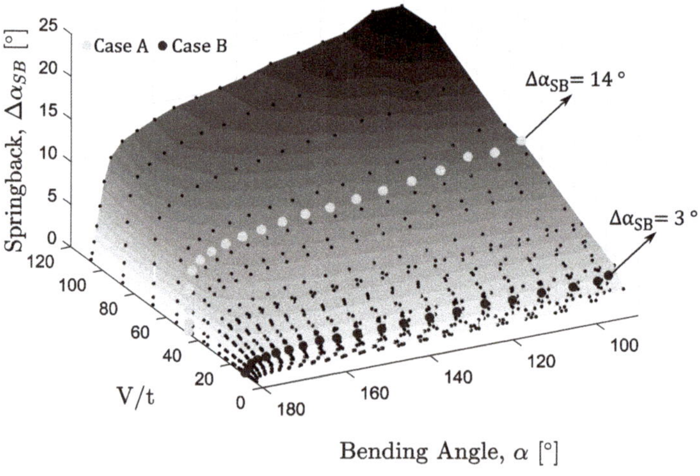


**Figure 6.** Springback angle values ($\Delta\alpha_{SB}$) for different bending conditions ($V/t$ ratios) and considering different required bending angles ($\alpha$).

To sum up, the air bending problem (b) of this work aims to develop a method capable of providing a fast and accurate estimate of the punch displacement, $y_p$, to obtain a desired angle $\alpha$ for the complete range of $V/t$ ratios presented in Figure 4. The developed neural networks should not only present the punch displacement for a desired bending angle but also an estimate of a springback angle for each case. For such purpose, only one material will be considered, a dual phase steel (DP590) with thicknesses between $0.5 < t < 6$ mm. The die opening values considered are defined between $10 < V < 50$ mm which translate in a "die opening-thickness" ratio between $1.6 < V/t < 100$. The selected die openings ($V$) correspond to dimensions for standard industrial press-brake bending dies and the sheet metal thicknesses represent the most common practical intended applications.

### 2.2. Proposed Approach Using ANN

Modeling and solving a problem using artificial neural networks (ANN) can be included within machine learning (ML) approaches, as its development is based on the available data that characterizes the problem. Most often, the ANN algorithms consist of a supervised learning method, although unsupervised ANN algorithms are also used. Considering its widespread interest and use since mainly the late 1980s [50,51], currently, there has been a renewed interest due to the latest developments associated with, which is normally referred to as deep learning, or deep neural networks (DL/DNN) [52]. These DNN entail an increased complexity of the models and have been successful in numerous pattern recognition problems. In the present work we will be concentrated on a conventional ANN approach, i.e., shallow ANN, rather than on DNN which we believe will be of interest when escalating the approach to a higher scale of generalization ability.

The main idea behind ANN, either shallow or deep models, is a structure of multiple simple processing elements (PE), or nodes, with a pattern of interconnections (weights) that process information at its input to provide a solution, to the problem, at its output. As such, it is usually characterized as being inspired in the human brain. In a supervised learning algorithm the process of learning consists on the adjustment of parameters (i.e., weight's values of the interconnections) in order to minimize an error function that represents a measure of the deviation between the known, or target response, and the ANN response, when examples, or instances of the problem are available in the data sets. A typical ANN architecture is represented in Figure 7 where the nodes (i.e., PE) are organized in successive layers from Input to Output, and without backwards connections, i.e., in a feedforward structure [50]. The use of nonlinear functions in the PE, combined with multiple PE, enables ANN to model highly nonlinear problems. The definition of the Input and Output layers defines how the problem is formulated and encoded in the ANN. This structure provides great flexibility by allowing the combination of different types of information we can give to a single ANN, as well as enabling multiple types of information a single ANN can provide to the user. Before an ANN can be ready to provide a solution in a use phase, its variable parameters must be adjusted in a training, or learning phase. This requires the selection of the performance function and the algorithm used to minimize this function by selecting the appropriate values for each adjustable parameter. Multiple algorithms can be used, being most common backpropagation or gradient descent based learning methods and the mean squared error (MSE) as the performance function, namely in function approximation type problems. Having the ANN architecture defined and the training phase completed, the ANN represents a well-defined mathematical function that delivers output values when input values are specified. The ANN developed in the present work for both problems were based on this architecture. As represented in Figure 8 for problem (a) the objective is to define and encode at the input layer, information relative to the force-displacement ($F_p$-$y_p$) bending test, in order to obtain at the output layer the parameters of the constitutive hardening law —Equation (1). In problem (b) the objective is to define at the ANN input layer the die opening ($V$), the material thickness ($t$) and the desired bending angle ($\alpha$) and having at the ANN output the punch displacement ($y_P$) that should be used and the magnitude of the springback angle ($\Delta\alpha_{SB}$). Both problems were considered as function approximation type problems. The required training and validation data sets were obtained through the use of FEA models, as described in the next section.

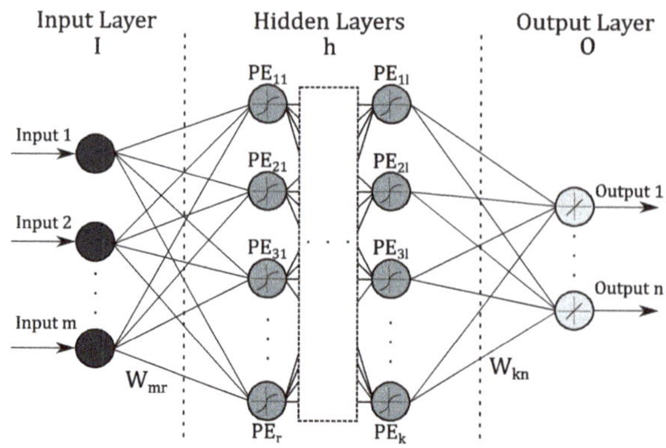


**Figure 7.** Neural network: example of a feedforward structure, with $m$ input nodes, several hidden layers with specific numbers of processing units (PE) each implementing a nonlinear function (i.e., sigmoid), $n$ output nodes implementing a linear function and interconnections weights (i.e., $W_{kn}$ and $W_{mr}$).

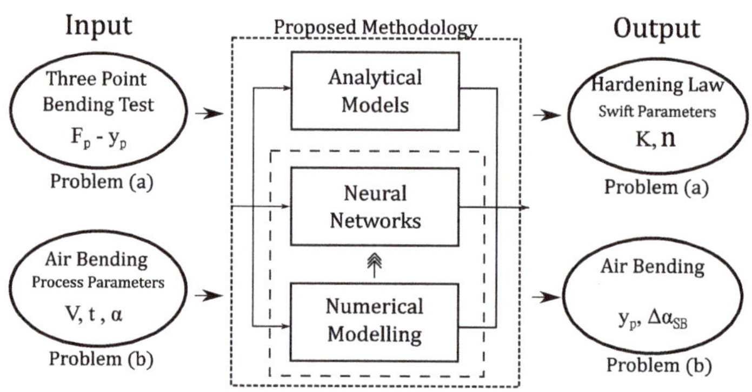


**Figure 8.** Schematic representation of the main objectives and the proposed methodology for solving problems (a) and (b).

### 2.3. Finite Element Model

The air bending process and the three point bending test can be defined as plane strain problems, in which the blank width is much larger than the blank thickness. Due to the symmetry of these two proposed bending processes, only half of real experimental setup was considered in the 2D finite element model. Figures 9 and 10 illustrates the fundamental geometry and variables defined for FE models, used for the bending problem (a) and press-brake bending problem (b), respectively. Both numerical models were validated by experiments in previous works [3,53].

#### 2.3.1. Problem (a)—Material Characterization

The three-point bending process simulation, problem (a) was performed using ABAQUS with implicit analysis (ABAQUS/ Standard). The blank material is modeled with an elasto-plastic behavior using the Swift law for the hardening curve. The selected materials and corresponding properties are presented in the Table 2. The sheet blank was discretized with 819 deformable four node solid elements (CPE4R type from ABAQUS Library) and nine layers through thickness. Mesh discretization is regularly spaced in both thickness and length (xx-axis) directions. Punch and die were modelled as analytical rigid surfaces and for Coulomb friction a value of 0.1 has been defined, which follows previous results obtained from experiments validating the numerical model [53].

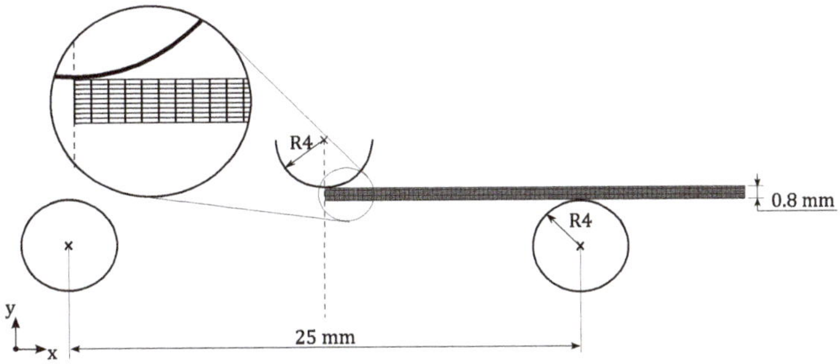


**Figure 9.** Three-point bending test geometry and variables defined for problem (a) FE model.

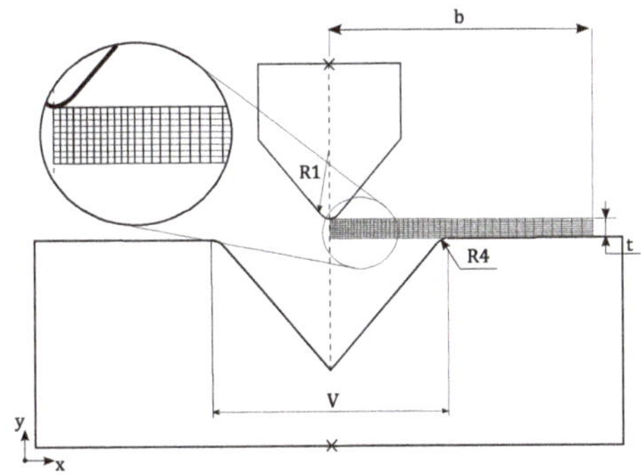


**Figure 10.** Geometry and variables defined for problem (b) FE model.

**Table 2.** Mechanical properties and Swift law hardening parameters of selected materials used in problem (a).

| Property | | |
|---|---|---|
| Elastic modulus | E [GPa] | 210 |
| Poisson coefficient | $\nu$ | 0.3 |
| | $K$ | 400–1600 |
| Swift parameters | $\varepsilon_0$ | 0.01 |
| $\sigma = K(\varepsilon_0 + \varepsilon)^n$ | $n$ | 0.05–0.35 |

2.3.2. Problem (b)—Air Bending: Forming and Springback Prediction

The press-brake air bending process simulation, problem (b) was performed using ABAQUS with implicit analysis (ABAQUS/ Standard) for this quasi-static problem. Therefore, both the bending process and the springback can be processed efficiently by using two steps for simulation. The blank material, dual-phase steel DP590, is modeled with an elastoplastic behavior using the Swift law for the hardening curve. The selected materials and corresponding properties are presented in the Table 3. The sheet blank was meshed with 450 deformable four node solid elements (CPE4R type from ABAQUS Library) and nine layers through thickness. Mesh discretization along xx direction is done with a bias ratio, so that several nodes accommodates to a small punch radius but also a similar ratio accommodates to a higher die radius, always having in mind a right balance with proportion for elements. Punch and die were modelled as analytical rigid surfaces. The friction has been considered for the interacting surfaces, with a Coulomb coefficient of 0.15, following previous results with experiments to validate the numerical model [3].

**Table 3.** Mechanical properties and Swift law hardening parameters of selected materials used in problem (b).

| Property | | DP590 |
|---|---|---|
| Elastic modulus | E [GPa] | 210 |
| Poisson coefficient | $\nu$ | 0.3 |
| Yield stress | $R_e$ [MPa] | 393 |
| Ultimate tensile strength | $R_m$ [MPa] | 641 |
| Total elongation | $e_t$ [%] | 26 |
| | $K$ | 1000 |
| Swift Parameters | $\varepsilon_0$ | 0.0024 |
| $\sigma = K(\varepsilon_0 + \varepsilon)^n$ | $n$ | 0.155 |

Python scripts were developed to create and modify automatically the parts of finite element model, for different bending conditions and also to submit the analysis, since a total of 740 analysis were considered. In order to acquire results from each of previous numerical simulations, an additional python script was written so that the fundamental data is retrieved for the ANN development such as punch displacement, bending angle, bending conditions and other variables.

## 3. Neural Networks Implementation

Once a specific architecture is selected, developing an ANN solution involves mainly an iterative process of specifying values for the variable parameters in the learning algorithms and evaluating the resulting performance, in order to select the best settings. Due to the high number of parameters involved, this development turns out to be a case dependent problem with few guidelines available to guarantee success in every case.

The first step is the generation and analysis of the data available for each problem. The higher the quantity of data available, the better are the expectations of developing a successful ANN solution. The data used should represent the variety of situations, i.e., instances, of the problem in order to obtain good generalization abilities. In order to promote these objectives data is normally divided in: data sets used during the learning phase for parameters adjustment (i.e., training sets) and data sets new to the ANN (i.e., test set). In some learning algorithms a third set is used (validation set), not for parameters adjustment, but to favor ANN generalization capability, by stopping training when performance deteriorates in the validation set (early stopping).

Analyzing the data is also crucial, especially in shallow ANN, as it may enable reducing the number of elements required to represent the information in the Input layer and therefore the size of the ANN. A good knowledge of the problem is also important to interpret and define the Output layer elements required. The next sections describe the specific problem formulation and respective ANN implementation using the Deep Learning Toolbox available in Matlab [54].

### *3.1. Problem (a)—Material Characterization*

In this problem, the input data, provided by FEA, consists of a punch force-displacement, $F_p$-$y_p$, curve comprising a total of 730 discrete points. In feedforward shallow ANN, each layer is fully connected to the next layer, which in this problem makes unpraticable to use all points as elements in the Input layer. So, in order to simplify the ANN structure and reduce the number of parameters to be adjusted, only five points were considered as illustrated in Figure 11a. P1 corresponds to the point that divides the elastic deformation zone, characterized by a linear $F_p$-$y_p$ trend, from the plastic deformation zone on the test curve. On the other hand, P2 represents the point of maximum force in each test. Similarly, P3, P4 and P5 are points with arbitrarily chosen fixed displacement values (15 mm, 17 mm, 20 mm) used in each curve. Therefore, as represented in Figure 11b, the feedforward ANN for this first problem will be characterized by ten processing elements corresponding to the five pairs of punch force-displacement values in the input layer and two output values corresponding to parameters $K$ and $n$ of the Swift law, represented in Equation (1).

A total of 91 curves were generated considering 13 different values for $K$ parameter and 7 values for the $n$ parameter. These curves are obtained from bending tests having the same conditions for punch displacement, from 0 mm (flat sheet blank) to 20 mm (bent specimen). Figure 12 shows the data sets (output values) used for neural network development. It can be seen that the 13 $K$ parameters were selected in the interval [400, 1600] with 100 units increments, and the 7 $n$ values with increments of 0.05 units in the interval [0.05, 0.35]. From these, 55 cases (60.4% of the total) were used as the training set. The remaining cases (39.6% of the total), corresponding to $K = 800$, $K = 1400$ and $n = 0.15$, $n = 0.25$ were randomly split into a testing and validating set, each with 18 cases.

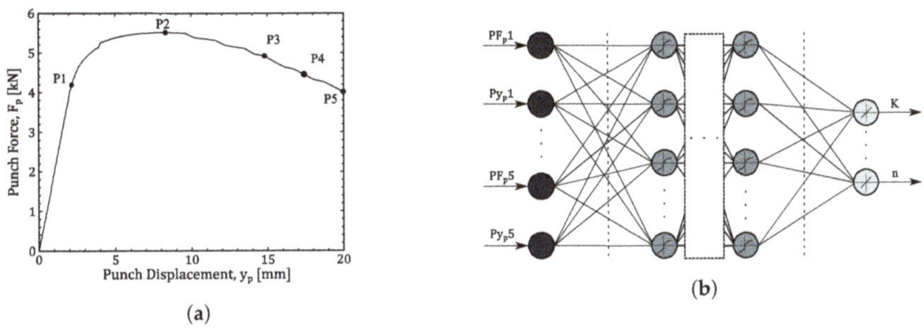


**Figure 11.** Implementation of material characterization problem: (**a**) punch force-displacement curve and the selected five pairs of values (P1–P5); (**b**) structure of the neural network with ten processing elements in the input layer corresponding to the five pairs of punch force-displacement values and two nodes in the output layer (Swift parameters — $K$ and $n$).

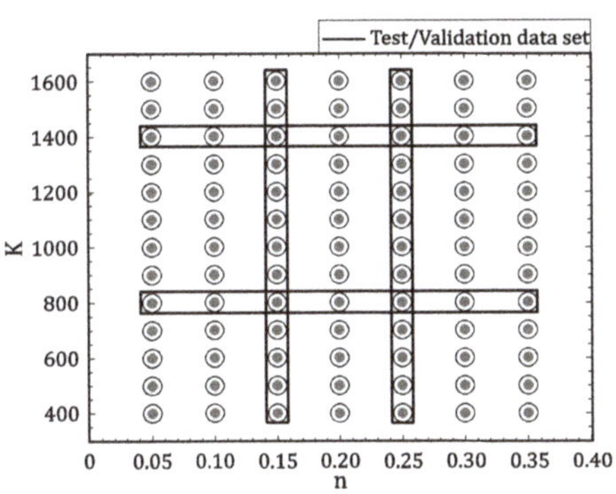


**Figure 12.** Selected data sets for problem (a).

Various combinations of hidden elements and layers were tested in multiple and repeated runs using the Levenberg–Marquardt learning algorithm and the mean squared error (MSE) performance function [55], in order to identify the size of the NN structure that seemed to fit better the available data. For the best performant cases, several runs (10) were made, starting with different initial weight values and initial learning rate parameters. The weight values were updated in a batch mode, i.e., after all data cases were presented to the ANN. The input and output values were normalized in the [−1,1] range, and several learning parameter combinations were tested. The condition to stop learning was based on the performance function on the values of the validation set (i.e., early stop). The best performance was obtained with five hidden layers, each with five elements. Following the training phase, the ANN performance were compared against the simulation solutions, after conversion of the NNs output to un-normalized values. The performance obtained can be observed through the error's histograms in Figure 13 , where the extreme error values appear on a reduced number of cases. Table 4 presents the performance of the developed ANN, after conversion to un-normalized values, for each of the three data sets (training, validating, and testing) and in terms of root mean squared error, RMSE—Equation (4), maximum and minimum extreme error values.

$$\text{RMSE} = \sqrt{\text{MSE}} = \sqrt{\frac{1}{N} \cdot \left[ \sum_{i=1}^{N} (target_i - output_i)^2 \right]} \tag{4}$$

In absolute terms it looks as the ANN performs worse in modeling the $K$ parameter comparatively to the $n$ parameter. However, when the performance measures, RMSE, are expressed in relation to the range values of each parameter (i.e., 1200 for $K$, 0.3 for $n$) and in relation to the $K$ and $n$ values for the maximum and minimum values, the performance is of similar magnitudes and behavior in the three data sets for both parameters. The overall performance (RMSE) is better in the training sets, less than 0.5% in both cases, with the validating and testing sets below 2% in the worst case. The extreme values are higher (i.e., −8.6%), but they occur in a few cases.

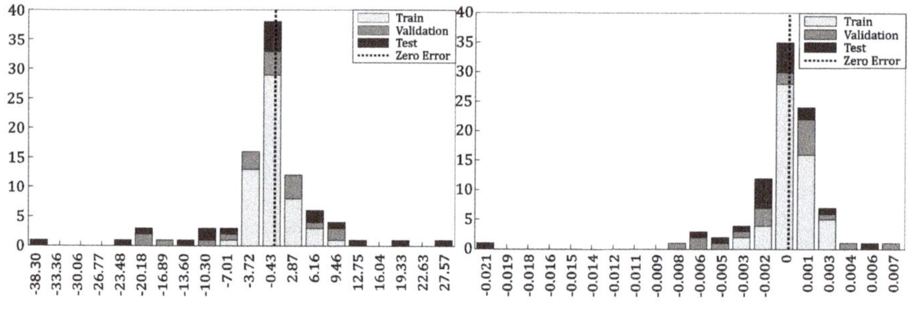


(**a**) $K$ error histogram (**b**) $n$ error histogram

**Figure 13.** Neural network histogram of errors for (**a**) $K$ output variable and (**b**) $n$ output variable considering un-normalized values, all data sets and considering grouping each of the 91 cases error in 20 classes (i.e., bins).

**Table 4.** Neural networks performance error for problem (a) in terms of RMSE, Max. and Min.

| | **Training Data Set** | | | **Validating Data Set** | | | **Testing Data Set** | | |
|---|---|---|---|---|---|---|---|---|---|
| | **RMSE** | **Max.** | **Min.** | **RMSE** | **Max.** | **Min.** | **RMSE** | **Max.** | **Min.** |
| $K$ | 2.959 | 9.508 | −7.938 | 8.955 | 10.201 | −21.225 | 15.376 | 27.567 | −38.296 |
| $n$ | 0.0013 | 0.0031 | −0.0034 | 0.0037 | 0.0078 | −0.0081 | 0.0057 | 0.0051 | −0.0216 |
| $K$ | 0.3% | 0.6% | −0.3% | 0.8% | 1.3% | −2.7% | 1.3% | 1.7% | −7.6% |
| $n$ | 0.4% | 0.9% | −3.4% | 1.2% | 5.2% | −3.4% | 1.9% | 3.4% | −8.6% |

### 3.2. *Problem (b)—Air Bending: Forming and Springback Prediction*

In this work the advantages of formulating specific ANN modeling of forming and springback in air bending are highlighted by the possibility of extending the use of press-brake variables ($V/t$ ratios) outside the limits of practical industry guidelines that are mainly supported by analytical and experience based knowledge [3]. Furthermore, it is desired to explore whether a multitask versus single task ANN [34] can prove to be more effective in increasing the generalization ability and reduce errors and outliers. The additional information in the multitask ANN can also be considered of interest for the air bending application. The input layer includes the die opening ($V$), the sheet thickness ($t$) and the desired bending angle ($\alpha$) (Figure 14) in both multitask and single task ANN. In one of the single task ANN the output layer consists on the punch displacement ($y_p$) required to obtain the desired bending angle, after removing the tool (i.e., after springback occurred). Another single task ANN provides as output, the springback angle ($\Delta\alpha_{SB}$). In the multitask ANN, the output includes, in addition to the punch displacement, the magnitude of the springback angle.

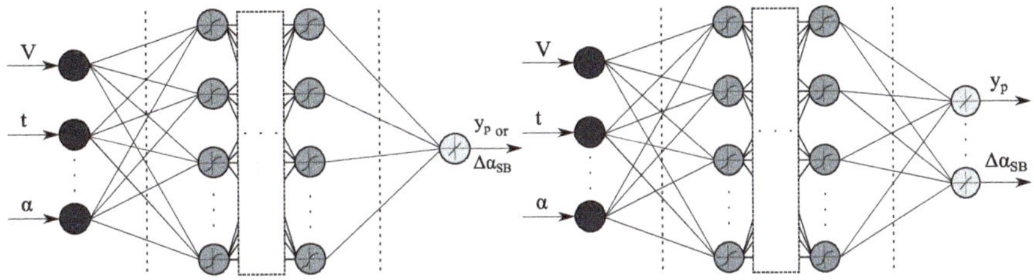


(**a**) Single-task (ST) ANN structure (**b**) Multitask (MT) ANN structure

**Figure 14.** Structure of the neural networks considering (**a**) single task (ST) and (**b**) multitask (MT) formulations.

A total number of 37 cases were used (Figure 15) resulting from different combinations of die opening and material thickness (Table 5). Two different separations of data (DD1, DD2) between training and validation/testing sets were used in the multiple runs of ANN development stages. In both separations, the same proportion of training (25/37) versus validation/testing (12/37) was used. An automated Bayesian regularization learning algorithm using the respective learning function available in Matlab ('trainbr') [55] combined with early stopping, and the mean squared error (MSE) performance function, as it could enhance reduction of extreme errors.

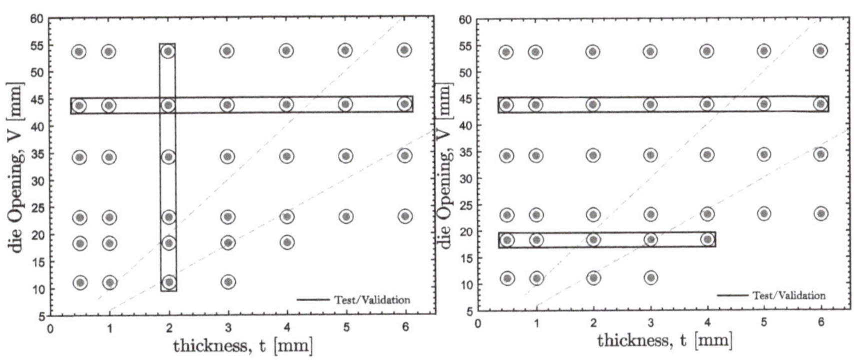


**Figure 15.** Selected bending processing conditions ($V/t$ ratios) for (**a**) data set division I (DD1) and (**b**) data set division II (DD2).

The results obtained with the considered best performant ANN, either ST-NN, MT-NN and considering the two data divisions (DD1, DD2), are represented in Figure 16. The results include an overall measure (RMSE), maximum positive and minimum negative errors, for each used data set (training, validation, testing). When considering all measures, it can be observed that, for the punch displacement ($y_p$), the performance on training, validation and testing data sets are in closer agreement for the MT-NN, although having slightly lower RMSE performance in the training set comparatively with the ST-NN. Combined with the general better performance on the outliers in all sets (training, validation, testing) it can be considered that a MT-NN generalizes better than ST-NN. When observing the results for the springback angle using the MT-NN, the same behavior is verified, with the MT-NN performing better and more homogeneously. However, the extreme error values occur in cases within the training set. These results will be further analyzed in the next section in order to evaluate the usefulness of the ANN models in both problems (a) and (b).

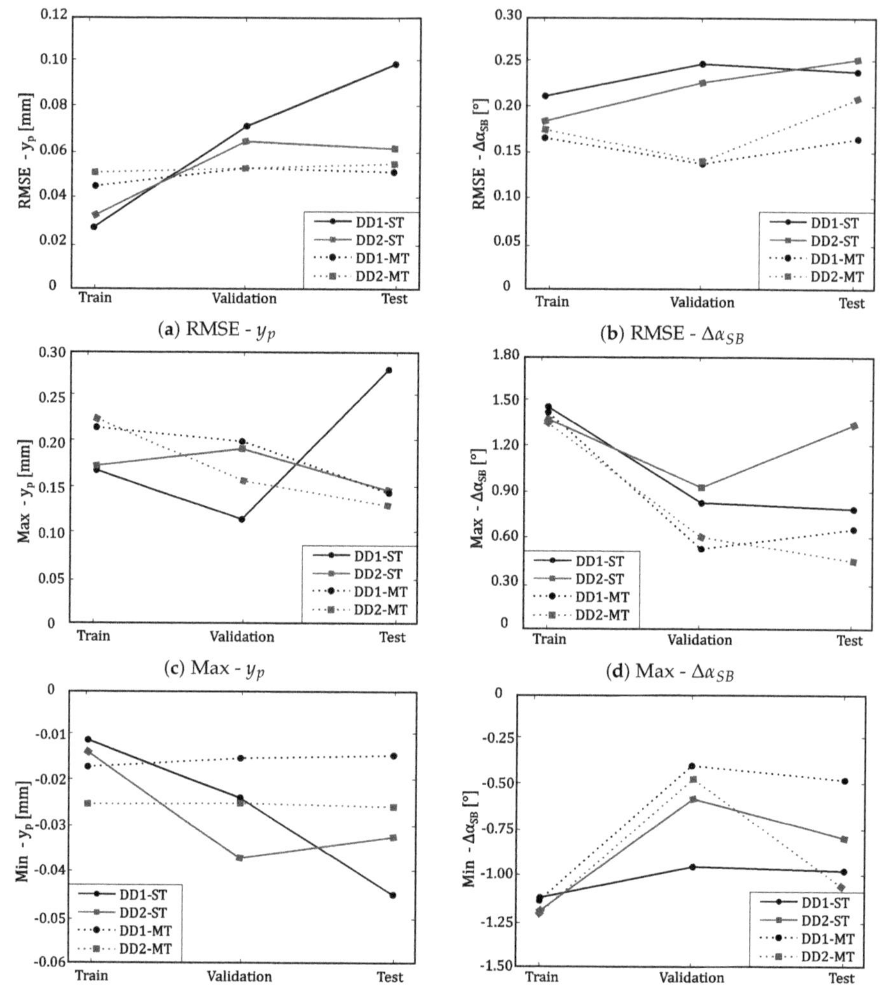


(**a**) RMSE - $y_p$ (**b**) RMSE - $\Delta\alpha_{SB}$ (**c**) Max - $y_p$ (**d**) Max - $\Delta\alpha_{SB}$ (**e**) Min - $y_p$ (**f**) Min - $\Delta\alpha_{SB}$

**Figure 16.** Performance measures of neural networks for the two different data sets (DD1 and DD2) and considering both single (ST) and multitask (MT) ANN structure; all performance measures are presented for the train, validation and test data sets; (**a**) punch displacement ($y_p$) RMSE value; (**b**) springback angle ($\Delta\alpha_{SB}$) RMSE value; (**c**) punch displacement ($y_p$) maximum value; (**d**) springback angle ($\Delta\alpha_{SB}$) maximum value; (**e**) punch displacement ($y_p$) minimum value; (**f**) springback angle ($\Delta\alpha_{SB}$) minimum value.

**Table 5.** Test combinations and dimensions for the tooling used in problem (b).

| V [mm] | $V_{nom}$ [mm] | $\alpha$ [mm] | $b$ [°] | Thickness [mm] 0.5 | 1 | 2 | 3 | 4 | 5 | 6 |
|---|---|---|---|---|---|---|---|---|---|---|
| 11.10 | 10 | 30 | 15 | · | · | · | · | | | |
| 18.30 | 16 | 30 | 25 | · | · | · | · | · | | |
| 23.06 | 20 | 30 | 35 | · | · | · | · | · | · | · |
| 34.21 | 30 | 40 | 50 | · | · | · | · | · | · | · |
| 43.73 | 40 | 80 | 50 | · | · | · | · | · | · | · |
| 53.73 | 50 | 80 | 35 | · | · | · | · | · | · | · |

## 4. Results

### *4.1. Problem (a)—Material Characterization*

In this section it will be presented, with detail, the results obtained for problem (a) in order to evaluate the influence of Swift parameter prediction error directly on a stress–strain curve. In this context, Figure 17 illustrates in a graphical format the 91 combinations of *K*-*n* used for the neural network development. This diagram shows not only the expected target (circle marker) but also the ANN output (star marker). The train data set (light color) and the validation/test data set (dark color) are all represented. These results support the performance analysis presented in Section 3.1 since, in general, the developed ANN can predict with accuracy the Swift parameters for the majority of the total cases. However, for some combinations, a substantial disagreement between output and target value is evident, especially for cases that belong to the validation and testing data set. Table 6 summarizes the differences between target-output values, for different cases $C_1$ to $C_4$, also identified in Figure 17, in terms of relative error for both Swift parameters. For case $C_1$, the relative errors are quite low (less than 0.5%) as expected, since this combination belongs to the train data set. On the contrary for cases belonging to validation/test data set ($C_2$, $C_3$, $C_4$) the errors are systematically higher, however always bellow 5%, except for one case (C4) which is bellow 9%. The higher error (C4) occur in 1 out of 91 cases, and therefore it can be considered an outlier for the test/validation data set.

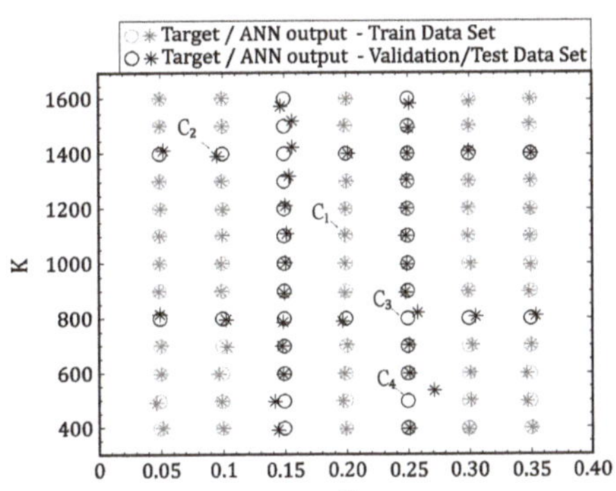


**Figure 17.** Graphical comparison between targets and ANN output values for the 91 *K*-*n* combinations.

In order to evaluate the influence of the Swift parameter error on a true stress–strain curve, Figure 18a compares the difference for cases $C_1$ to $C_4$ when considering the target Swift parameters (solid line) and the parameters obtained using the developed ANN (dotted line). From the graph it can be noted that case $C_4$ presents the largest gap between the two $\sigma$-$\varepsilon$ curves especially for higher true strain values. However, as seen in Figure 18b, the true stress error, represented by the absolute difference ($|\sigma_{ANN} - \sigma_{target}|$) for each value of true strain, is less than 15 MPa, which in this context can be considered completely acceptable.

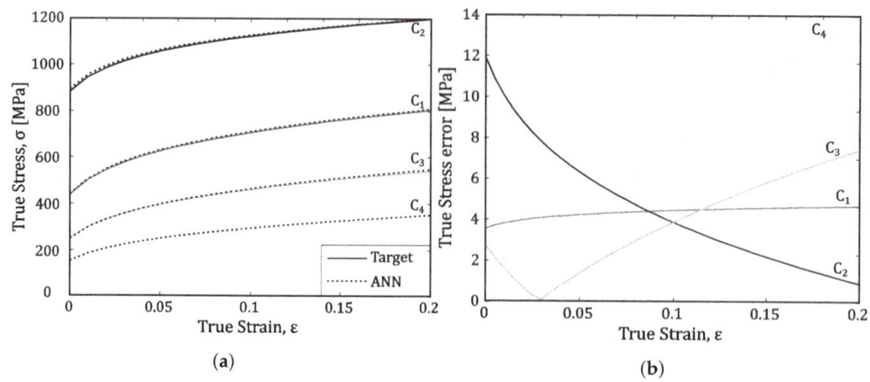


**Figure 18.** True stress–strain curves (**a**) for cases $C_1$ to $C_4$ considering the Swift parameters obtained using ANN and its comparison with the expected target results; (**b**) corresponding true stress error ($|\sigma_{Target} - \sigma_{ANN}|$) for each case and true strain values between [0, 0.2].

**Table 6.** Relative error [%] between the target reference and the ANN output values of four different cases ($C_1$ to $C_4$) and considering two Swift parameters ($K$ and $n$).

| | $K_{Target}$ | $K_{ANN}$ | $\lvert K_{Relative\ Error}\rvert$ | $n_{Target}$ | $n_{ANN}$ | $\lvert n_{Relative\ Error}\rvert$ |
|---|---|---|---|---|---|---|
| Case 1-$C_1$ | 1100 | 1105 | 0.5% | 0.200 | 0.199 | 0.5% |
| Case 2-$C_2$ | 1400 | 1392 | 0.6% | 0.100 | 0.096 | 4.0% |
| Case 3-$C_3$ | 800 | 821 | 2.6% | 0.250 | 0.258 | 3.2% |
| Case 4-$C_4$ | 500 | 538 | 7.6% | 0.250 | 0.272 | 8.8% |

Complementary Test

In order to confirm the results prediction, complementary tests were performed. For that purpose, an additional database was created, using FEA, considering Swift parameters that were not used in the development of the ANN. These new target parameters ($K_{target}$ and $n_{target}$) are summarized in Table 7 for six different extra tests. Additionally, the numerical punch force-displacement curves are represented in Figure 19a for each extra case. In this point it is important to note that the extra cases two and five have similar $F_p$-$y_p$ curves, and this proximity is especially evident in the elastic-plastic transition zone. However, the Swift parameters are completely distinct. Regarding the ANN prediction for these extra cases, the Swift parameters obtained using the developed neural network ($K_{ANN}$ and $n_{ANN}$) are in closer agreement with the targets, with relative errors below 2.5% for $ExT_{2\text{-}6}$ and 5.8% for $ExT_1$. Figure 19b represents the resulting hardening Swift curves (Equation (1)) for these considered extra cases with the parameters predicted using ANN ($K_{ANN}$, $n_{ANN}$) and the corresponding reference values ($K_{target}$, $n_{target}$). As previously observed, the true stress–strain curves (Swift) for both cases are in a good agreement for strain values between 0–0.2. Among these new cases, the $ExT_2$ and $ExT_5$ are of particular interest as, although having closer three-point bending test curves, the respective stress–strain curves are quite different. Observing the ANN performance in these two cases, Figure 19b, it can be verified that the ANN performs equally well. Therefore, it can be concluded that the five point selection, that represents the information from the bending test curves given to the ANN, it was usefully used by the trained ANN.

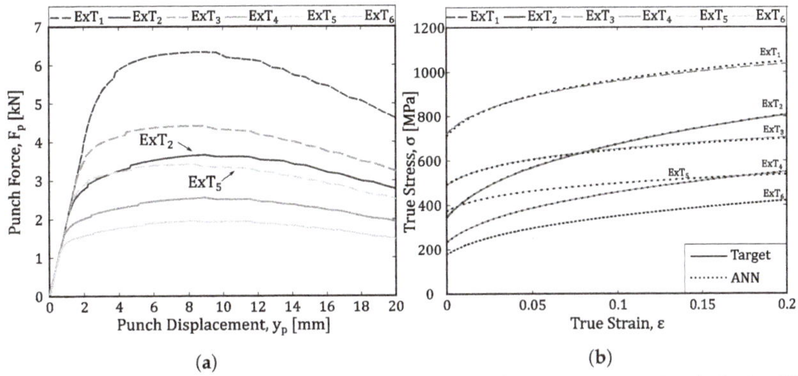


**Figure 19.** Three point bending test (**a**) punch force-displacement curves ($F_p$-$y_p$) obtained by finite element analysis for extra cases $ExT_1$ to $ExT_6$, and (**b**) the corresponding true stress–strain curves ($\sigma$-$\varepsilon$).

**Table 7.** Relative Error [%] between the target reference and the ANN output values for six different extra cases ($ExT_1$ to $ExT_6$) and considering two Swift parameters ($K$ and $n$).

| | $K_{Target}$ | $K_{ANN}$ | $\lvert K_{Relative\ Error}\rvert$ | $n_{Target}$ | $n_{ANN}$ | $\lvert n_{Relative\ Error}\rvert$ |
|---|---|---|---|---|---|---|
| $ExT_1$ | 1250 | 1278 | 2.3% | 0.120 | 0.127 | 5.8% |
| $ExT_2$ | 1250 | 1250 | ~0.0% | 0.280 | 0.279 | 0.4% |
| $ExT_3$ | 850 | 840 | 1.2% | 0.120 | 0.117 | 2.5% |
| $ExT_4$ | 850 | 851 | 0.1% | 0.280 | 0.280 | ~0.0% |
| $ExT_5$ | 650 | 648 | 0.3% | 0.120 | 0.120 | ~0.0% |
| $ExT_6$ | 650 | 651 | 0.2% | 0.280 | 0.280 | ~0.0% |

### *4.2. Problem (b)—Air Bending: Forming and Springback Prediction*

Turning now to problem (b), in this section it will be presented the results obtained by the multitask ANN developed in Section 3.2, using the DD1 division set (MT-ANN, DD1). In this context, Figure 20a,b details, respectively, the punch displacement ($y_p$) and the springback angle ($\Delta\alpha_{SB}$) curves as a function of required bending angle ($\alpha$), for the 37 $V/t$ combinations. This representation illustrates the overall capability of the developed ANN to model, in the same structure, two different, although related, functions ($y_p$ and $\Delta\alpha_{SB}$), and in a significant range of input parameter values ($V$, $t$, bending angle). It can also be observed that in the prediction of the springback angle, higher oscillations occur in relation to the reference curves.

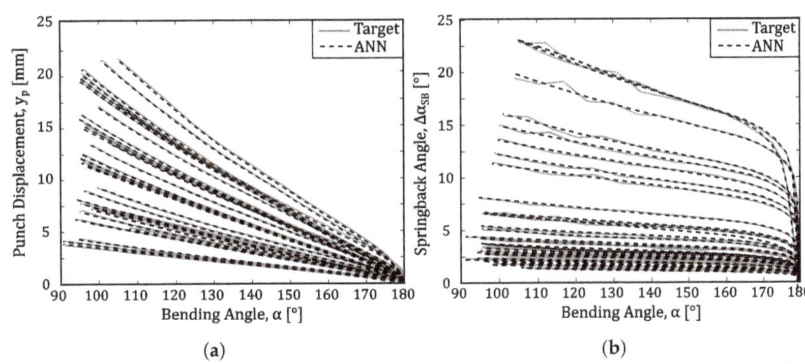


**Figure 20.** (**a**) Punch displacement ($y_p$) graph and (**b**) springback angle ($\Delta\alpha_{SB}$) graph as a function of required bending angle ($\alpha$) for the total $V/t$ combinations used in problem (b).

In order to evaluate the results of this problem in more detail, the three cases (A, B and C), presented in Section 2.1.2 (Figures 4–6), will be studied in this analysis. As already mentioned, these three cases correspond to different bending conditions: case A ($V/t$ = 43.7) is characterized by an excessive springback angle, case C ($V/t$ = 8.7) is characterized by indentation, and finally case B ($V/t$ = 4.6) where none of these phenomena occur and the bending conditions are considered appropriate, in closer agreement with analytical models and industry guidelines. Additionally, it can be noted that cases A and B belong to the test and validation data set, while case C belongs to the training data set, following the proposed data sets division (DD1), as represented in Figure 15a. It can be observed (Figure 21) that the ANN performs equally well in the three cases, which means it can provide a solution to a wide extension of the bending angles, that can be obtained with the same set of tools, in addition to the industry guidelines.

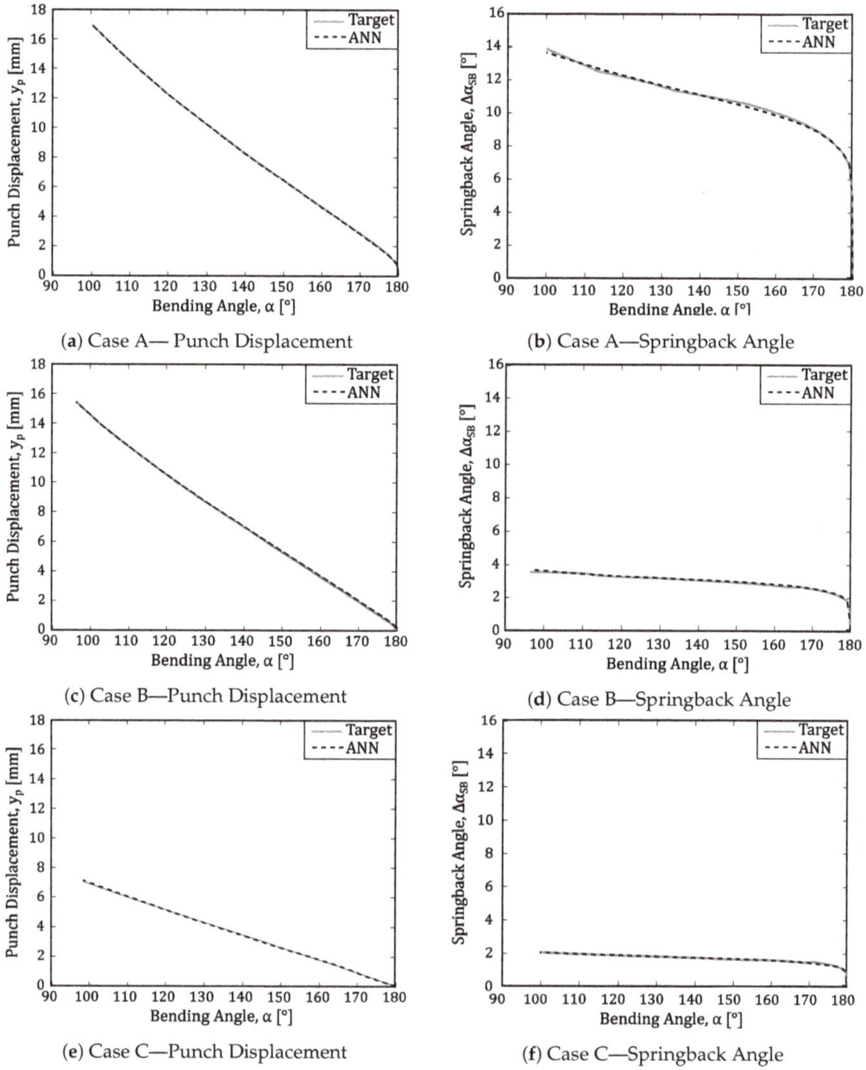


(**a**) Case A— Punch Displacement (**b**) Case A—Springback Angle

(**c**) Case B—Punch Displacement (**d**) Case B—Springback Angle

(**e**) Case C—Punch Displacement (**f**) Case C—Springback Angle

**Figure 21.** Punch displacement ($y_p$) graphs and springback angle ($\Delta\alpha_{SB}$) graphs as a function of required bending angle ($\alpha$) for cases A, B, C.

In order to have a closer look at the capability of ANN to model the punch displacement, Figure 22 includes both the $y_p$ and error graphs in case B, which favors the analytical or geometrical based solutions. It contains the $y_p$ reference curves obtained from simulation, after ($y\text{Sim}_{ASB}$) and before ($y\text{Sim}_{BSB}$) springback, the analytical solution ($y_{JBP}$) and the ANN solution ($y_{ANN}$). Although the $y_p$ curves seem to have a similar behavior, the error curves clearly differentiate the analytical solution when compared to the reference simulation curves, before springback and even more when springback is taken into account. Comparatively the ANN solution presents a lower error relative to reference simulation curve after springback, demonstrating that it provides an adequate solution even when springback is considered.

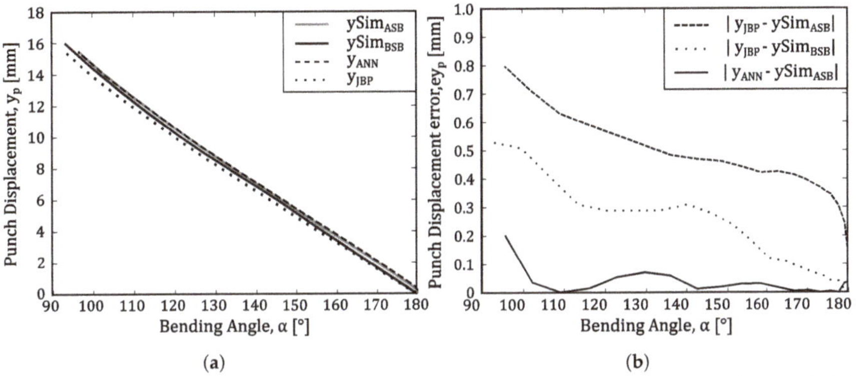


**Figure 22.** Punch displacement errors ($ey_p$), for case B, as a function of required bending angle ($\alpha$) considering ANN output values and JBP analytical results.

In order to evaluate the error of the ANN solution when predicting the springback angle ($\Delta\alpha_{SB}$), Figure 23 represents, for the previously described three cases (A, B, C) the evolution of springback angle ($\Delta\alpha_{SB}$) and the respective error curves relative to the simulation reference springback angle. In the overall, evolution of ($\Delta\alpha_{SB}$) angles follow a similar behavior to the reference curves and the errors are clearly higher and more irregular in case A, than B and C. This indicates higher difficulty in this area of the tools usage ($V/t > 10$). However, taking into account the objective of having an indicator of the magnitude of the springback angle, rather than a precise value for the user, the ANN solution can be considered to fulfill its purpose adequately.

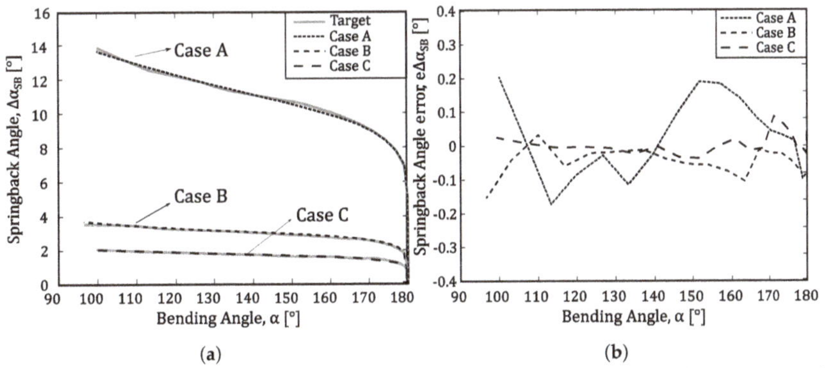


**Figure 23.** Sprinback analysis for cases A, B, C: (**a**) springback angle ($\Delta\alpha_{SB}$) and the corresponding expected value (Target) as a function of required bending angle ($\alpha$); (**b**) springback angle error ($e\Delta\alpha_{SB}$) for each case as a function of required bending angle ($\alpha$).

## 5. Conclusions

In this work, the use of machine learning algorithms was explored in the form of artificial neural networks to model different problems associated with sheet metal processing and material characterization. The ANN methods have the advantage of an efficient modeling of the complexity and nonlinearities associated with these problems. However, it must be considered that they do not provide by themselves an explicative solution for the problems nor a confidence level of the results obtained. Therefore, its usage must be carefully considered and other ML tools and methods could provide a complementary solution to overcome some of these limitations. First, it was intended to use the results of a simple and standard test (three-point bending) to perform the mechanical characterization of a metallic sheet material and finding the corresponding parameters for the Swift hardening curve. Second, in a different but related problem, it was intended to model the sheet metal press brake, air-bending process, to predict the required punch displacement corresponding to the desired bending angle, after removing the tools (i.e., after springback). In both cases, simulation results obtained with FEA models were used. The obtained results show that ANN can be a valuable tool to model these problems.

In the first problem, mechanical characterization with three-point bending, the results show a good agreement with the simulation and reference models, being able to closely predict the material $K$ and $n$ parameters in the ranges 400 to 1600, and 0.05 to 0.35 respectively, and characterizing adequately the strain–stress curves in the range of interest, i.e., up to 0.2 strain values. In the second problem, press-brake bending, it can be concluded that a single structure ANN was efficient in predicting simultaneously the required punching displacement and the springback angle. It was also proved beneficially to include a second learning task to better predict the punch displacement.

In spite of having more than two hidden layers, the ANN developed can be considered within a simpler shallow ANN classification, rather than included in deep learning ANN structures. Shallow ANN has the advantage of being faster to train and requiring less data. However, when expanding the solution to include other materials, tools or process parameters it can be expected that a deep learning structure will be beneficial. One envisaged future work will be to use convolution neural networks (CNN) in a deep learning structure. This approach allows for the use of more information about a problem to be included in the Input layer, without having the corresponding increase of the ANN size as in fully connected feedforward structures. Regarding the first problem, it may be explored whether the complete curve would provide better generalization. As for the second problem, it is intended to include the material force–displacement behavior in the learning tasks, to evaluate at what level the precision and generalization ability of the ANN can be guided using this information. In general, it is intended to expand the solutions for these problems by comparing ANN solutions with other methods in an attempt to include more materials and associated increased data sets.

**Author Contributions:** Data curation, R.L.A.; Investigation, D.J.C. and S.S.M.; Methodology, D.J.C., M.R.B. and A.D.S.; Software, D.J.C., A.D.S. and R.L.A.; Supervision, M.R.B.; Validation, A.D.S.; Visualization, D.J.C.; Writing—original draft, D.J.C.; Writing—review & editing, M.R.B. and A.D.S. All authors have read and agreed to the published version of the manuscript.

**Funding:** This research was funded by Project POCI-01-0145-FEDER-031243 —RDFORMING— Robust Design of Sheet Metal Forming Processes to Reduce Productivity Losses, cofinanced by Programa Operacional Competitividade e Internacionalização (Compete2020), through Fundo Europeu de Desenvolvimento Regional (FEDER) and by Fundação para a Ciência e Tecnologia through its component of the state budget. The fourth author is also grateful to the FCT for the Doctoral grant SFRH/BD/146083/2019 under the program POCH, cofinanced by the European Social Fund (FSE) and Portuguese National Funds from MCTES.

**Institutional Review Board Statement:** Not applicable.

**Informed Consent Statement:** Not applicable.

**Data Availability Statement:** Not applicable.

**Acknowledgments:** The authors highlight their appreciation and gratitude for the availability and fruitful discussions with our colleague and industry expert J. Bessa Pacheco.

**Conflicts of Interest:** The authors declare no conflict of interest.

## References

1. Zheng, K.; Politis, D.J.; Wang, L.; Lin, J. A review on forming techniques for manufacturing lightweight complex—Shaped aluminium panel components. *Int. J. Lightweight Mater. Manuf.* **2018**, *1*, 55–80. [CrossRef]
2. Pacheco, J.B.; Santos, A.D. A Study on the Nose Radius Influence in Press Brake Bending Operations by Finite Element Analysis. *Key Eng. Mater.* **2013**, *554–557*, 1432–1442. [CrossRef]
3. Miranda, S.S.; Barbosa, M.R.; Santos, A.D.; Pacheco, J.B.; Amaral, R.L. Forming and springback prediction in press brake air bending combining finite element analysis and neural networks. *J. Strain Anal. Eng. Des.* **2018**, *53*, 584–601. [CrossRef]
4. Rodrigues, J.; Martins, P. *Tecnologia Mecanica (Mechanical Technology)*; Escolar Editora: Lisboa, Portugal, 2005; Volume I.
5. de Vin, L. Computer Aided Process Planning for the Bending of Sheet Metal Components. Ph.D. Thesis, University of Twente, Enschede, The Netherlands, 1994.
6. Vin, L.D.; Streppel, A.H.; Singh, U.; Kals, H. A process model for air bending. *J. Mater. Process. Technol.* **1996**, *57*, 48–54. [CrossRef]
7. De Vin, L. Curvature prediction in air bending of metal sheet. *J. Mater. Process. Technol.* **2000**, *100*, 257–261. [CrossRef]
8. Pacheco, J.B.; Santos, A.D. Numerical Simulation on thevalidation of a press brake design criteria to minimizeangle deviations. In Proceedings of the CMNE 2011—Congresso de Métodos Numéricos em Engenharia, Conference Proceedings, Coimbra, Portugal, 14–17 June 2011.
9. Wang, C.; Kinzel, G.; Altan, T. Mathematical modeling of plane-strain bending of sheet and plate. *J. Mater. Process. Technol.* **1993**, *39*, 279–304. [CrossRef]
10. Tekkaya, A. State-of-the-art of simulation of sheet metal forming. *J. Mater. Process. Technol.* **2000**, *103*, 14–22. [CrossRef]
11. Tisza, M. Numerical modelling and simulation in sheet metal forming. *J. Mater. Process. Technol.* **2004**, *151*, 58–62. [CrossRef]
12. Vorkov, V.; Aerens, R.; Vandepitte, D.; Duflou, J.R. Springback Prediction of High-strength Steels in Large Radius Air Bending Using Finite Element Modeling Approach. *Procedia Eng.* **2014**, *81*, 1005–1010. [CrossRef]
13. Lingbeek, R.; Huétink, J.; Ohnimus, S.; Petzoldt, M.; Weiher, J. Thedevelopment of a finite elements based springback compensation toolfor sheet metal products. *J. Mater. Process. Technol.* **2005**, *169*, 115–125. [CrossRef]
14. Li, Y.; Rong, Q.; Shi, Z.; Sun, X.; Meng, L.; Lin, J. An accelerated springback compensation method for creep age forming. *Int. J. Adv. Manuf. Technol.* **2018**, *102*, 121–134. [CrossRef]
15. Ingarao, G.; Lorenzo, R.D. Design of Complex Sheet Metal Forming Processes: A New Computer Aided Progressive Approach. *Int. J. Mater. Form.* **2010**, *3*, 21–24. [CrossRef]
16. Katsikopoulos, K.V.; Durbach, I.N.; Stewart, T.J. When should we use simple decision models? A synthesis of various research strands. *Omega* **2018**, *81*, 17–25. [CrossRef]
17. Bock, F.E.; Aydin, R.C.; Cyron, C.J.; Huber, N.; Kalidindi, S.R.; Klusemann, B. A Review of the Application of Machine Learning and Data Mining Approaches in Continuum Materials Mechanics. *Front. Mater.* **2019**, *6*, 110. [CrossRef]
18. Cao, J.; Brinksmeier, E.; Fu, M.; Gao, R.X.; Liang, B.; Merklein, M.; Schmidt, M.; Yanagimoto, J. Manufacturing of advanced smart tooling for metal forming. *CIRP Ann.* **2019**, *68*, 605–628. [CrossRef]
19. Jaremenko, C.; Ravikumar, N.; Affronti, E.; Merklein, M.; Maier, A. Determination of Forming Limits in Sheet Metal Forming Using Deep Learning. *Materials* **2019**, *12*, 1051. [CrossRef]
20. Liu, S.; Shi, Z.; Lin, J.; Li, Z. Reinforcement learning in free-form stamping of sheet-metals. *Procedia Manuf.* **2020**, *50*, 444–449. [CrossRef]
21. Liu, X.; Du, Y.; Lu, X.; Zhao, S. Springback Prediction and Forming Accuracy Control of Micro W-bending Using Support Vector Machine. In Proceedings of the 2019 6th International Conference on Frontiers of Industrial Engineering (ICFIE), London, UK, 10–12 September 2019; pp. 23–27.
22. Dib, M.A.; Oliveira, N.J.; Marques, A.E.; Oliveira, M.C.; Fernandes, J.V.; Ribeiro, B.M.; Prates, P.A. Single and ensemble classifiers for defect prediction in sheet metal forming under variability. *Neural Comput. Appl.* **2019**, *32*, 12335–12349. [CrossRef]
23. Pathak, K.; Panthi, S.; Ramakrishnan, N. Application of Neural Network in Sheet Metal Bending Process. *Def. Sci. J.* **2005**, *55*, 125–131. [CrossRef]
24. Forcellese, A.; Gabrielli, F.; Ruffini, R. Effect of the training set size on springback control by neural network in an air bending process. *J. Mater. Process. Technol.* **1998**, *80–81*, 493–500. [CrossRef]
25. Spathopoulos, S.C.; Stavroulakis, G.E. Springback Prediction in Sheet Metal Forming, Based on Finite Element Analysis and Artificial Neural Network Approach. *Appl. Mech.* **2020**, *1*, 97–110. [CrossRef]
26. Trzepieciński, T.; Lemu, H.G. Improving Prediction of Springback in Sheet Metal Forming Using Multilayer Perceptron-Based Genetic Algorithm. *Materials* **2020**, *13*, 3129. [CrossRef] [PubMed]
27. Baseri, H.; Bakhshi-Jooybari, M.; Rahmani, B. Modeling of spring-back in V-die bending process by using fuzzy learning back-propagation algorithm. *Expert Syst. Appl.* **2011**, *38*, 8894–8900. [CrossRef]

28. Inamdar, M.; Date, P.; Desai, U. Studies on the prediction of springback in air vee bending of metallic sheets using an artificial neural network. *J. Mater. Process. Technol.* **2000**, *108*, 45–54. [CrossRef]
29. Viswanathan, V.; Kinsey, B.; Cao, J. Experimental Implementation of Neural Network Springback Control for Sheet Metal Forming. *J. Eng. Mater. Technol.* **2003**, *125*, 141–147. [CrossRef]
30. Fu, Z.; Mo, J.; Chen, L.; Chen, W. Using genetic algorithm-back propagation neural network prediction and finite-element model simulation to optimize the process of multiple-step incremental air-bending forming of sheet metal. *Mater. Des.* **2010**, *31*, 267–277. [CrossRef]
31. Narayanasamy, R.; Padmanabhan, P. Comparison of regression and artificial neural network model for the prediction of springback during air bending process of interstitial free steel sheet. *J. Intell. Manuf.* **2010**, *23*, 357–364. [CrossRef]
32. Özgü Şenol.; Esat, V.; Darendeliler, H. Springback Analysis in Air Bending Process through Experiment based Artificial Neural Networks. *Procedia Eng.* **2014**, *81*, 999–1004.
33. Guo, Z.; Tang, W. Bending Angle Prediction Model Based on BPNN-Spline in Air Bending Springback Process. *Math. Probl. Eng.* **2017**, *2017*, 1–11. [CrossRef]
34. Liu, S.; Xia, Y.; Shi, Z.; Yu, H.; Li, Z.; Lin, J. Deep Learning in Sheet Metal Bending with a Novel Theory-Guided Deep Neural Network. *IEEE/CAA J. Autom. Sin.* **2021**, *8*, 565–581. [CrossRef]
35. Eggertsen, P.A.; Mattiasson, K. An efficient inverse approach for material hardening parameter identification from a three-point bending test. *Eng. Comput.* **2009**, *26*, 159–170. [CrossRef]
36. Gorji, M.B.; Mozaffar, M.; Heidenreich, J.N.; Cao, J.; Mohr, D. On the potential of recurrent neural networks for modeling path dependent plasticity. *J. Mech. Phys. Solids* **2020**, *143*, 103972. [CrossRef]
37. Jenab, A.; Sari Sarraf, I.; Green, D.E.; Rahmaan, T.; Worswick, M.J. The Use of genetic algorithm and neural network to predict rate-dependent tensile flow behaviour of AA5182-O sheets. *Mater. Des.* **2016**, *94*, 262–273. [CrossRef]
38. Lin, Y.; Zhang, J.; Zhong, J. Application of neural networks to predict the elevated temperature flow behavior of a low alloy steel. *Comput. Mater. Sci.* **2008**, *43*, 752–758. [CrossRef]
39. Zhang, A.; Mohr, D. Using neural networks to represent von Mises plasticity with isotropic hardening. *Int. J. Plast.* **2020**, *132*, 102732. [CrossRef]
40. Gorji, M.B.; Mohr, D. Towards neural network models for describing the large deformation behavior of sheet metal. *IOP Conf. Ser. Mater. Sci. Eng.* **2019**, *651*, 012102. [CrossRef]
41. Li, X.; Roth, C.C.; Mohr, D. Machine-learning based temperature- and rate-dependent plasticity model: Application to analysis of fracture experiments on DP steel. *Int. J. Plast.* **2019**, *118*, 320–344. [CrossRef]
42. Garcia-Romeu, M.; Ciurana, J.; Ferrer, I. Springback determination of sheet metals in an air bending process based on an experimental work. *J. Mater. Process. Technol.* **2007**, *191*, 174–177. [CrossRef]
43. Raabe, D.; Roters, F.; Barlat, F.; Chen, L.Q. *Continuum Scale Simulation of Engineering Materials: Fundamentals, Microstructures, Process Applications*; Wiley-VCH: Weinheim, Germany, 2004.
44. *Standard Test Methods For Bend Testing Of Material For Ductility*; ASTM International: West Conshohocken, PA, USA, 2014.
45. Naseem, S.; Perdahcıoğlu, E.S.; Geijselaers, H.J.M.; van den Boogaard, A.H. A New in-Plane Bending Test to Determine Flow Curves for Materials with Low Uniform Elongation. *Exp. Mech.* **2020**, *60*, 1225–1238. [CrossRef]
46. Kim, Y.J.; Lee, C.H.; Kim, J.H.; Lim, J.H. Numerical modeling of shape memory alloy plates considering tension/compression asymmetry and its verification under pure bending. *Int. J. Solids Struct.* **2018**, *136–137*, 77–88. [CrossRef]
47. Badr, O.M.; Rolfe, B.; Zhang, P.; Weiss, M. Applying a new constitutive model to analyse the springback behaviour of titanium in bending and roll forming. *Int. J. Mech. Sci.* **2017**, *128–129*, 389–400. [CrossRef]
48. Capilla, G.; Hamasaki, H.; Yoshida, F. Determination of uniaxial large-strain workhardening of high-strength steel sheets from in-plane stretch-bending testing. *J. Mater. Process. Technol.* **2017**, *243*, 152–169. [CrossRef]
49. Nádai, Á.; Wahl, A.M.; Bingham, E.C. Plasticity, A Mechanics of the Plastic State of Matter. *J. Rheol.* **1931**, *2*, 455–456. [CrossRef]
50. Rumelhart, D.E.; Hinton, G.E.; Williams, R.J. Learning representations by back-propagating errors. *Nature* **1986**, *323*, 533–536. [CrossRef]
51. Hertz, J.; Krogh, A.; Palmer, R.G. *Introduction to the Theory of Neural Computation*; CRC Press: Boca Raton, FL, USA, 2018.
52. Goodfellow, I.; Bengio, Y.; Courville, A. *Deep Learning*; MIT Press: Cambridge, MA, USA, 2016.
53. Ferreira, B. Integração de Metodologias de Análise do Processo de Quinagem. Master's Thesis, Faculdade de Engenharia da Universidade do Porto, Porto, Portugal, 2015.
54. MATLAB. *Version 9.10.0 (R2021a)*; The MathWorks Inc.: Natick, MA, USA, 2010.
55. *MATLAB Deep Learning Toolbox™ User's Guide: PDF Documentation for Release R2021a*; The MathWorks Inc.: Natick, MA, USA, 2021.

*metals*

MDPI

*Article*

# Robust Optimization and Kriging Metamodeling of Deep-Drawing Process to Obtain a Regulation Curve of Blank Holder Force

**Maria Emanuela Palmieri *, Vincenzo Domenico Lorusso and Luigi Tricarico**

Department of Mechanics, Mathematics and Management, Polytechnic University of Bari, Via Orabona 4, 70125, Bari, Italy; vincenzodomenico.lorusso@poliba.it (V.D.L.); luigi.tricarico@poliba.it (L.T.)
* Correspondence: mariaemanuela.palmieri@poliba.it; Tel.: +39-0805-963-723

**Abstract:** In recent decades, the automotive industry has had a constant evolution with consequent enhancement of products quality. In industrial applications, quality may be defined as conformance to product specifications and repeatability of manufacturing process. Moreover, in the modern era of Industry 4.0, research on technological innovation has made the real-time control of manufacturing process possible. Moving from the above context, a method is proposed to perform real-time control of a deep-drawing process, using the stamping of the upper front cross member of a car chassis as industrial case study. In particular, it is proposed to calibrate the force acting on the blank holder, defining a regulation curve that considers the material yield stress and the friction coefficient as the main noise variables of the process. Firstly, deep-drawing process was modeled by using commercial Finite Element (FE) software AutoForm. By means of AutoForm Sigma tool, the stability and capability of deep-drawing process were analyzed. Numerical results were then exploited to create metamodels, by using the kriging technique, which shows the relationships between the process parameters and appropriate quality indices. Multi-objective optimization with a desirability function was carried out to identify the optimal values of input parameters for deep-drawing process. Finally, the desired regulation curve was obtained by maximizing total desirability. The resulting regulation curve can be exploited as a useful tool for real-time control of the force acting on the blank holder.

**Keywords:** sheet metal forming; deep-drawing; kriging metamodeling; multi-objective optimization; FE (Finite Element) AutoForm robust analysis; defect prediction

**Citation:** Palmieri, M.E.; Lorusso, V.D.; Tricarico, L. Robust Optimization and Kriging Metamodeling of Deep-Drawing Process to Obtain a Regulation Curve of Blank Holder Force. *Metals* **2021**, *11*, 319. https://doi.org/10.3390/met11020319

Academic Editors: Pedro Prates and André Pereira

Received: 5 January 2021
Accepted: 9 February 2021
Published: 12 February 2021



## 1. Introduction

Sheet metal cold forming processes, or deep-drawing processes, play an important role in modern industry, since components of complex geometry can be produced. However, there are some aspects that must be taken into consideration such as: (i) the influence of sheet anisotropy; (ii) the formability limits; and (iii) the spring-back phenomenon that is not negligible. The cold forming process involves plastic deformation, and it should not involve alterations in the thickness of the starting sheet. Actually, the thickness of the blank may have considerable variations during stamping. Cold forming consists of pressing the blank on a punch, by means of a die. The correct execution of the process is ensured by the blank holder, which, by exerting a force on the blank edges, allows the correct material draw-in in the die, avoiding part defects such as wrinkles, thickening, thinning and cracks. It is possible to identify the following main phases of the process: (i) gravity, during which the sheet, resting on the tool, undergoes a first deformation due to its weight; (ii) holding, during which the sheet metal is closed between the die and the blank holder; (iii) stamping, during which die-blank-blank holder system moves towards the punch for the plastic deformation; and (iv) trimming, during which the excess metal is removed while spring-back occurs in the finished part

Deep-drawing process is very effective especially for symmetrical pieces, but, if the component is non-axisymmetric, it is important to have a uniform material flux in the die; in fact, in sheet metal forming, it is fundamental to control the rate of material flow into the die cavity [1]. To control material flow during drawing operation in order to achieve the optimal forming of a part without cracks and wrinkles, it is generally necessary to slow down the sliding of the blank regions that flow more easily. This can be achieved by calibrating the blank holder force and eventually by drawbeads, which are rib-like projections mounted on the binder and designed with the aim of improving the metal flow control [2]. For better effectiveness, during the stamping phase, the action of the drawbeads and the force on the blank holder can be differentiated in the different regions of the sheet. This means having active drawbeads and active blank holder control system (different forces on the different segments of the blank holder). A good optimization leads to a better distribution of thickness on the formed part reducing the occurrence of defects such as fracture and wrinkling [3].

In general, defects on stamped components can be multiple:

- Wrinkles are generally caused by insufficient force on blank holder.
- Cracks occur when the applied load exceeds the maximum resistance of the material.
- Spring-back is due to the deformations of the component in the elastic–plastic field. Spring-back is a critical aspect of drawing processes especially in the automotive industry, where high-dimensional accuracy is often required. Many parameters influence this phenomenon such as Young modulus, yield stress, punch radius and sheet thickness.

The good performances of the deep-drawing process depend on the correct setting of the process parameters that govern the phenomenon such as pressure on the blank holder, gap between die and punch, radius of the tools, lubrication and initial geometry of the blank. In the literature, in fact, several studies evaluate the effects of different process parameters on the quality of the final product [4,5].

In this perspective, to support and facilitate the analysis of the criticalities of the process, there is various simulation software such as AutoForm and PamStamp, which is becoming increasingly widespread. The success of these software packages is due to the ability to conduct a preliminary analysis of the process to identify critical parameters, thus reducing the costs of the experimentation (in terms of time, material, energy resources, etc.).

In the present work, the cold forming process for the production of an upper cross member was modeled using the finite element (FE) commercial software AutoForm. Once the process was modeled, it was decided to investigate how some input parameters affect the quality of the final product. The input parameters considered are the blank holder force, friction coefficient and yield stress of blank material. The first was considered a design parameter, while the other two were considered noise parameters. Instead, the quality of the final product was assessed by optimizing at the end of drawing phase the output responses: thickening, insufficient stretch, safe zone, potential splits and thinning.

It is important to consider the noise variables in addition to design variables, because in everyday production it is possible for parts to be produced safely one day, and the next day problems arise even though production condition have apparently not changed. This is probably due to noise and variation during forming process. Therefore, the robustness analysis was indispensable. It can be verified whether a forming process provides stable results under the influence of the noise parameters. Therefore, in this work, after robustness analysis, thanks to the numerical results, metamodels were built with the kriging technique for each quality criterion considered. The combination of finite element analysis with metamodeling techniques is a consolidated methodology in the literature [6–8]. Metamodeling is a powerful tool that allows deriving the mathematical relationship between inputs and outputs even when the analysis is based on a deterministic computer experiments.

In the present work, after the metamodeling phase, a multi-objective optimization with a desirability approach was carried out. The innovative aspect of this work is linked to the need to find a regulation curve of the force to be imparted to the blank holder as a function of the yield stress of the material in order to control the process online from the perspective of Industry 4.0.

## 2. Materials and Methods

The component studied in this work is the upper front cross member of a car currently being produced at Tiberina company (Sangro – Atessa (CH), Italy); Figure 1 shows the image of this component realized in HR 440Y580T-FB-UC steel (2 mm thick) that is common in the automotive field for cold forming of structural components. It is a hot-rolled steel strip; in particular, it belongs to the family of ferritic-bainitic steel. This microstructure offers a particularly attractive combination of high strength and good cold workability.

**Figure 1.** Upper cross member B-Suv.

Table 1 shows chemical composition of investigated steel and Figure 2 shows the mechanical characteristics of the material considered for the studied component.

**Table 1.** Chemical composition, heat analysis in mass%.

| C Max | Si Max | Mn Max | P Max | S Max | Al | Ti+Nb Max | Cr+Mo Max | B Max | Cu Max |
|---|---|---|---|---|---|---|---|---|---|
| 0.18 | 0.50 | 2 | 0.05 | 0.010 | 0.015–2 | 0.15 | 1 | 0.01 | 0.2 |

Specifically, Figure 2a shows the hardening curve. AutoForm requires the true stress as a function of true plastic strain measured in the direction of rolling. In this image, the values of the uniform elongation ($Ag$), yield stress ($\sigma_0$), tensile strength ($Rm$) and strain hardening exponent ($n$) are highlighted.

Figure 2b shows the yield surface defined with the BBC criterion (Banabic et al.) in order to take into account material anisotropy [9]. The main values of this model are illustrated in this figure: $r_m$ is the average of plastic strain ratio at 0°, 45° and 90° of rolling direction; $r_b$ is plastic strain ratio at biaxial stress, which is defined as the ratio of strains $\varepsilon_2$ and $\varepsilon_1$; $\sigma_b/\sigma_0$ is the ratio between onset of yielding at equi-biaxial stress and yield stress; $\sigma_{ps0}/\sigma_0$ is the ratio between plane strain stress at 0° of rolling direction and yield stress; $\sigma_{ps90}/\sigma_0$ is the ratio between plane strain stress at 90° of rolling direction and yield stress; and $\sigma_{shear}/\sigma_0$ is the ratio between shear stress and yield stress.

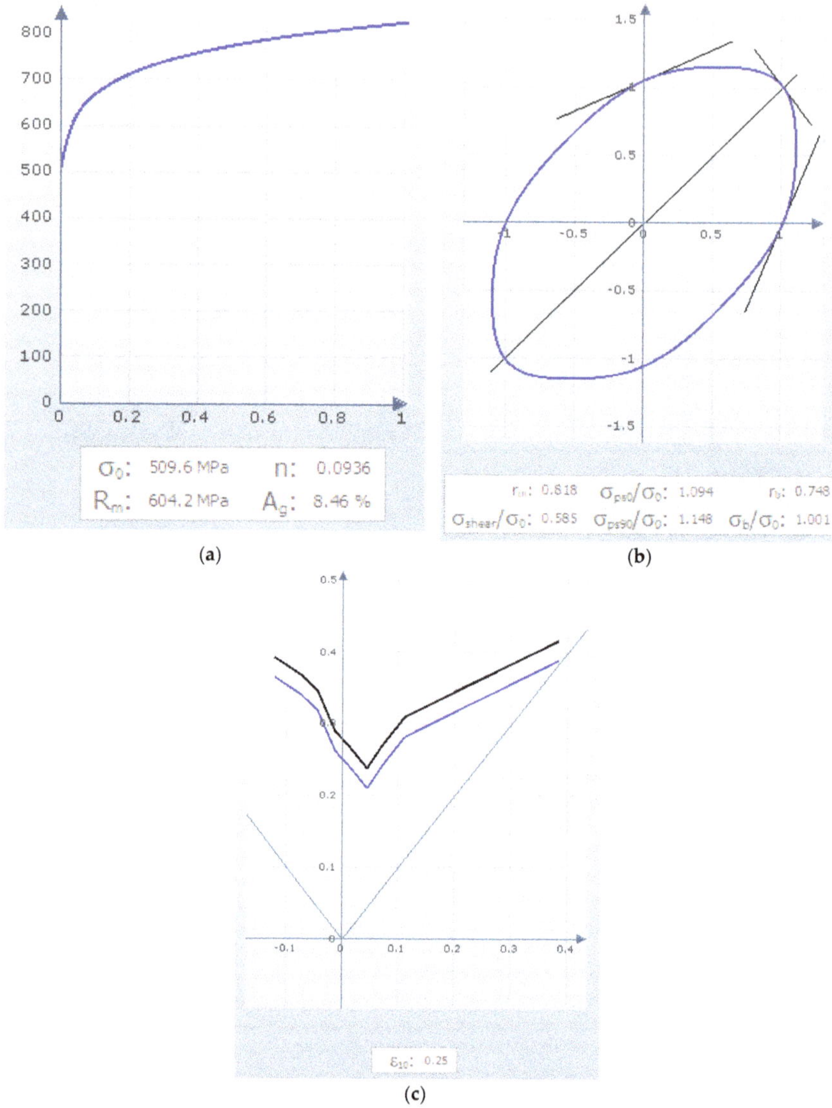


**Figure 2.** (**a**) Hardening curve; (**b**) yield surface with BBC criterion; and (**c**) Formability Limit Curve (FLC).

Figure 2c shows the Formability Limit Curve (FLC). The curve represents the maximum values of the principal strains $\varepsilon_1$ and $\varepsilon_2$ measured at the onset of material failure.

The goal of industrial digitization is to increase production efficiency and improve the quality of the final product. In fact, aiming at zero defect production, the number of scrap products is reduced and consequently the production costs are reduced. Therefore, it is necessary to optimize and design the process correctly. However, it must be taken into account that unwanted system changes may occur during a production process. In this work, for the examined deep-drawing process, possible fluctuations of the material in a coil (yield stress) and a variation of the lubrication conditions (friction coefficient) were considered. These two parameters are called noise factors; this means that they cannot be controlled. A possible controllable design parameter is the force on the blank holder, which can be adjusted in the production line.

The objective of this work, in fact, is the robust optimization of the process investigated. Moreover, once the process has been optimized, the goal is to find a regulation curve that allows, once it is implemented in the process through an algorithm, to identify how to adjust the force on the blank holder as the yield stress varies for different values of the friction coefficient. In the article by P. Fischer et al. [10], the control based on the feed-forward algorithm for the force on the blank holder is studied considering the fluctuations of the yield stress measured through the eddy currents.

The methodology adopted to derive the regulation curves is shown in Figure 3.

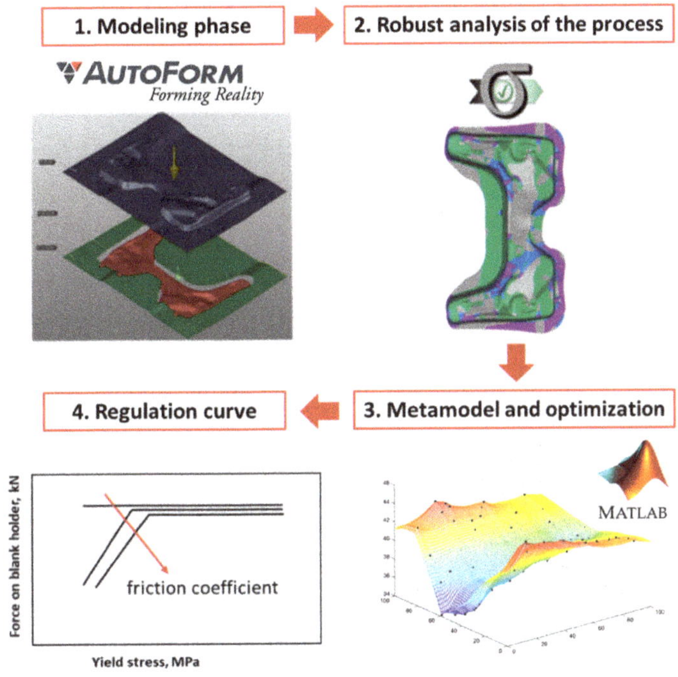


**Figure 3.** Scheme of adopted methodology.

In particular: (1) The component and the process phases were modeled in the AutoForm (R8, GmbH, Zurich, Switzerland) environment. (2) Thanks to the AutoForm Sigma module, the process was simulated and studied as the process parameters changed. In particular, the parameters that were changed are the force on the blank holder, the friction coefficient and the yield stress. These last two parameters were considered noise parameters. The target value and a standard deviation to all parameters considered were assigned. The software, then, proceeded to a sampling according to the Latin Hypercube statistical method (Latin Hypercube Sampling, LHS), generating a near-random samples of parameter values. In total, 81 numerical simulations were performed. Once the results of the numerical simulations were obtained, a robust analysis was carried out to analyze the influence of the noise variables on the forming process. The quality indices taken into consideration for this study were thickening, insufficient stretch, safe zone, potential splits and thinning. The results of robust analysis were used to predict the stability and capability of the process. (3) Given the dataset obtained with the numerical simulations of AutoForm Sigma, thanks to the Matlab DACE toolbox (2.0, Technical University of Denmark DK-2800 Kgs, Lyngby, Denmark), the metamodels were obtained. These metamodels allow predicting a new (untried) site [11]. These sites on which to evaluate the predictor were generated thanks to a definition of a grid points. We chose a 39 × 39 mesh of points distributed equidistantly in the area $[0, 100]^2$ covered by the design sites. After kriging meta-modeling phase, there was the multi-objective optimization phase. The approach used for optimization is that of desirability. With the optimization phase, the combination of input parameters (force on the blank holder, friction coefficient and yield stress) which guarantees a stamped component without defects (wrinkles, thinning, thickening and breakage) was identified. (4) Considering combination of input parameters that give high desirability, force regulation curves on the blank holder were obtained as a function of the yield stress of the material for three different values of friction coefficient. Finally, by comparing an optimized solution with a non-optimized one, the draw-in of metal sheet was evaluated, and it was observed that the sheet has different sliding in the two conditions.

## 3. Results

### *3.1. Design of Stamping Process Using Finite Element Model (FEM)*

The process was first numerically modeled using commercial Finite Element (FE) software AutoForm. The numerical model provides for the definition of tools geometries (die, punch and blank holder), the initial blank and their reference systems and material characteristics and production plan (defining the individual operations of the production process). Figure 4 shows the tools modeled in AutoForm.

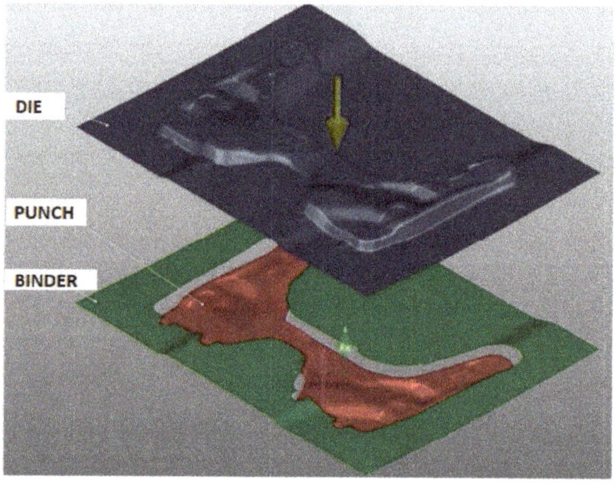


**Figure 4.** Tools geometry (die, punch and binder or blank holder).

The die and the punch were defined as rigid tools. The blank holder was defined as a force-controlled tool, which means that the assigned force is automatically increased; if the reaction force acting on the binder exceeds the defined force, the binder always remains closed [12].

For model construction, a membrane element, extended by an approximate bending stiffness (Bending Enhanced Membrane, BEM) is chosen. AutoForm software uses an adaptive mesh and an implicit solver.

Once the component and the process were modeled, robust analysis was carried out by means of AutoForm-Sigma, an AutoForm module. This tool allows analyzing and improving the robustness of sheet metal products and processes; in fact, it enables identifying which design and noise parameters influence part quality and to what extent. It also supports in determining appropriate correction measures during tryout and production. In addition, it identifies the correction measures that have no effect at all, as well as those that offer a real chance of resolving the particular problem at hand. By analyzing process performance and, in particular, process capability, it is possible to validate the stamping process, minimize part rejects and maximize production efficiency.

The friction coefficient and the yield stress, as well as the force on the blank holder, are the considered input parameters. The force on the blank holder was considered as a design variable and a variability of 25% was imposed with respect to the nominal value of 1470.5 kN; instead, the friction coefficient and the yield stress were considered noise variables. Variabilities of 10% and 15% were set for the friction coefficient (0.15 as nominal value) and for the yield stress (509.61 MPa as nominal value), respectively. The value of the force on the blank holder was chosen thanks to the design data of the press present in Tiberina Sangro company and the value of yield stress was set according to material datasheet. Instead, the software default value was chosen for the friction coefficient, specifying a mill oil for the lubrication condition. The Coulomb lubrication model was selected for the numerical simulation of deep-drawing process. However, the Coulomb model is only an approximation of the real friction behavior. In fact, the friction coefficient is not a constant in reality, but is dependent on multiple factors such as contact pressure.

The need to distinguish the parameters into two categories (controllable or design variables and non-controllable or noise variables) arises from the need to evaluate the process robustness before production phase. The variability of noise and controllable parameters, in fact, will lead to a response variation, thus causing changes in the product characteristics. If the response differs too much from the expected characteristics, the product may be unacceptable. However, while controllable variables can be corrected in the design phase, or even in-process, the noise variables must be carefully defined and studied; therefore, the process must be developed so as that these variations do not lead to worsening product quality.

The qualities of the final product were assessed on the basis of the percentage of thickened area, the percentage of the area with insufficient elongation, the percentage of safe zones, the percentage of area with potential cracks and the thinning at the two Critical Points A and B indicated in Figure 5. These two points are critical because at Point A the minimum global thinning value of 25.2% is reached, while at Point B a value of 23.6% is reached. The different maximum thinning values at Points A and B are justified by the lack of symmetry of the part.

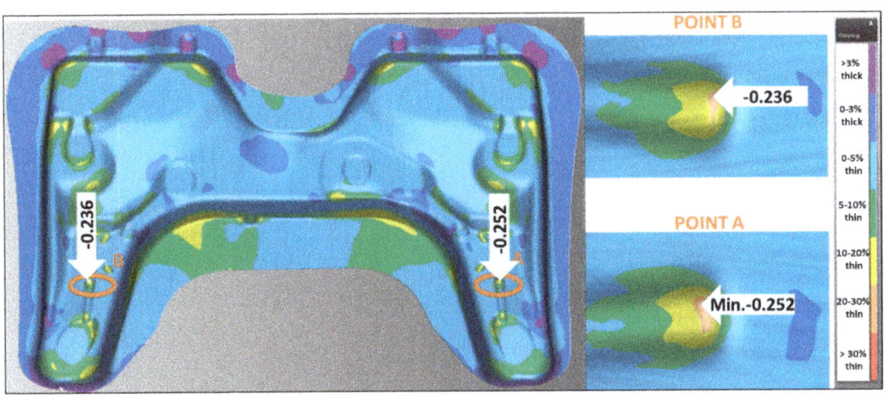


**Figure 5.** Critical Points A and B for the evaluation of thinning.

The output variables were evaluated at the end of the drawing phase, before the trimming operation in order to optimize the drawing phase.

These issues can be determined by means of the Formability Limit Diagram (FLD).

Thanks to the FLC imported during the material definition phase, the software is able to represent the FLD showing the strain state of all elements for each time step.

As an example, Figure 6 shows the formability limit diagram at the end of drawing phase for the nominal case (process parameters set at the nominal values); the different colored areas represent the behavior of the material during the deformation process. Figure 7 shows the formability map on the component.

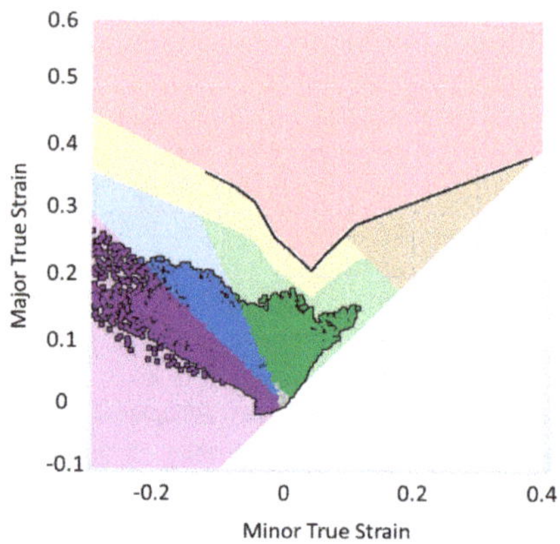


**Figure 6.** Formability limit diagram (FLD).

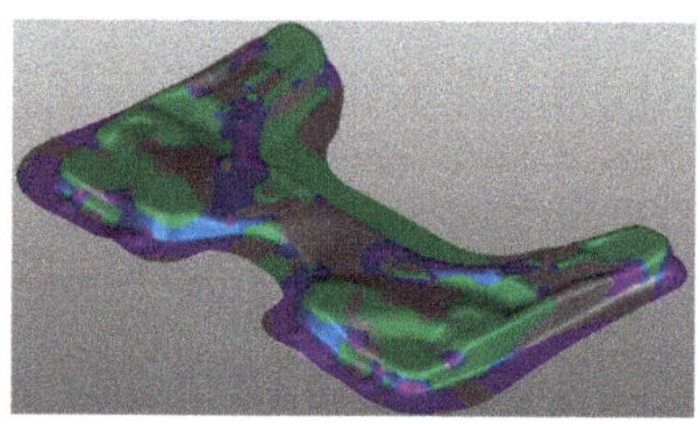

**Figure 7.** Formability map on component at the end of drawing phase.

The red region, above the formability limit curve, represents the area of points subject to splits. The area of the points subject to risk of splits is shown in yellow. The regions in which the deformation occurs in an optimal way are highlighted in green. When the material is not sufficiently deformed, it is called insufficient stretch, and this region is colored in gray. The other two regions in blue and purple represent, respectively, the compression zone and the thickening zone, where there may be a greater tendency to wrinkle.

These areas are calculated as the ratio between the area of the finite elements that present critical issues and total area of the component.

The thinning issue shows the thickness variation of the blank during the process. It is important to predict which areas are subjected to excessive thinning because it is more likely that the ruptures occur there.

### *3.2. Robust Analysis*

Fluctuations of lubrication conditions are noise factors in forming process, and yield strength, which can vary from coil to coil and supplier to supplier, is a noise in material properties. These are some of important but unavoidable and uncontrollable variations during deep-drawing process under real production conditions. In this work, the friction coefficient and the yield stress were taken into account to verify whether the process provides stable results under the influence of these most common noise parameters.

The robustness was analyzed with index process capability ($cp_k$), which indicates controllability of the process around the defined specification limits. This index is calculated as:

$$cp_k = \min\left(\frac{USL - \mu}{3\sigma}, \frac{\mu - LSL}{3\sigma}\right) \quad (1)$$

where $USL$ is the Upper Specification Limit, $LSL$ is the Lower Specification Limit, $\mu$ is the mean value and $\sigma$ is the standard deviation.

Upper and lower specification limits are defined thanks to an evaluation standard present in AutoForm that, being a commercial software, ensures to industrial companies that results are always evaluated in the same way. However, these standards can be modified by the user.

The discrete representation provided by the FE software AutoForm reports the following results [12]:

- $cp_k < 0.67$: The process is not acceptable, as more than 2.25% of the results do not meet the specification limits.
- $cp_k \in [0.67 : 1)$: The process is unreliable, as 0.14–2.25% of results do not meet the specification limits.
- $cp_k \in [1 : 1.33)$ : The process may be acceptable because the results fall within limit imposed. However, a check is required.
- $cp_k \geq 1.33$: The process is reliable, as less than 0.004% of the results exceed the limits.

For splitting and wrinkling, parameters to evaluate the quality of the final product, typically specification limits, are defined as: (i) thinning, for which only the lower $cp_k$ is relevant; (ii) maximum failure (defined as the ratio between the maximum major strain computed at an element and the major strain of the strain-based FLC for the same minor strain), for which only the upper $cp_k$ is relevant; and (iii) potential wrinkles, for which only the upper $cp_k$ is relevant [13].

By evaluating the lower $cp_k$ for thinning and the upper $cp_k$ for maximum failure after drawing phase, a value of $cp_k$ greater than 1.33 was obtained. The upper $cp_k$ for potential wrinkles is shown in Figure 8.

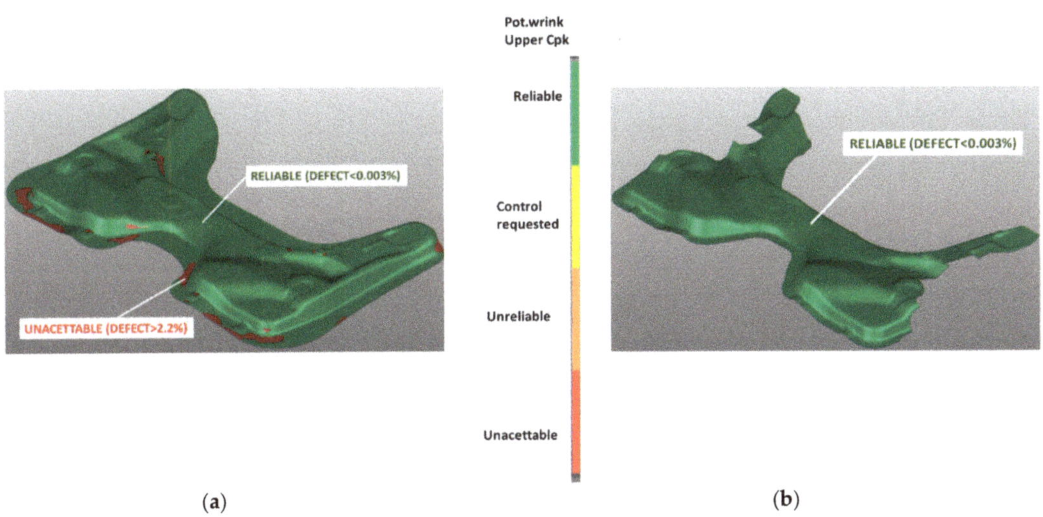


**Figure 8.** (**a**) Upper $cp_k$ for potential wrinkles after drawing phase; and (**b**) upper $cp_k$ for potential wrinkles at the end of the process.

Figure 8a shows that the variable potential wrinkles produces unacceptable results if the end of drawing phase is considered. However, with the subsequent trimming step and thanks to the spring-back phenomenon, the regions with defects are eliminated (Figure 8b). Therefore, it can be assumed that the process is reliable.

### 3.3. Metamodeling with Kriging Methodology

To study the relationship of the process parameters and material parameters with forming quality indices, a kriging method was used. For this purpose, a Matlab toolbox called DACE (Design and Analysis of Computer Experiments) was employed. This software allows constructing a kriging approximation model based on data from computer experiments and using this approximation model as a surrogate for the computer model [11]. The advantage of using this technique is that kriging models are accurate because they interpolate the sampled points and they are not limited by the type of function chosen, unlike other polynomial regression models. Moreover, kriging models are chosen to interpolate the data and are fit using maximum likelihood estimation [14]; for this reason, the surfaces may not be perfectly smooth, unlike the surfaces that could be obtained with response surface modeling that typically employs least squares regression to fit a polynomial model to the sampled data.

Figures 9–13 show the metamodels obtained in correspondence with the nominal force value (1470.5 kN). These metamodels represent, respectively, how the percentage of thickened zone, the percentage of area with insufficient stretching, the percentage of the safe zone, the percentage of zone with potential splits and the percentage of thinning at Critical Points A and B vary as the two noise variables (friction coefficient and yield stress) vary.

In these graphs, it is possible to observe that the noise variables greatly influence the quality indices of the final product. In particular, the figures show that an increase in the yield stress involves an increase in the percentage of thickened areas, areas with insufficient stretch, areas with potential splits (only for low friction coefficient) and an increase in thinning at Critical Points A and B. Consequently, increasing this parameter, there is a reduction in the percentage of the safe zone. Furthermore, if the blank lubrication conditions are such that they have a reduction in the friction coefficient, there is an increase in the percentage of the thickened area, a reduction in the safe area, a reduction in areas with potential splits and a reduction in thinning of the part region near Point B.

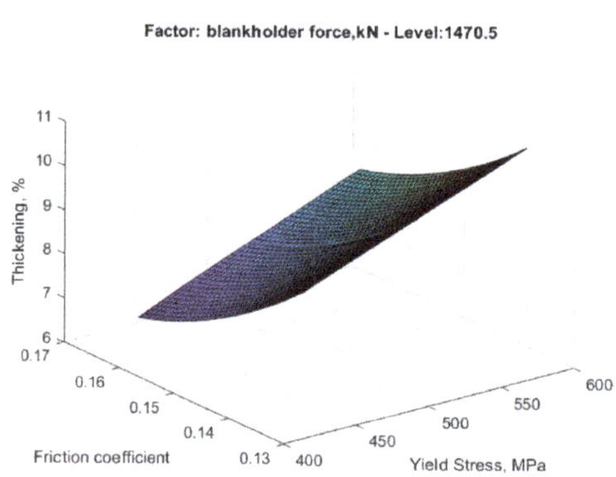


**Figure 9.** Metamodel of the percentage of thickened zone as friction coefficient and yield stress vary.

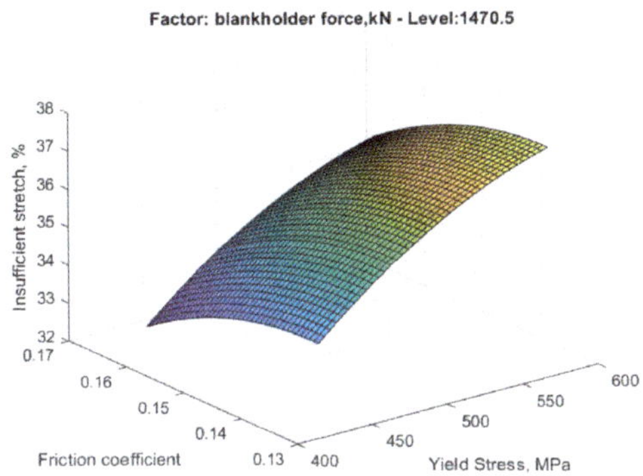


**Figure 10.** Metamodel of the percentage of area with insufficient stretching as friction coefficient and yield stress vary.

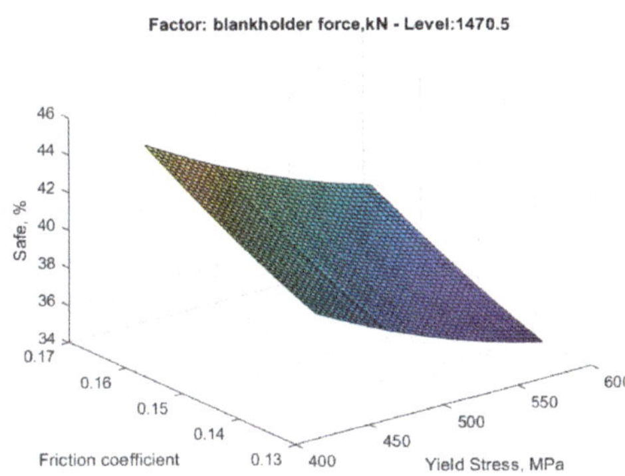


**Figure 11.** Metamodel of the percentage of safe zone as friction coefficient and yield stress vary.

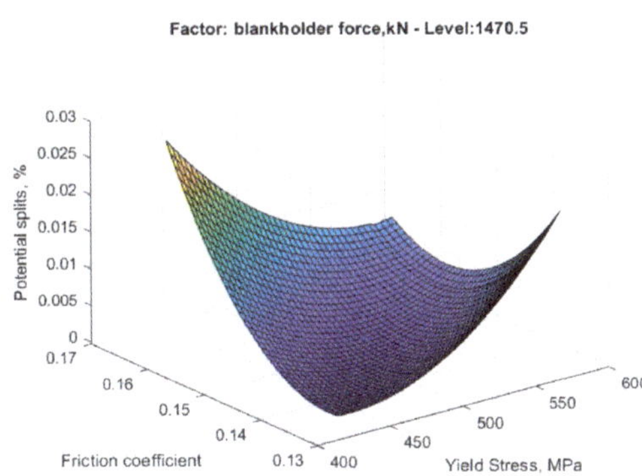


**Figure 12.** Metamodel of the percentage of area with potential splits as friction coefficient and yield stress vary.

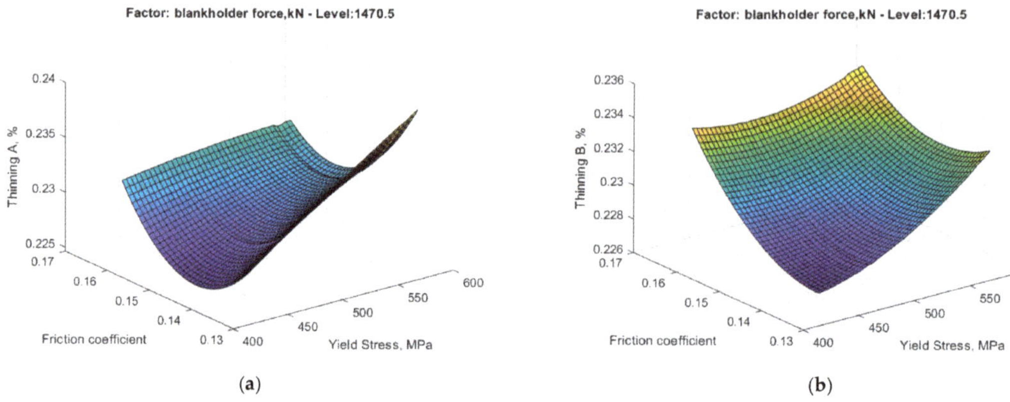


**Figure 13.** (**a**) Metamodel of the percentage of thinning at Critical Point A as friction coefficient and yield stress vary; and (**b**) metamodel of the percentage of thinning at Critical Point B as friction coefficient and yield stress vary.

In the part region near Point A, there is a minimum value for the friction coefficient beyond which the percentage of thinning in this region begins to increase. The percentage of the zone with insufficient stretch increases as the friction coefficient decreases up to a maximum value, beyond which it begins to decrease.

### *3.4. Multi-Objective Optimization and Regulation Curve*

To obtain an upper cross member free of defects, it is necessary to minimize the percentage of thickening, the percentage of areas with insufficient stretch, the percentage of areas with risk of splits and the percentage of thinning. Moreover, it is necessary to maximize the percentage of safe zones as well. It is clear that these characteristics cannot simultaneously assume the best possible values; therefore, the need to identify a compromise solution arises. This explains the multi-objective nature of optimization.

In this study, the Desirability Function Approach (DFA) was chosen. This approach is one of the most common methods for the optimization of multiple response processes in the industry field. According to this method, when the quality of a product or a process depends on several characteristics, if even one of them exceeds the imposed limits, the product or process is not acceptable. The DFA identifies the operating conditions that return the "most desirable" response values [15].

For each output variable, the criterion to be followed for optimization was chosen. In particular for the response relating to the percentage of safe zone, the criterion "the larger the better" was chosen, and, for all the other responses, the criterion was "the smaller the better".

If a "the larger the better" response is desired, i.e., if the response is to be maximized, the desirability function is of the type:

$$d_i(Y_i) = \begin{cases} 0 \quad \text{if} \quad Y_i(x) < L_i \\ \left(\frac{Y_i(x)-L_i}{T_i-L_i}\right)^s \quad \text{if} \quad L_i \leq Y_i(x) \leq T_i \\ 1 \quad \text{if} \quad Y_i(x) > T_i \end{cases} \tag{2}$$

If a "the smaller the better" response is desired, instead, i.e., if the response is to be minimized, the desirability function is defined as follows:

$$d_i(Y_i) = \begin{cases} 1 \quad \text{if} \quad Y_i(x) < T_i \\ \left(\frac{Y_i(x)-U_i}{T_i-U_i}\right)^s \quad \text{if} \quad T_i \leq Y_i(x) \leq U_i \\ 0 \quad \text{if} \quad Y_i(x) > U_i \end{cases} \tag{3}$$

According to this approach, total desirability is defined as:

$$D = \prod_{i=1}^{k} [d_i(Y_i)]^{1/k} \quad (4)$$

Table 2 presents the notation related to the equations.

**Table 2.** Notation for the definition of desirability functions.

| Symbol | Definition |
|---|---|
| $Y_i(x)$ | *i*th response as a function of *x* parameter |
| $d_i(Y_i)$ | *i*th desirability function correlated to the response parameter $(Y_i)$ |
| $L_i$ | Minimum *i*th value |
| $U_i$ | Maximum *i*th value |
| $T_i$ | Target *i*th value |
| $s$ | Exponent that defines the shape of the function ($s$ = 0.1 convex function) |
| $D$ | Total desirability |
| $k$ | Number of responses |

This optimization was a useful tool to obtain the regulation curves shown in Figure 14. These curves, in fact, were obtained considering the points of maximum desirability ($D > 0.9$). In this figure, the desirability curves for each value of the friction coefficient are also shown. In particular, regulation curves of the force on the blank holder as a function of yield stress for all friction coefficient (f) considered are highlighted with solid lines. The curves of the total desirability as a function of the yield stress for all values of the friction coefficient are highlighted with dashed lines.

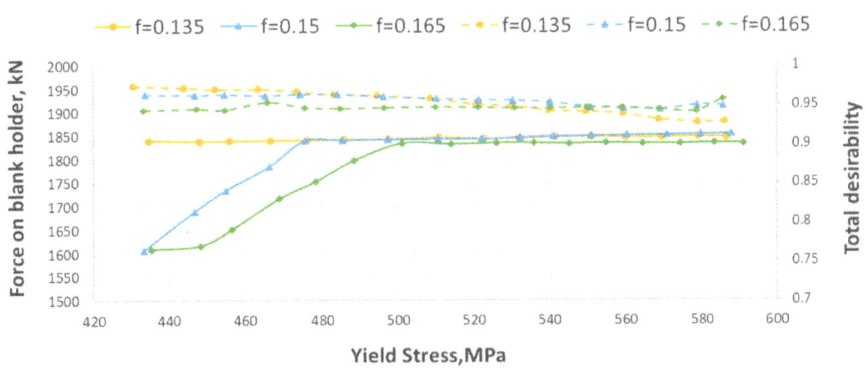


**Figure 14.** Regulation and total desirability curves.

Figure 14 shows that, at low values of the friction coefficient, the process requires higher force values, while, for the other levels of friction considered, the force initially increases, and then it stabilizes at a constant value. In particular, for a blank with a yield stress lower than 500 MPa and high friction (>0.135), it is necessary to reduce the force on the blank holder. Moreover, at low friction coefficient values, the process is insensitive and the maximum desirability curve is horizontal.

The regulation curves identified provide the control of the deep-drawing process in the case of random variations of the material mechanical characteristics (yield stress and friction coefficient). These noise parameters affect the sheet draw-in. In fact, in Figure 15, taking some reference points, the draw-in is compared as a function of the punch stroke for one of the conditions with maximum desirability (safe) and for a generic non-optimized condition (cracks).

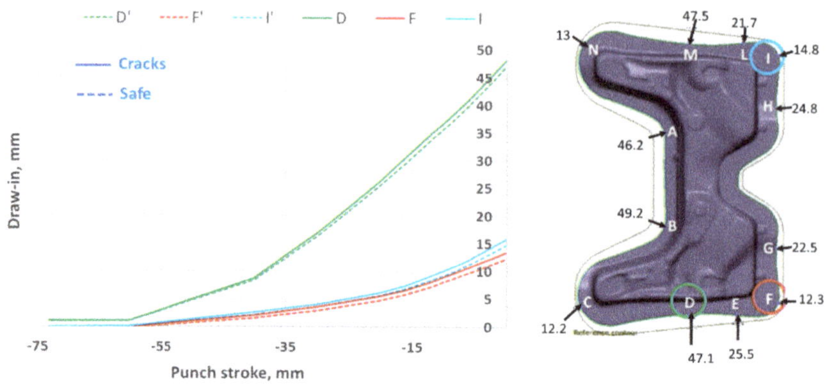


**Figure 15.** Draw-in as a function of the punch stroke at Points D, F and I of the sheet in the optimized condition (safe) and in the non-optimal condition (cracks).

From the comparison, the different sliding of the sheet is observed in the non-optimized case compared to the optimal case; this leads to excessive thinning or rupture. The comparison in-process of draw-in with the safe condition proves to be a promising strategy for online monitoring of the stamping process. In fact, through laser sensors placed on the most critical points, it is possible to evaluate sliding of the sheet by comparing it with the optimal case. Thus, if there is no correspondence, a signal is sent to the piezoelectric actuators on the blank holder which acts on the force, modifying it. Neugebauer [16] used piezoelectric actuators for manipulating the blank holder force. The used state variable was the edge draw-in, which was measured by a laser displacement sensor developed in the work of Bräunlich [17].

## 4. Conclusions

The results of the numerical simulations show that the factors considered for the evaluation of the quality of the final product (thickening, insufficient stretch, safe zone, potential splits and thinning in the part region near Points A and B) are strongly influenced by the causal variation of the yield stress and the coefficient of friction. Therefore, these disturbing factors should be taken into consideration when designing the process.

The main result of this work shows that numerical simulation using AutoForm FE software, meta-modeling using kriging technique and multi-objective optimization with a desirability approach are support tools for obtaining regulation curves that can be implemented by means of some control algorithms in the stamping process investigated. In this work, the regulation curve of the force on the blank holder was obtained as a function of the yield stress for different lubrication conditions.

These curves could allow regulating in process the force on the blank holder, in view of Industry 4.0, avoiding defects at the end of the process, when there are random variations in the yield stress of the material coil or in the lubrication conditions.

From the results of the draw-in as a function of the punch stroke for some points, it emerges that, to have a safe stamped component, it is possible to monitoring the sheet sliding online by correcting the force on the blank holder in process if the draw-in differs from that in the optimal case.

**Author Contributions:** Conceptualization, M.E.P., V.D.L. and L.T.; methodology, M.E.P., V.D.L. and L.T.; software, M.E.P. and V.D.L.; validation, M.E.P. and V.D.L.; formal analysis, M.E.P.,V.D.L. and L.T.; investigation, M.E.P.,V.D.L. and L.T.; resources, M.E.P.,V.D.L. and L.T; data curation, M.E.P.,V.D.L. and L.T; writing—original draft preparation, M.E.P. and L.T.; writing—review and editing, M.E.P. and L.T.; visualization, M.E.P.,V.D.L. and L.T; supervision, L.T; project administration, L.T.; funding acquisition, L.T. All authors have read and agreed to the published version of the manuscript.

**Funding:** This research was funded by MIUR thanks to the project PICO&PRO.

**Institutional Review Board Statement:** Not applicable.

**Informed Consent Statement:** Not applicable.

**Data Availability Statement:** Not applicable.

**Acknowledgments:** The authors are grateful to thank MIUR (PICO&PRO project) for funding this research. Moreover, the authors thank Tiberina Sangro company and AutoForm for the technical support.

**Conflicts of Interest:** The authors declare no conflict of interest.

## References

1. Bruschi, S.; Altan, T.; Banabic, D.; Bariani, P.F.; Brosius, A.; Cao, J.; Ghiotti, A.; Khraisheh, M.; Merklein, M.; Tekkaya, A.E. Testing and modelling of material behaviour and formability in sheet metal forming. *CIRP Ann. Manuf. Technol.* **2014**, *63*, 727–749. [CrossRef]
2. Schmid, H.; Merklein, M. Study of the mechanical properties of sheet metals drawn through drawbeads. *Manuf. Rev.* **2019**, *6*, 14. [CrossRef]
3. Liu, G.; Lin, Z.; Bao, Y. Optimization design of drawbead in drawing tools of autobody cover panel. *J. Eng. Mater. Technol.* **2002**, *124*, 278–285. [CrossRef]
4. Su, C.; Su, C.; Lou, S.; Xu, T.; Zhang, K.; Wang, Q. Effects of variable blank holder forces and a controllable drawbead on the springback of shallow-drawn TA2M titanium alloy boxes. *Int. J. Adv. Manuf. Technol.* **2017**, *93*, 1627–1635. [CrossRef]
5. Kardan, M.; Parvizi, A.; Askari, A. Influence of process parameters on residual stresses in deep-drawing process with FEM and experimental evaluations. *J. Braz. Soc. Mech. Sci. Eng.* **2018**, *40*, 157. [CrossRef]
6. Prates, P.; Marques, A.E.; Oliveira, M.C.; Fernandes, J.V. Comparing metamodeling techniques for variability analysis in sheet metal forming processes. In *AIP Conference Proceedings*; AIP Publishing LLC: Coimbra, Portugal, 2019; p. 120003.
7. Fischer, P.; Harsch, D.; Heingärtner, J. Influence of scattering material properties on the robustness of deep drawing processes. In Proceedings of the Forming Technology Forum 2017, Enschede, The Netherlands, 12–13 October 2017.
8. Van-Tuan, D.; Labergère, C.; Lafon, P. Adaptive metamodel-assisted shape optimization for springback in metal forming processes. *Int. J. Mater. Form.* **2019**, *12*, 535–552.
9. Banabic, D. *Sheet Metal Forming Processes: Constitutive Modelling and Numerical Simulation*; Springer Science & Business Media: Berlin/Heidelberg, Germany, 2010.
10. Fischer, P.; Harsch, D.; Heingärtner, J. Approaches for control in deep drawing. In Proceedings of the Forming Technology Forum, Enschede, The Netherlands, 12–13 October 2017; pp. 12–13.
11. Lophaven, S.N.; Nielsen, H.B.; Søndergaard, J. *Dace A Matlab Kriging Toolbox Version 2.0, August 1, 2002*; Imm Informatics and Mathematical Modelling; Technical University of Denmark DK-2800 Kgs: Lyngby, Denmark, 2002.
12. Available online: https://www.autoform.com/en/ (accessed on 5 January 2021).
13. Ling, D.; Williams, D.; Carleer, B.; Stippak, M. Systematic Process Improvement of Stamping Processes. In *Key Engineering Materials*; Trans Tech Publications Ltd: Dortmund, Germany, 2013; pp. 13–20.
14. Koehler, J.R.; Owen, A.B. *Computer Experiments*; Handbook of Statistics; Ghosh, S., Rao, C.R., Eds.; Elsevier Science: New York, NY, USA, 1996.
15. Bhaskar, P.; Sahoo, S.K. Optimization of Machining Process by Desirability Function Analysis (DFA): A Review. *CVR J. Sci. Technol.* **2020**, *18*, 138–143. [CrossRef]
16. Neugebauer, R.; Hoffmann, M.; Roscher, H.; Scheffler, S.; Wolf, K. Control of sheet-metal forming processes with piezoactuators in smart structures. In Proceedings of the Smart Structures and Materials 2006: Industrial and Commercial Applications of Smart Structures Technologies, San Diego, CA, USA, 30 March 2006.
17. Bräunlich, H. *Blecheinzugsregelung Beim Tiefziehen Mit Niederhalter-ein Beitrag zur Erhöhung der Prozessstabilität*; Verlag Wiss. Scripten: Chemnitz, Germany, 2002.

metals

Article

# Strip Steel Surface Defects Classification Based on Generative Adversarial Network and Attention Mechanism

**Zhuangzhuang Hao [1,2], Zhiyang Li [1], Fuji Ren [2], Shuaishuai Lv [1] and Hongjun Ni [1,*]**

1 School of Mechanical Engineering, Nantong University, Nantong 226019, China; hao_zhuangzhuang@163.com (Z.H.); zyli023@aliyun.com (Z.L.); lvshuaishuai@ntu.edu.cn (S.L.)
2 Graduate School of Advanced Technology and Science, Tokushima University, Tokushima 770-8506, Japan; ren@is.tokushima-u.ac.jp
* Correspondence: ni.hj@ntu.edu.cn

**Abstract:** In a complex industrial environment, it is difficult to obtain hot rolled strip steel surface defect images. Moreover, there is a lack of effective identification methods. In response to this, this paper implements accurate classification of strip steel surface defects based on generative adversarial network and attention mechanism. Firstly, a novel WGAN model is proposed to generate new surface defect images from random noises. By expanding the number of samples from 1360 to 3773, the generated images can be further used for training classification algorithm. Secondly, a Multi-SE-ResNet34 model integrating attention mechanism is proposed to identify defects. The accuracy rate on the test set is 99.20%, which is 6.71%, 4.56%, 1.88%, 0.54% and 1.34% higher than AlexNet, VGG16, ShuffleNet v2 1×, ResNet34, and ResNet50, respectively. Finally, a visual comparison of the features extracted by different models using Grad-CAM reveals that the proposed model is more calibrated for feature extraction. Therefore, it can be concluded that the proposed methods provide a significant reference for data augmentation and classification of strip steel surface defects.

**Keywords:** hot rolled strip steel; defect classification; generative adversarial network; attention mechanism; deep learning

**Citation:** Hao, Z.; Li, Z.; Ren, F.; Lv, S.; Ni, H. Strip Steel Surface Defects Classification Based on Generative Adversarial Network and Attention Mechanism. *Metals* **2022**, *12*, 311. https://doi.org/10.3390/met12020311

Academic Editor: Pedro Prates

Received: 20 January 2022
Accepted: 7 February 2022
Published: 10 February 2022



## 1. Introduction

As one of the main products of the steel industry, hot rolled strip steel is widely used in automobile manufacturing, aerospace and light industry [1]. Surface quality is one of the key indicators of strip steel's market competitiveness. Due to the influence of raw materials, rolling process and external environment, the strip steel surface will inevitably appear oxide scale, inclusion, scratch and other defects in the production process, which not only seriously affects the appearance, but also reduces the fatigue resistance. At the same time, these shortcomings cannot be completely overcome by improving the process [2,3]. Therefore, the classification of surface defects can provide an important reference for the production process. Through the corresponding tuning, the purpose of further improving the yield rate and reducing production costs is achieved.

The traditional surface defect detection mainly relies on manual visual inspection [4]. Although the implementation of this method is relatively simple, it is difficult to detect small defects with the continuous acceleration of the production line. In addition, long-term manual work will lead to visual fatigue and affect physical and mental health. Many researchers have used machine learning algorithms to overcome the drawbacks of manual visual inspection. Kim et al. [5] developed a K-Nearest Neighbor (KNN) classifier for eight defects with a classification performance of about 85%. Karthikeyan et al. [6] proposed a texture-based approach, where discrete wavelet transform based local configuration pattern features were given as input to a KNN classifier with an overall accuracy of 96.7%. Martins et al. [7] adopted principal component analysis to extract features from the defect images and used self-organizing maps to classify six types of defects obtained in the ArcelorMittal mill with an overall accuracy of 87%. Bulnes et al. [8] proposed a non-invasive system based on computer vision, which uses a neural network for classification and a genetic algorithm to determine the optimal values of the parameters. This method improves flexibility and the whole process can be executed quickly. Hu et al. [9] extracted geometric features, shape features, texture features and grey-scale features from defect images and their corresponding binary images. A classification model was developed by combining a hybrid chromosome genetic algorithm and a support vector machine (SVM) classifier, achieving a higher average prediction accuracy than that of the traditional SVM-based model. Jiang et al. [10] proposed an adaptive classifier with Bayesian kernel. Firstly, abundant features were introduced to cover detailed information of defects, and then a series of SVMs were constructed by using the random subspace of features. Finally, an improved Bayesian classifier was trained by fusing the results of basic SVMs, which has a strong adaptive capability. Zaghdoudi et al. [11] proposed an efficient system which for the first time used binary Gabor pattern feature descriptors to extract local texture features, and experimental results on the NEU defect database demonstrated the effectiveness of the method. The defect classification scheme based on machine learning has achieved certain results, which can guide the actual production. However, the expression ability of defect features extracted by the above method is limited and vulnerable to subjective experience, which often leads to low classification accuracy. In addition, new detection tasks need to redesign new algorithms, which is difficult to realize the migration of algorithms.

In the past few years, with the improvement of computing power and the establishment of large-scale datasets, deep learning-based classification methods have shown better performance compared to traditional recognition methods. Yi et al. [12] proposed an end-to-end recognition system based on symmetric surround saliency map and deep convolutional neural network (CNN). The excellent detection performance for seven types of strip steel surface defects is demonstrated. Fu et al. [13] proposed a compact and effective CNN model using pre-trained SqueezeNet as the backbone to achieve high accuracy on a diversity-enhanced steel surface defect dataset containing severe nonuniform illumination, camera noise and motion blur. Liu et al. [14] proposed a classification method based on deep CNN, adding an identity mapping to GoogLeNet and using this network to detect defects (such as scar, burrs, inclusion) with an accuracy of 98.57%. Konovalenko et al. [15] proposed an automated method based on ResNet50, which allows inspection with specific efficiency and speed parameters. The overall accuracy on the test set was 96.91%, proving that the residual neural network has excellent recognition performance and can be used as an effective tool. Wang et al. [16] proposed a VGG16-ADB network. Using VGG16 as the benchmark model, reducing system consumption and memory usage by decreasing the depth and width of the network structure, and adding a batch normalization layer to speed up convergence, which outperformed other classification models in terms of accuracy and speed. Wan et al. [17] proposed a complete process based on improved gray-scale projection algorithm, ROI image enhancement algorithm, and transfer learning. The fast screening, feature extraction, category balancing, and classification of defect images was achieved, and the recognition accuracy reached 97.8%. The deep learning-based classification algorithms for strip steel surface defects has been effective, but there are still shortcomings in the current research. On the one hand, the performance of deep learning model mainly depends on the size and quality of training samples [18]. Nevertheless, it is difficult to obtain sufficient number of defect samples in complex industrial scenes, so expanding the data set has become an urgent problem to be solved. On the other hand, attention mechanism has been proved to enable the model to focus on more valuable information, which is conducive to improving the recognition accuracy [19,20]. However, the current research rarely introduces attention mechanism into the classification algorithm of strip steel surface defects.

Based on Generative Adversarial Network(GAN) and attention mechanism, accurate classification of strip steel surface defects is realized. Firstly, a novel Wasserstein GAN(WGAN) model is proposed for data augmentation. Secondly, a Multi-SE-ResNet34 model is proposed and used for defect classification. Comparative experiments verify the excellent performance of the proposed model. Finally, the features extracted by the proposed model are visualised, demonstrating robustness and calibration for the identification of multiple defects. Our methods provide a reference for solving the small sample and classification problems of strip steel surface defects.

The rest of this paper is structured as follows. The second part introduces related theories and proposed methods. The third part gives the experimental results. The fourth part explains the proposed method. The fifth part summarizes the full text.

## 2. Methodologies

### 2.1. GAN

The GAN [21] is an unsupervised deep learning model that can learn the distribution of samples and generate new sample data without relying on prior assumptions. The typical structure is shown in Figure 1. GAN optimizes generator and discriminator by alternate iteration. G(z) tries to satisfy the probability distribution of the real sample x, while discriminator D tries to distinguish between x and G(z). Through continuous confrontation training, the generator and discriminator finally reach Nash equilibrium.

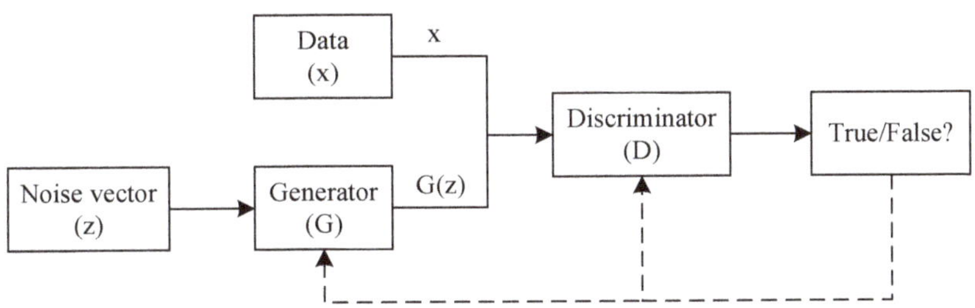


**Figure 1.** GAN structure.

For the original GAN, Jensen-Shannon (JS) divergence is used to measure the gap between the generated sample and the real sample. In the process of seeking Nash equilibrium, model collapse or gradient disappearance will lead to the non-convergence of the neural network. In WGAN, JS distance is replaced by Wasserstein distance [22]. The replacement of loss function brings the following advantages: the problem of unstable GAN training is completely solved, and it is no longer necessary to carefully balance the training degree of generator and discriminator; the problem of collapse mode is solved to ensure the diversity of generated samples; the design of network architecture becomes simple, which is conducive to the combination with CNN to realize image generation. The Wasserstein distance is defined as:

$$W(P_r, P_g) = \inf_{\delta \in \Pi(P_r, P_g)} E_{(x,y)\sim\delta}[\|x - y\|] \quad (1)$$

where $P_r$ and $P_g$ represent the data distribution of the real sample and the generated sample; $\Pi(P_r, P_g)$ represents the set of joint probability distribution $\delta$ with $P_r$ and $P_g$ as the marginal distribution; $W(P_r, P_g)$ represents the distance of $x$ to $y$ required to fit $P_g$ to $P_r$. The Kantorovich-Rubinstein dual form of $W(P_r, P_g)$ is adopted in the actual calculation, as shown in Equation (2).

$$W(P_r, P_g) = \sup_{\|f\|_L \leq 1} E_{x\sim P_r}[f(x)] - E_{x\sim P_g}[f(x)] \quad (2)$$

$\|f\|_L \leq 1$ means that $f(x)$ satisfies the 1-Lipschitz condition. WGAN uses weight clipping to limit the weight of the discriminator network to a fixed range to approximate the Wasserstein distance. The generator network is optimized to minimize the Wasserstein distance, thereby effectively narrowing the distribution of generated samples and real samples. The loss functions of generator and discriminator are defined as $\mathrm{Loss}_G$ and $\mathrm{Loss}_D$, respectively, as shown in Equations (3) and (4).

$$\mathrm{Loss}_G = -E_{x \sim P_g}[D(x)] \quad (3)$$

$$\mathrm{Loss}_D = E_{x \sim P_g}[D(x)] - E_{x \sim P_r}[D(x)] \quad (4)$$

### 2.2. Squeeze-and-Excitation Block

Squeeze-and-excitation block (SE block) [23] is shown in Figure 2. By learning the weights of the feature maps, effective channels are amplified and invalid or less effective channels are suppressed, thereby achieving the purpose of improving the accuracy of the model.

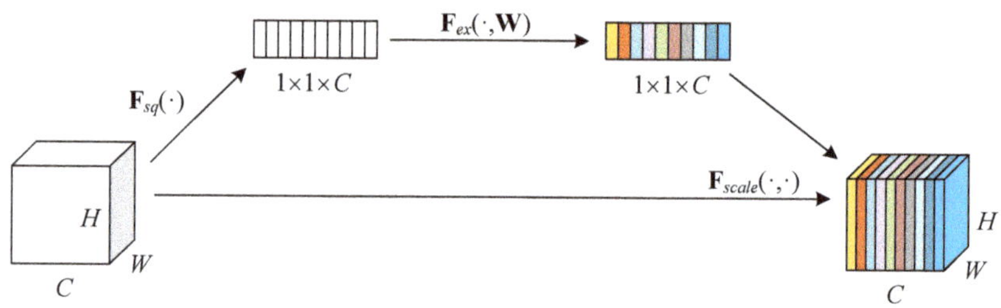


**Figure 2.** SE block.

The height, width, and channel number of the input feature map $u_c$ are $H$, $W$ and $C$, respectively. Through squeeze and global average pooling algorithm, the output feature map is transformed from $H \times W \times C$ to $1 \times 1 \times C$, as shown in Equation (5).

$$z_c = \boldsymbol{F}_{sq}(u_c) = \frac{1}{W \times H} \sum_{i=1}^{W} \sum_{j=1}^{H} u_c(i,j) \quad (5)$$

where $Z_c$ represents the output feature map, and $(i,j)$ represents the coordinate position on the feature map. Through excitation, two fully connected layers $\mathbf{W}_1$ and $\mathbf{W}_2$ are utilised to merge the information of the channels. The dimension of $\mathbf{W}_1$ is set to $1 \times 1 \times \frac{C}{r}$ to reduce the computational effort, where $r$ represents reduction ratio. The dimension of $\mathbf{W}_2$ is restored to $1 \times 1 \times C$. Finally, the channel weight $v$ is obtained, as shown in Equation (6).

$$v = \boldsymbol{F}_{ex}(z_c, \mathbf{W}_i) = \delta(\mathbf{W}_2 \sigma(\mathbf{W}_1 z_c)) \quad (6)$$

where $\sigma$ is ReLU activation function and $\delta$ is Sigmoid activation function. The adjustment parameters between the channels are multiplied by the original feature map to realize the recalibration, as shown in Equation (7).

$$\boldsymbol{X}_c = \boldsymbol{F}_{\text{scale}}(u_c, v_c) = u_c v_c \quad (7)$$

where $v_c$ represents the weight parameter of the $c$ th feature map, $\boldsymbol{X}_c$ represents the adjusted feature map.

### 2.3. Feature Visualization

The features extracted by deep convolutional networks are highly abstract, which is difficult to visually display the information of interest. With the deepening of research, Gradient-weighted Class Activation Mapping (Grad-CAM) [24] has gradually become a powerful visualization tool. Grad-CAM is able to present the features of most interest to the model in the form of a heat map, which calculates the weights of the features primarily by employing a global average of the gradients.

The gradient of the model score for category $C$ is first calculated for a particular convolutional layer, while for the gradient information obtained by the above process, the importance weights of the neurons are obtained by averaging the pixel values over each channel dimension, as shown in Equation (8).

$$\alpha_i^c = \frac{1}{Z}\sum_{k=1}^{c_1}\sum_{j=1}^{c_2}\frac{\partial S_c}{\partial A_{kj}^i} \tag{8}$$

where $Z$ is the number of pixels in the feature map, $S_c$ is the classification score for category $C$. $c_1 \times c_2$ represents the dimension of the feature map. $A_{kj}^i$ represents the pixel value of the $k$ th row and $j$ th column of the $i$ th feature map, and $\alpha_i^c$ is the weight of class C relative to the $i$ th channel of the feature map output by the last convolution layer. The weighted average is executed and then passed through the ReLU function to obtain the Grad-CAM feature map. The formula is shown in Equation (9).

$$L^c = \mathrm{ReLU}\left(\sum_i \alpha_i^c A^i\right) \tag{9}$$

where $L^c$ represents the activated heat map of class $C$ and $A^i$ represents the $i$ th feature map.

### 2.4. Our Methods

#### 2.4.1. A Novel WGAN Model

A novel WGAN model is proposed and used for data augmentation of strip steel surface defect images, as shown in Figure 3. The implementation of the discriminator is similar to that of a general CNN [25]. The activation functions between discriminator convolutional layers all use LeakyReLU. It should be noted that the Sigmoid function is not used in the last layer. The input of the generator is a 128-dimensional random noise vector conforming to the standard normal distribution. Between levels, batch normalization is used to accelerate convergence and slow down overfitting. The tanh function is used to activate the output layer, and the ReLU function is used to activate the remaining layers. With the transposed convolution, the number of channels gradually decreases and the dimensions continue to increase, so that the three-channel pseudo image is finally generated.

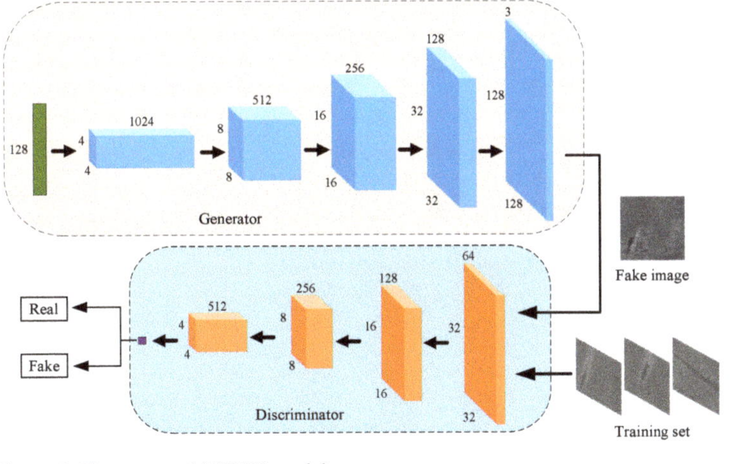


**Figure 3.** The proposed WGAN model.

By modifying the dimension of the last layer of the generator to 128 × 128, the generated image can directly maintain the same size as the original image, which facilitates subsequent classification research.

#### 2.4.2. Multi-SE-ResNet34 Model

Based on current experience, increasing the depth of network can improve network performance. However, the degradation phenomenon that occurs during the back propagation of the error gradient may cause difficulties in network convergence. In the deep residual network (ResNet) proposed by He et al. [26] in 2015, the addition of identity mapping solves the problem that deep network models are difficult to train. In the last few years, ResNet has been widely used in various classification tasks [27–30] with strong capabilities. On this basis, a Multi-SE-ResNet34 model combined with the attention mechanism is proposed, and the structure is shown in Figure 4.

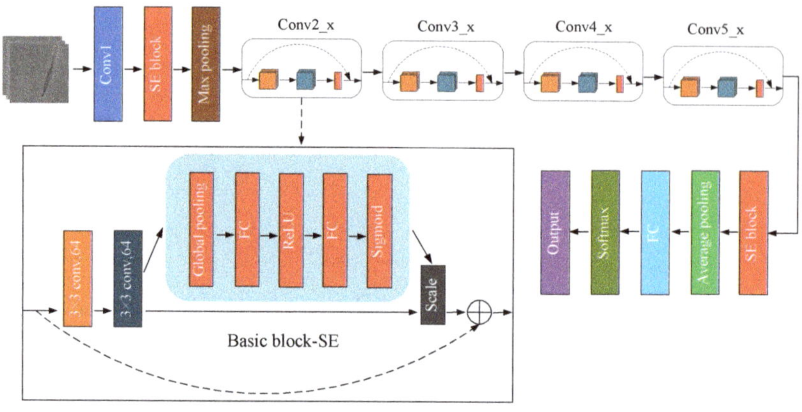


**Figure 4.** The proposed Multi-SE-ResNet34 model.

Multi-SE-ResNet34 is an improvement of ResNet34, which is mainly composed of four different types of Basic block-SE modules. This module embeds SE block in each residual unit. From Conv2_x to Conv5_x, there are 3, 4, 6, and 3 Basic block-SEs, and all Basic block-SEs use a 3 × 3 convolution kernel. As the depth of the model increases, the number of convolution kernels keeps consistent with ResNet34. Moreover, two additional SE blocks are added outside the residual structure, which are located after the first convolutional layer and before the average pooling layer. Due to the attention mechanism, the performance of the proposed model is better than that of the basic ResNet34, which will give support in the discussion.

#### 2.4.3. Overall Process

The overall process of our methods is shown in Figure 5. First, the WGAN model is constructed for data augmentation. The generated image and the original image together form a new data set. Second, the enhanced data set is divided into training set, validation set and test set. The function of the test set lies in the evaluation of performance and the output of classification results.

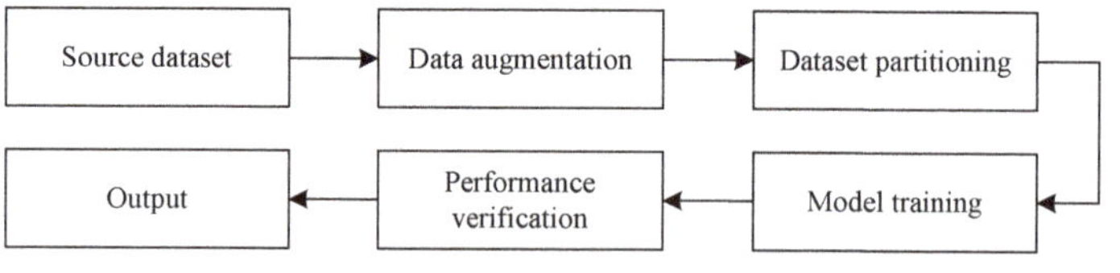


**Figure 5.** Overall flow of the proposed method.

## 3. Experiments and Results

The experiment is based on the following hardware and software environment: Windows10 operating system of Microsoft, Intel(R) Core (TM) i7-11800H CPU, NVIDIA GeForce RTX 3060 Laptop GPU, NVIDIA CUDA-11.1.1 and cuDNN-11.2, Pytorch v1.8.0 deep learning framework.

### *3.1. Introduction to the Data Set*

The X-SDD data set [31] contains 1360 strip steel surface defect images in 7 categories. The size of each image is 128 × 128 pixels, and the format is 3-channel JPG. Several samples of each defect are shown in Figure 6. For the convenience of description, the 7 types of images are marked with tags of 0, 1, 2, 3, 4, 5, and 6.

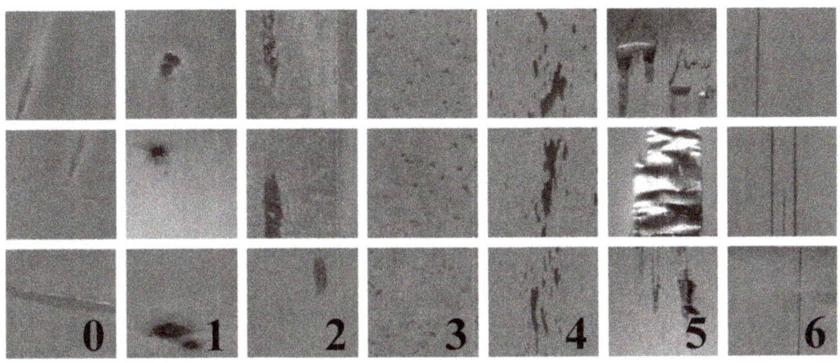


**Figure 6.** Seven kinds of strip steel surface defect image samples in X-SDD data set, including (0) finishing roll printing; (1) iron sheet ash; (2) oxide scale of plate system; (3) oxide scale of temperature system; (4) red iron; (5) slag inclusion; (6) surface scratch.

### *3.2. Image Generation*

After training the discriminator five times, the generator is trained once. Both the generation network and the discriminant network use RMSProp algorithm to update parameters, including learning rate of 0.00005, clipping parameter of 0.01, batch size of 32, and epoch of 7000. The strip steel surface defect images generated by the proposed WGAN model at different stages are shown in Figure 7.

It can be seen that when the number of iterations is 500, the generated image contains more meaningless information. At this point, the discriminator can easily distinguish false samples. When the number of iterations reaches 2000, the generator gradually learns the data distribution of the real image. At this point, the generated image has a rough outline of the defect. However, a lot of texture information is lost and blurred visually. After 7000 epochs, the generated image is close to the real image, with clear outline and distribution of defects. Unlike linear transformations such as rotation and scaling, the generated image guarantees the diversity of features. The total number of samples increases from 1360 to 3773 after data augmentation. The specific number of each type of defect is shown in Table 1.

**Table 1.** Number of defective images.

| Category | 0 | 1 | 2 | 3 | 4 | 5 | 6 |
|---|---|---|---|---|---|---|---|
| Original | 203 | 122 | 63 | 203 | 397 | 238 | 134 |
| Enhanced | 589 | 517 | 498 | 530 | 595 | 488 | 556 |

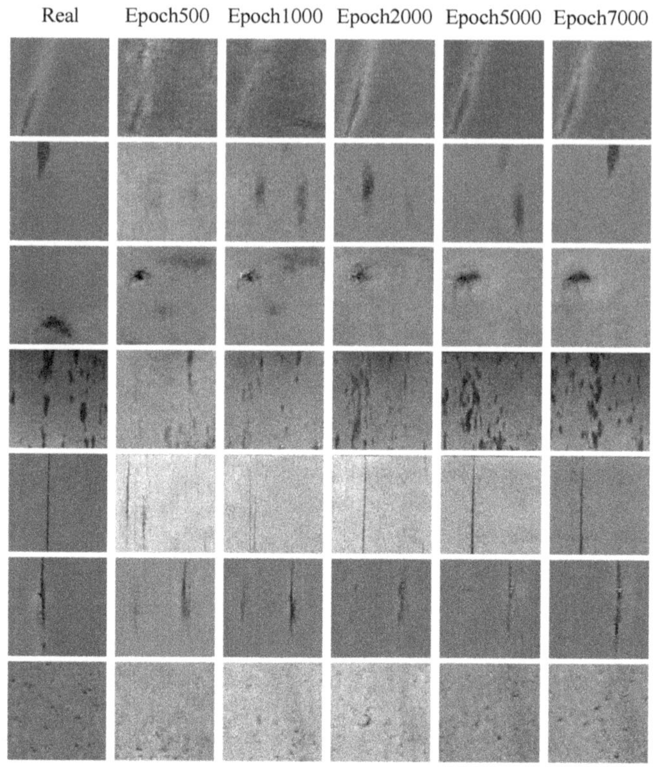


**Figure 7.** Strip steel surface defect image samples generated by WGAN in different iterations.

### 3.3. *Defect Classification*

In the classification experiment, the data set after data augmentation is divided. First, 10% sample is randomly sampled to form a testing set. Then, the remaining images are divided into training set and validation set with the ratio of 8:2. The number of images in the training set, validation set, and testing set are 2722, 678 and 373, respectively. The input image of Multi-SE-ResNet34 is set to a size of 224 $\times$ 224 and normalized with batch size of 16. The reduction ratio of SE block is set to 16. Stochastic gradient descent with momentum is used for parameter update with the momentum factor of 0.9 and initial learning rate of 0.001. The learning rate is reduced to one-tenth of the original after 20 epochs. Moreover, L2 regularization is used to prevent overfitting, with the weight decay coefficient of 0.0001. Figure 8 shows the loss and accuracy curves. During the first 10 iterations, the loss drops rapidly and the accuracy rises. As the learning rate decreases, the model tends to stabilize. The loss approaches 0 after the iteration is completed.

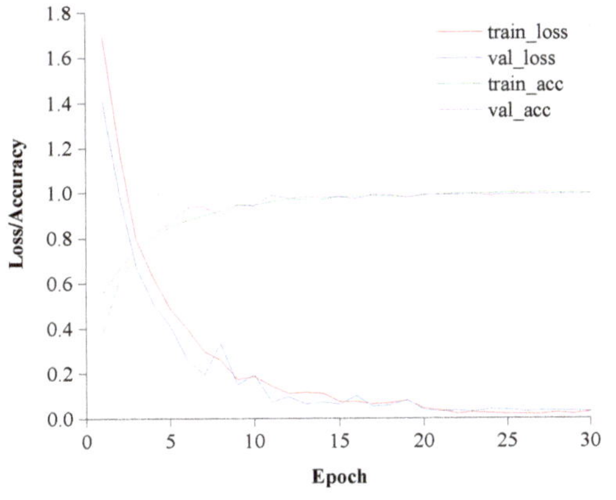


**Figure 8.** Curves of loss and accuracy during training.

In the test set, the classification performance of the model is evaluated. We chose indicators such as Accuracy, Macro-Precision, Macro-Recall and Macro-F1. The above indicators are given by Equations (10)–(13).

$$Accuracy = \frac{n\_correct}{n\_total} \tag{10}$$

$$Macro - Precision = \frac{1}{N}\sum_{i=1}^{N}\frac{TP_i}{TP_i + FP_i} \tag{11}$$

$$Macro - Recall = \frac{1}{N}\sum_{i=1}^{N}\frac{TP_i}{TP_i + FN_i} \tag{12}$$

$$Macro - F_1 = \frac{1}{N}\sum_{i=1}^{N}\frac{2 \times P_i \times R_i}{P_i + R_i} \tag{13}$$

where, $n\_correct$ is the number of samples correctly classified by the model; $n\_total$ is the total number of samples; $TP$, $FP$, $TN$ and $FN$ represent true positive, false positive, true negative, and false negative, respectively. $N$ is the number of defect types. $P$ and $R$ represent precision and recall.

The classification results are shown in Table 2. The generated confusion matrix is shown in Figure 9. The accuracy of Multi-SE-ResNet34 is 99.20%, demonstrating the robustness of our method for feature recognition of a wide range of strip steel surface defects. According to the confusion matrix, defects 0, 1, 2, 4, and 5 can be identified 100%. The accuracy of defect 6 is relatively low, and two images are classified as defect 4. Some of the defects 4 have a slender distribution, which is similar to that of defects 6, which leads to an increase in the difficulty of classification. On the whole, our method can accurately classify 7 kinds of strip steel surface defects.

**Table 2.** Classification results.

| Accuracy (%) | Macro-Precision (%) | Macro-Recall (%) | Macro-F1 (%) |
|---|---|---|---|
| 99.20 | 99.29 | 99.21 | 99.24 |

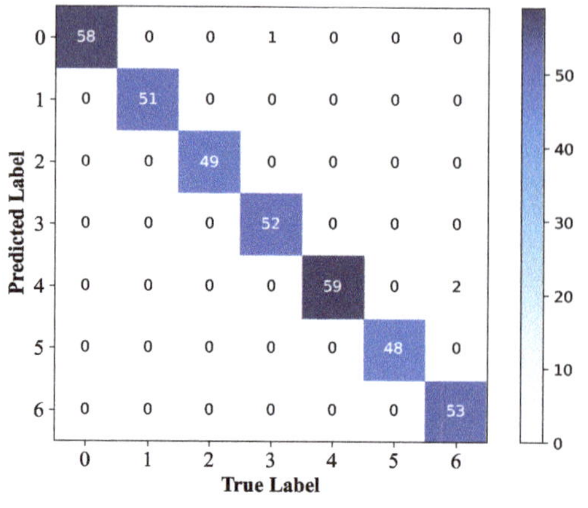


**Figure 9.** Confusion matrix.

### *3.4. Grad-CAM Visualization*

Seven defect images are randomly selected and used to generate visual heat maps of each layer of Multi-SE-ResNet34, as shown in Figure 10. It can be clearly seen that the number of layers in the network at the end of Conv1 is very shallow and the model extracts few features. As the number of convolutional layers increases, the feature recognition capability is enhanced, and the features learned by the model becomes rich at the end of Conv4_x, but still insufficient to cover the whole defect. The model extracts enough features at the end of Conv5_x, and at the same time, the area of interest is exactly where the defects are located due to the addition of the attention module. It can be concluded that our model has excellent recognition performance for all seven strip surface defects features.

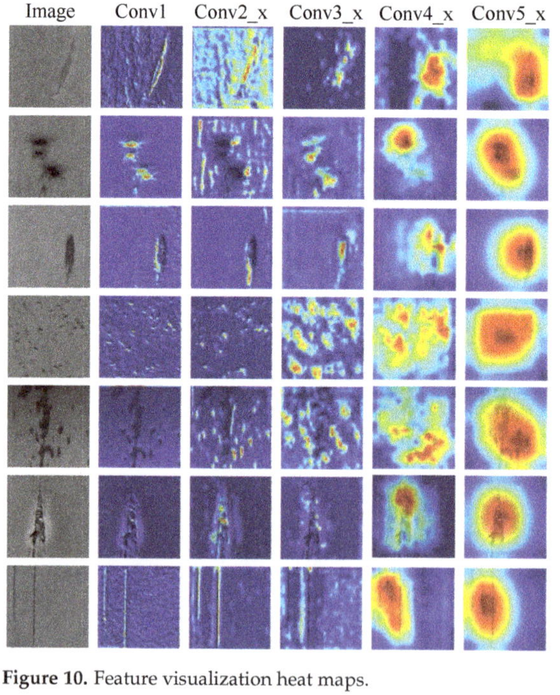


**Figure 10.** Feature visualization heat maps.

## 4. Discussions

### 4.1. The Impact of Sample Size on Classification Results

Classification using Multi-SE-ResNet34 on the source dataset yielded an accuracy of 93.98%. Nevertheless, the accuracy is improved by 5.22% after data augmentation, i.e., 99.20%, which shows the classification performance is closely related to the number of samples. Although studies have pointed this out [32,33], there are few complete identification cases. Therefore, our method generates realistic images and improves recognition accuracy, providing an effective solution for the small sample size of strip steel surface defect images.

### 4.2. Comparison with Other Models

In order to further verify the remarkable performance of our method, the classical models of AlexNet [34], VGG16 [35], ShuffleNet v2 1× [36], ResNet34 and ResNet50 [26] are selected for comparison using the enhanced dataset with the same hyperparameters. The classification results of each model on the test set are shown in Table 3. It can be seen that our method obtains the highest accuracy rate, which is 6.71%, 4.56%, 1.88%, 0.54% and 1.34% higher than AlexNet, VGG16, ShuffleNet v2 1×, ResNet34, and ResNet50, respectively. At the same time, our model is also optimal on three other evaluation indicators.

**Table 3.** Comparison of different models.

| Model | Accuracy (%) | Macro-Precision (%) | Macro-Recall (%) | Macro-F1 (%) |
|---|---|---|---|---|
| AlexNet | 92.49 | 92.82 | 92.15 | 92.19 |
| VGG16 | 94.64 | 95.06 | 94.30 | 94.45 |
| ShuffleNet v2 1× | 97.32 | 97.38 | 97.26 | 97.30 |
| ResNet34 | 98.66 | 98.68 | 98.59 | 98.61 |
| ResNet50 | 97.86 | 97.88 | 97.72 | 97.77 |
| Our method | 99.20 | 99.29 | 99.21 | 99.24 |

Figure 11 shows the accuracy curves for the training set of each model. It can be seen that after 10 iterations of training, the accuracy of all models except AlexNet exceeds 90%, with AlexNet having the lowest accuracy due to its shallow network layers. The accuracy of each model increases over the first 20 epochs, reaching its maximum value and stabilising after the learning rate is reduced; after the completion of iterations, all models except AlexNet obtain an accuracy of over 99.41%. In terms of convergence speed, AlexNet is the slowest, in contrast to ResNet34. The lower convergence speed of ShuffleNet than VGG16 is attributed to the reduction in the number of parameters due to the lightweight implementation, where the recognition ability is diminished. Our method achieves a satisfactory convergence rate, comparable to that of ResNet50, but lower than that of ResNet34. One possible reason is that the number of parameters increased with the addition of multiple SE blocks, and fewer iterations are not sufficient to extract enough features. However, our method has the highest accuracy and achieves a balance between recognition effectiveness and number of parameters, which can be considered more advantageous.

The loss curves in the validation set of each model are shown in Figure 12. It can be seen that both AlexNet and VGG16 have large fluctuations and are less stable. The curve of ShuffleNet is the smoothest. There are several fluctuations in ResNet34 and ResNet50 where stability is compromised. The curve of our method is relatively smooth overall, with only a few minor fluctuations that do not affect the decreasing course of loss. All models converge after 20 iterations. At the end of training, the loss of our method is the lowest, maintaining at 0.029. On the whole, a stable training process, the lowest loss value and the highest accuracy have been obtained, therefore our method is optimal for the classification of strip surface defects.

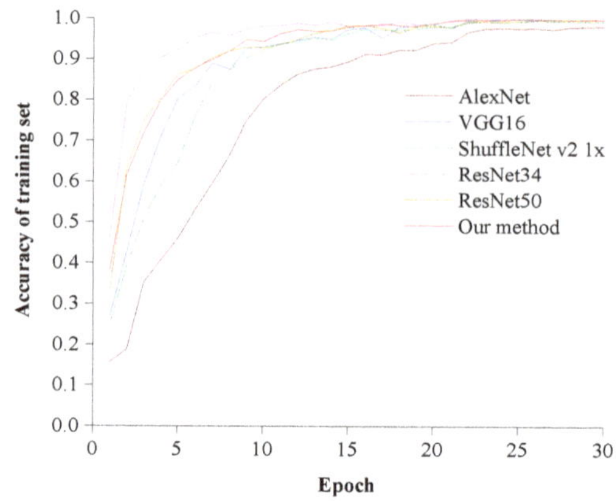


**Figure 11.** Comparison of accuracy of each model training set.

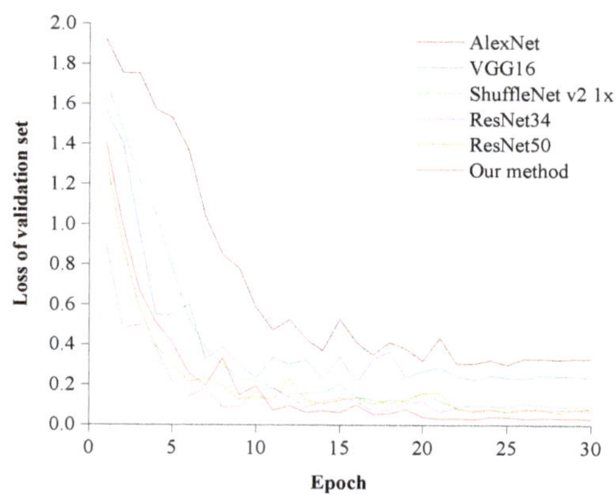


**Figure 12.** Comparison of loss of each model validation set.

### *4.3. Influence of Attention Mechanism on Feature Extraction*

Heat maps of the strip surface defect features extracted by the last convolutional layer of each model are generated to explore the influence of attention mechanism on feature extraction, as shown in Figure 13. It can be seen that AlexNet struggles to extract features effectively due to its shallow network layers. VGG16 simply stacks convolutional layers, with no obvious improvement in feature extraction capability compared to AlexNet. The features extracted by ShuffleNet increased but with a large amount of useless information. In particular, despite the relatively deep depth of the ResNet50 network, it failed to accurately extract features of defect 0 and defect 4. The performance of ResNet34 is outstanding with an excellent feature extraction capability. Nevertheless, in comparison, our method not only extracts sufficient features, but also reduces invalid information in the background and locates feature regions more precisely, which verifies the comparison results in Section 4.2. In other words, benefiting from the attention mechanism, our method is more calibrated in terms of feature extraction.

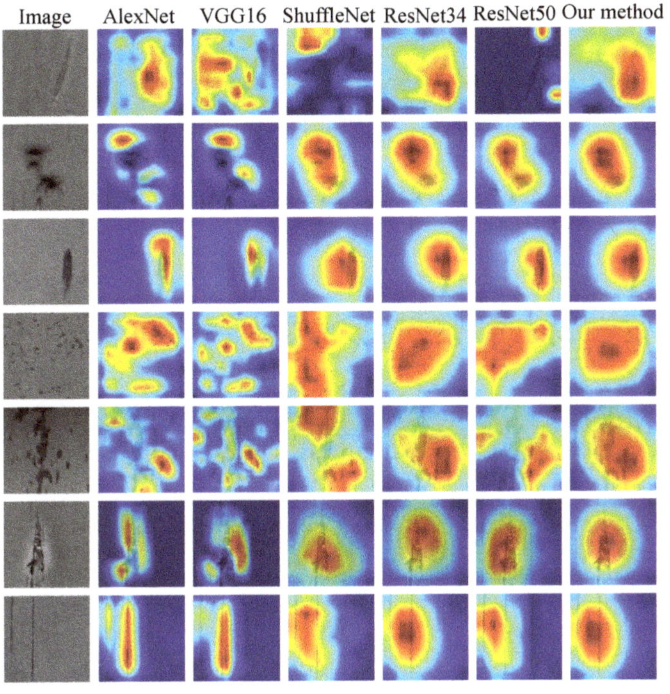


**Figure 13.** Visualization of feature extraction in the last convolution layer of each model.

## 5. Conclusions

1. For the small sample size of strip steel surface defect images, a novel WGAN model is proposed and used for data augmentation. The generated image has a resolution of 128 × 128 and the appearance is close to the real image, which can be directly used to expand the original data set.
2. A Multi-SE-ResNet34 model combining channel attention mechanism is proposed and used for defect classification with 99.20% accuracy. In addition, Multi-SE-ResNet34 outperforms the other models in terms of Macro-Precision, Macro-Recall and Macro-F1. The training process of Multi-SE-ResNet34 is stable, and the validation set loss tends to 0. Furthermore, there is no over-fitting phenomenon.
3. The Grad-CAM method is used to visually analyze the defect features extracted by different models, which shows that the attention mechanism can make the model pay attention to more valuable information and improve the classification accuracy. The advantages of our method are further demonstrated.

In the future, we have the expectation of combining spatial attention and channel attention to further improve the recognition rate and realize the lightweight of the network.

**Author Contributions:** Conceptualization, Z.L. and Z.H.; methodology, Z.H. and F.R.; software, Z.H.; validation, Z.H.; formal analysis, Z.H. and F.R.; investigation, Z.L.; resources, H.N.; data curation, F.R.; writing—original draft preparation, Z.H.; writing—review and editing, S.L.; visualization, F.R.; supervision, S.L.; project administration, H.N.; funding acquisition, H.N. All authors have read and agreed to the published version of the manuscript.

**Funding:** This research was funded by the Priority Academic Program Development of Jiangsu Higher Education Institutions, grant number PAPD; Jiangsu Province Policy Guidance Program (International Science and Technology Cooperation) Project, grant number BZ2021045; Nantong Applied Research Project, grant number JCZ21066, JCZ21043, JCZ21013; Key R&D Projects of Jiangsu

Province, grant number BE2019060; University-Industry Collaborative Education Program, grant number 202102236001.

**Institutional Review Board Statement:** Not applicable.

**Informed Consent Statement:** Not applicable.

**Data Availability Statement:** Not applicable.

**Conflicts of Interest:** The authors declare no conflict of interest.

## References


1. Tang, W.; Liong, S.; Chen, C.; Tsai, M.; Hsieh, P.; Tsai, Y.; Chen, S.; Wang, K. Design of Multi-Receptive Field Fusion-Based Network for Surface Defect Inspection on Hot-Rolled Steel Strip Using Lightweight Dataset. *Appl. Sci.* **2021**, *11*, 9473. [CrossRef]
2. Sun, J.; Peng, W.; Ding, J.; Li, X.; Zhang, D. Key intelligent technology of steel strip production through process. *Metals* **2018**, *8*, 597. [CrossRef]
3. Feng, X.; Gao, X.; Luo, L. A ResNet50-Based Method for Classifying Surface Defects in Hot-Rolled Strip Steel. *Mathematics* **2021**, *9*, 2359. [CrossRef]
4. Zhang, J.; Kang, X.; Ni, H.; Ren, F. Surface defect detection of steel strips based on classification priority YOLOv3-dense network. *Ironmak. Steelmak.* **2021**, *48*, 547–558. [CrossRef]
5. Kim, C.; Choi, S.; Kim, G.; Joo, W. Classification of surface defect on steel strip by KNN classifier. *J. Korean Soc. Precis. Eng.* **2006**, *23*, 80–88.
6. Karthikeyan, S.; Pravin, M.C.; Sathyabama, B.; Mareeswari, M. DWT Based LCP Features for the Classification of Steel Surface Defects in SEM Images with KNN Classifier. *Aust. J. Basic Appl. Sci.* **2016**, *10*. Available online: https://ssrn.com/abstract=2792637 (accessed on 17 April 2021).
7. Martins, L.A.; Pádua, F.L.; Almeida, P.E. Automatic detection of surface defects on rolled steel using computer vision and artificial neural networks. In Proceedings of the IECON 2010-36th Annual Conference on IEEE Industrial Electronics Society, Glendale, AZ, USA, 7–10 November 2010; pp. 1081–1086.
8. Bulnes, F.G.; García, D.F.; De la Calle, F.J.; Usamentiaga, R.; Molleda, J. A non-invasive technique for online defect detection on steel strip surfaces. *J. Nondestruct. Eval.* **2016**, *35*, 1–18. [CrossRef]
9. Hu, H.; Liu, Y.; Liu, M.; Nie, L. Surface defect classification in large-scale strip steel image collection via hybrid chromosome genetic algorithm. *Neurocomputing* **2016**, *181*, 86–95. [CrossRef]
10. Jiang, M.; Li, G.; Xie, L.; Xiao, M.; Yi, L. Adaptive classifier for steel strip surface defects. *J. Phys.* **2017**, *787*, 012019. [CrossRef]
11. Zaghdoudi, R.; Seridi, H.; Ziani, S. Binary Gabor pattern (BGP) descriptor and principal component analysis (PCA) for steel surface defects classification. In Proceedings of the 2020 International Conference on Advanced Aspects of Software Engineering (ICAASE), Constantine, Algeria, 28–30 November 2020; pp. 1–7.
12. Yi, L.; Li, G.; Jiang, M. An end-to-end steel strip surface defects recognition system based on convolutional neural networks. *Steel Res. Int.* **2017**, *88*, 1600068. [CrossRef]
13. Fu, G.; Sun, P.; Zhu, W.; Yang, J.; Cao, Y.; Yang, M.Y.; Cao, Y. A deep-learning-based approach for fast and robust steel surface defects classification. *Opt. Laser Eng.* **2019**, *121*, 397–405. [CrossRef]
14. Liu, Y.; Geng, J.; Su, Z.; Zhang, W.; Li, J. Real-time classification of steel strip surface defects based on deep CNNs. In Proceedings of the 2018 Chinese Intelligent Systems Conference, Wenzhou, China, 10–13 March 2019; pp. 257–266.
15. Konovalenko, I.; Maruschak, P.; Brezinová, J.; Viňáš, J.; Brezina, J. Steel surface defect classification using deep residual neural network. *Metals* **2020**, *10*, 846. [CrossRef]
16. Wang, W.; Lu, K.; Wu, Z.; Long, H.; Zhang, J.; Chen, P.; Wang, B. Surface Defects Classification of Hot Rolled Strip Based on Improved Convolutional Neural Network. *ISIJ Int.* **2021**, *61*, 1579–1583. [CrossRef]
17. Wan, X.; Zhang, X.; Liu, L. An Improved VGG19 Transfer Learning Strip Steel Surface Defect Recognition Deep Neural Network Based on Few Samples and Imbalanced Datasets. *Appl. Sci.* **2021**, *11*, 2606. [CrossRef]
18. Xu, L.; Tian, G.; Zhang, L.; Zheng, X. Research of Surface Defect Detection Method of Hot Rolled Strip Steel Based on Generative Adversarial Network. In Proceedings of the 2019 Chinese Automation Congress (CAC), Hangzhou, China, 22–24 November 2019; pp. 401–404.
19. Cao, Y.; Xu, J.; Lin, S.; Wei, F.; Hu, H. Gcnet: Non-local networks meet squeeze-excitation networks and beyond. In Proceedings of the IEEE/CVF International Conference on Computer Vision Workshops, Seoul, Korea, 27–28 October 2019.
20. Wang, F.; Jiang, M.; Qian, C.; Yang, S.; Li, C.; Zhang, H.; Wang, X.; Tang, X. Residual attention network for image classification. In Proceedings of the IEEE Conference on Computer Vision and Pattern Recognition, Honolulu, HI, USA, 21–26 July 2017; pp. 3156–3164.
21. Goodfellow, I.J.; Pouget-Abadie, J.; Mirza, M.; Xu, B.; Warde-Farley, D.; Ozair, S.; Courville, A.; Bengio, Y. Generative adversarial networks. *Adv. Neural Inf. Process. Syst.* **2014**, *3*, 2672–2680. [CrossRef]
22. Jiao, Z.; Ren, F. WRGAN: Improvement of RelGAN with Wasserstein Loss for Text Generation. *Electronics* **2021**, *10*, 275. [CrossRef]
23. Hu, J.; Shen, L.; Sun, G. Squeeze-and-excitation networks. In Proceedings of the IEEE Conference on Computer Vision and Pattern Recognition, Salt Lake City, UT, USA, 18–23 June 2018; pp. 7132–7141.

24. Selvaraju, R.R.; Cogswell, M.; Das, A.; Vedantam, R.; Parikh, D.; Batra, D. Grad-cam: Visual explanations from deep networks via gradient-based localization. In Proceedings of the IEEE International Conference on Computer Vision, Venice, Italy, 22–29 October 2017; pp. 618–626.
25. Wang, K.; Zhang, J.; Ni, H.; Ren, F. Thermal Defect Detection for Substation Equipment Based on Infrared Image Using Convolutional Neural Network. *Electronics* **2021**, *10*, 1986. [CrossRef]
26. He, K.; Zhang, X.; Ren, S.; Sun, J. Deep residual learning for image recognition. In Proceedings of the IEEE Conference on Computer Vision and Pattern Recognition, Las Vegas, NV, USA, 27–30 June 2016; pp. 770–778.
27. Li, B.; He, Y. An improved ResNet based on the adjustable shortcut connections. *IEEE Access* **2018**, *6*, 18967–18974. [CrossRef]
28. Ren, F.; Liu, W.; Wu, G. Feature reuse residual networks for insect pest recognition. *IEEE Access* **2019**, *7*, 122758–122768. [CrossRef]
29. Zhang, Y.; Wa, S.; Sun, P.; Wang, Y. Pear Defect Detection Method Based on ResNet and DCGAN. *Information* **2021**, *12*, 397. [CrossRef]
30. Yang, Y.; Wang, H.; Jiang, D.; Hu, Z. Surface Detection of Solid Wood Defects Based on SSD Improved with ResNet. *Forests* **2021**, *12*, 1419. [CrossRef]
31. Feng, X.; Gao, X.; Luo, L. X-SDD: A New Benchmark for Hot Rolled Steel Strip Surface Defects Detection. *Symmetry* **2021**, *13*, 706. [CrossRef]
32. Antoniou, A.; Storkey, A.; Edwards, H. Augmenting image classifiers using data augmentation generative adversarial networks. In Proceedings of the International Conference on Artificial Neural Networks, Rhodes, Greece, 4–7 October 2018; pp. 594–603.
33. Liu, K.; Li, A.; Wen, X.; Chen, H.; Yang, P. Steel surface defect detection using GAN and one-class classifier. In Proceedings of the 2019 25th International Conference on Automation and Computing (ICAC), Lancaster, UK, 5–7 September 2019; pp. 1–6.
34. Krizhevsky, A.; Sutskever, I.; Hinton, G.E. Imagenet classification with deep convolutional neural networks. *Adv. Neural Inf. Process. Syst.* **2012**, *25*, 1097–1105. [CrossRef]
35. Simonyan, K.; Zisserman, A. Very deep convolutional networks for large-scale image recognition. *arXiv* **2014**, arXiv:1409.1556.
36. Zhang, X.; Zhou, X.; Lin, M.; Sun, J. Shufflenet: An extremely efficient convolutional neural network for mobile devices. In Proceedings of the IEEE Conference on Computer Vision and Pattern Recognition, Salt Lake City, UT, USA, 18–23 June 2018; pp. 6848–6856.

metals

Article

# Intelligent Recognition Model of Hot Rolling Strip Edge Defects Based on Deep Learning

Dongcheng Wang [1,2,*], Yanghuan Xu [1], Bowei Duan [1], Yongmei Wang [1], Mingming Song [1], Huaxin Yu [1] and Hongmin Liu [1,2]

1 National Engineering Research Center for Equipment and Technology of Cold Rolling Strip, Yanshan University, Qinhuangdao 066004, China; xu_yanghuan@163.com (Y.X.); bwduan1@163.com (B.D.); wangym1213@163.com (Y.W.); smm19891123@126.com (M.S.); zyl19950114@126.com (H.Y.); zyl18833989190@163.com (H.L.)

2 State Key Laboratory of Metastable Materials Science and Technology, Yanshan University, Qinhuangdao 066004, China

* Correspondence: wangdongcheng@ysu.edu.cn

**Abstract:** The edge of a hot rolling strip corresponds to the area where surface defects often occur. The morphologies of several common edge defects are similar to one another, thereby leading to easy error detection. To improve the detection accuracy of edge defects, the authors of this paper first classified the common edge defects and then made a dataset of edge defect images on this basis. Subsequently, edge defect recognition models were established on the basis of LeNet-5, AlexNet, and VggNet-16 by using a convolutional neural network as the core. Through multiple groups of training and recognition experiments, the model's accuracy and recognition time of a single defect image were analyzed and compared with recognition models with different learning rates and sample batches. The experimental results showed that the recognition model based on the AlexNet had a maximum accuracy of 93.5%, and the average recognition time of a single defect image was 0.0035 s, which could meet the industry requirement. The research results in this paper provide a new method and thought for the fine detection of edge defects in hot rolling strips and have practical significance for improving the surface quality of hot rolling strips.

**Keywords:** hot rolling strip; edge defects; intelligent recognition; convolutional neural networks

**Citation:** Wang, D.; Xu, Y.; Duan, B.; Wang, Y.; Song, M.; Yu, H.; Liu, H. Intelligent Recognition Model of Hot Rolling Strip Edge Defects Based on Deep Learning. *Metals* **2021**, *11*, 223. https://doi.org/10.3390/met11020223

Academic Editor: Leszek Adam Dobrzanski

Received: 16 December 2020
Accepted: 22 January 2021
Published: 27 January 2021



## 1. Introduction

Surface quality is an important indicator of hot rolling strip products. Surface defects not only have an influence on product appearance and rolling yield, but also have a harmful effect on the production of downstream processes [1,2]. Surface defects can be detected quickly and accurately through a surface quality detection system, which has practical significance for improving the surface quality of a strip. A new direction for strip surface quality detection has been provided with the rapid development of artificial intelligence, machine vision theory, and technology [3–6]. Many scholars have conducted related research.

Xu et al. [7] used eight 1024-pixel linear CCD (Charge Coupled Device) cameras as an image acquisition device and proposed the procedure of defect detection, and a recognition algorithm based on the surface features of a hot rolling strip, which was applied to a 1700 mm hot rolling strip production line. Later, a new method based on Tetrolet transform and kernel locality preserving projection for dimension reduction was proposed to detect the surface defects of hot rolling strips [8], and the recognition accuracy on the defect sample database was 97.3846%. He et al. [9] developed a long-distance and super-bright LED light and solved the problem of inhomogeneous illumination from a long distance at a high temperature. It simultaneously met the illumination request of line scan camera and plane scan camera imaging, and the real-time recognition of the strip surface defects was carried out using the decision tree classification model. Gan et al. [10] used a decision tree and expert experience classification model to recognize silicon steel surface defects. Zhao [11] used the VggNet16 model to recognize silicon steel surface defects, and the accuracies of the model on training and test sets were 97.5% and 92.27%, respectively. Han et al. [12] proposed a surface defect recognition method based on a BP (Back Propagation) neural network, and its accuracy reached more than 84%. Chu [13] used a least square twin support vector machine to recognize strip surface defects with an accuracy of 95.4% while greatly improving the recognition efficiency. Xing [14] improved the AlexNet model constructure, which was used to classify the surface defects of hot rolling strips by adjusting the size and number of convolution kernels with a recognition accuracy of 93.4%. Subsequently, a defect classification and index system based on MATLAB was developed.

Song et al. [15] established an NEU (Northeastern University) surface defect database, including the six kinds of typical surface defects of hot rolling strips: rolled-in scale (RS), patches (Pa), crazing (Cr), pitted surface (PS), inclusion (In), and scratches (Sc). Subsequently, a robust feature descriptor against noise named the adjacent evaluation completed local binary patterns (AECLBP) was proposed for defect recognition, and its recognition accuracy reached 97.89%. The NEU surface defect database was a significant contribution to the research of strip defect recognition. For instance, Hu [16] compared the recognition performance between AdaBoost and AdaBoost.BK on the NEU surface defect database, and their results showed that AdaBoost.BK had the highest classification accuracy when the backward step was F = 2. Though a long training time was needed, the recognition time of a single defect image was only 0.002132 s. Xie [17] used a deep residual network and transfer learning method, and its recognition accuracy on the NEU surface defect database reached 98.54%, which was better than that of VggNet16. Gao et al. [18,19] adopted a deep residual network and semi-supervised learning method to research the NEU surface defect database. The recognition accuracy of the deep residual network reached 99.889%, and the recognition accuracy of semi-supervised learning reached 86.72%. However, the data label can be omitted in the semi-supervised learning method, which improves the efficiency. This method is more suitable for the defect recognition mission that has a limitation on labeling. Saiz et al. [20] combined traditional machine learning techniques with convolutional neural networks and proposed an automatic classification method of strip surface defects based on a deep learning method. The best classifier parameters were obtained through plenty of experiments, and the robustness of the classifier was verified. The classification time of a single image was only 0.019 s on the NEU surface defect database, thereby achieving a classification accuracy of 99.95%. Lee et al. [21] proposed a relative approach for diagnosing steel defects using a deep structured neural network with class activation maps, and it achieved detection performance values of 99.44% and 0.99 in terms of accuracy and the F-1 score metric, respectively. He et al. [22,23] proposed a generative adversarial network and fused multiple hierarchical feature network defect classification method, respectively, and the recognition accuracy on the NEU surface defect database reached values of 99.56% and 99.67%. Dong et al. [24] proposed a pyramid feature fusion and global context attention network for the pixel-wise detection of surface defect named PGA-Net. The average pixel accuracy of the proposed method on the four defect datasets was 92.15% for NEU-Seg, 74.78% for DAGM 2007, 71.31% for MT_defect, and 79.54% for Road_defect, all of which were better than existing methods. Guan et al. [25] evaluated the image quality of the potential feature information of a strip defect through visualization and proposed a strip surface defect classification algorithm. Compared with VggNet19 and ResNet, the proposed method had a better performance in terms of prediction accuracy and speed on the NEU surface defect database. Fu et al. [26] proposed a lightweight convolutional neural network (CNN) defect recognition model that emphasized the training of low-level features and incorporated multiple receptive fields to achieve fast and accurate classification on the NEU surface defect database. The model could realize real-time detection, and its running speed could reach 100 fps in computer equipped with single NVIDIA TITAN X and 12G RAM (Random Access Memory).

In summary, the existing detection equipment, theories, and techniques of hot rolling strip surface defects have basically achieved a satisfactory performance for typical defects with obvious features (e.g., crazing, inclusion, patches, scratches, rolled-in scale, and pitted surface). However, the strip edges (approximately 50 mm on both sides) are frequent areas of surface defects in production practice. Common defects include upwarps, black lines, cracks, slag inclusions, and gas holes. The generation mechanisms and corresponding solutions of these defects vary, but the macroscopic features are relatively similar, and the online surface quality detection system often recognizes them as the same type of defect. To eliminate concrete defects, the production line generally needs to further subdivide defects through manual detection, which seriously reduces production efficiency and increases labor intensity. To this end, the authors of this article subdivided the edge defects of a hot rolling strip into five types and the intelligent recognition model of edge defects was investigated.

## 2. Convolutional Neural Network Recognition Model for Edge Defects

### 2.1. Characteristics of Edge Defects

As shown in Figure 1, during the actual production process of hot rolling strips, surface defects often appear. There are many types of defects, and these defects often occur in the head, tail, and both sides of the strip. Frequently, the feature difference between a perfect image and a defect image is obvious. Traditional machine vision theory can be used to solve this binary classification problem, and it is not difficult for a surface quality detection system to complete this task. However, there are some problems when classifying defect images. Because the feature difference among the various defects is not obvious, traditional machine vision theory cannot perform well to complete this multi classification problem. For this reason, many scholars have tried to solve this problem by using deep learning neural networks [15,22–24]. The edge defects are more special in a defect image set. The various edge defects generally have similar linear features, and a surface quality detection system often classifies these different types of defects into one classification, which is not conducive to the further analysis of the defect generation mechanism and the proposition of corresponding solutions. The authors of this paper took the edge defect set as the research object and studied the recognition model of edge defect image based on a convolutional neural network. The purpose was to improve the recognition accuracy of edge defects.

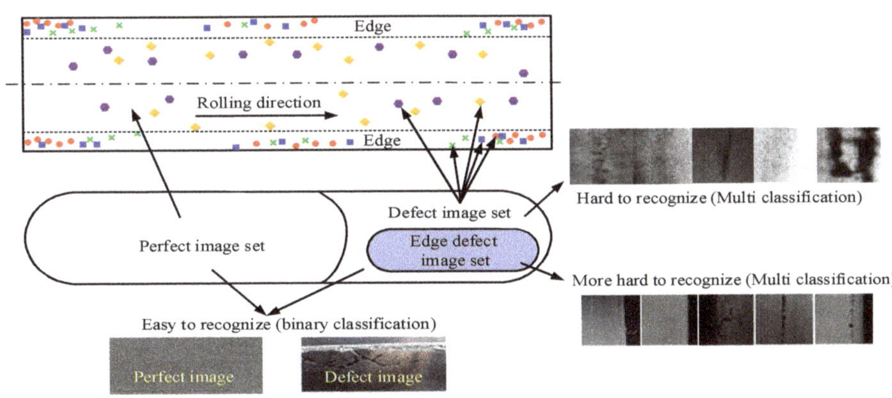


**Figure 1.** Relationship between the edge defect image set and the perfect image set.

The edge defects of a hot rolling strip occur on the operation and drive sides of the strip. The defects are detected by cameras on both edge sides of the surface quality detection system (SQDS). The detection position is located between the exit of the finishing mill's seventh stand and laminar cooling areas (Figure 2). These edge defects are evolved by heating, rough rolling, finishing rolling, and other processes. The evolution process is shown in Figure 3. In practical production, each defect must be accurately detected, and effective control method must be carried out.

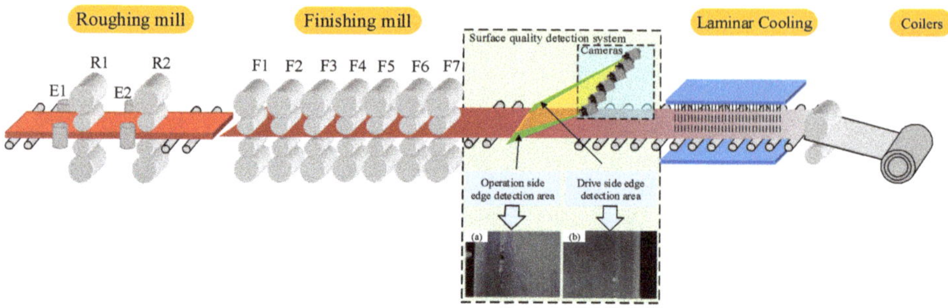


**Figure 2.** Edge defect detection of a hot rolling strip.

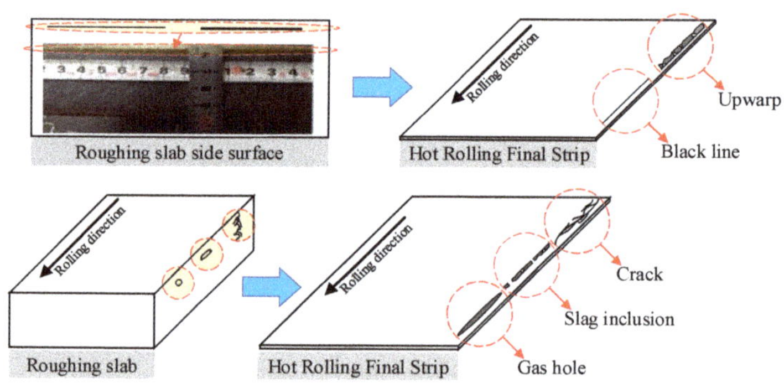


**Figure 3.** Evolution process of edge defects.

Take the upwarp as an example. This defect often appears in IF (Interstitial-Free) steel. The generation mechanism of the defect is the temperatures in the edge and corner of the intermediate slab drop too fast during the hot rolling process, so the $\gamma \rightarrow \alpha$ phase transformation is likely to occur in advance, thus resulting in an uneven distribution of flow stress and transverse flow in the thickness direction of the intermediate slab. The side of intermediate slab forms a large fold. As the rolling process continues, this large fold flips to the surface of the strip and forms edge upwarp [27,28]. In actual production, once the edge upwarp occurs, the temperature of the heating furnace should be appropriately increased, and an edge heater should be turned on at the same time so that the large folds in edge and corner of intermediate slab of subsequent products can be eliminated to avoid the occurrence of edge upwarp.

In this paper, after long-term tracking, sampling analysis, and technique exchanges for a 2250 mm hot rolling production line, the edge defects were divided into five types, namely upwarp, black line, crack, slag inclusion, and gas hole. The length, width, and specific features of these five types of defects are shown in Table 1. Table 1 shows that, except for the crack, the features of the four other types of defects presented a certain linear feature, but the line's width, length, color, and specific texture features were not completely consistent. Among them, the upwarp and black line were found to have the same generation mechanism, and their features reflected a certain similarity. Thus, they both belong to the edge seam defects [27,28]. However, because of the different severities of the two types of defects, they were divided into two types during the recognition process. Meanwhile, the images of slag inclusion and gas hole easily caused confusion, and detection errors of these two types of defects often occurred. Compared with other typical surface defects of hot rolling strips (e.g., crazing, inclusion, patches, scratches, rolled-in scale, and pitted surface) (Figure 4 [15]), the detection of edge defects is relatively difficult.

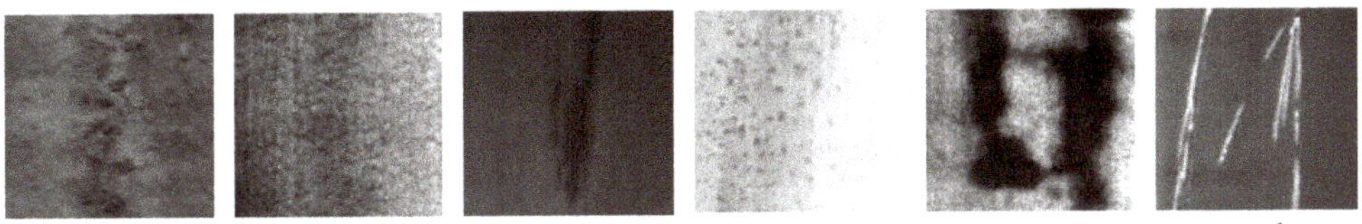


**Figure 4.** Typical surface defects of a hot rolling strip.

**Table 1.** Morphological features of edge defects. SQDS: surface quality detection system.

| Feature | Upwarp | Black Line | Crack | Slag Inclusion | Gas Hole |
|---|---|---|---|---|---|
| Length | Continuous or discontinuous with different lengths. | Continuous or discontinuous with different lengths. | 20~100 mm | 100~1000 mm | 150~500 mm |
| Width | 1~3 mm | <1 mm | 5~20 mm | 2~5 mm | 1~5 mm |
| Shape | A straight and uniform-width line accompanied with upwarp. | A straight and uniform-width line. | Branch. | strip with uniformwidth. | strip with texture. |
| Morphology (Physical) | 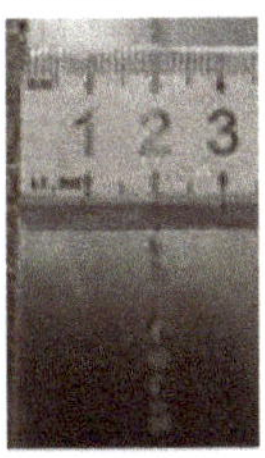  |  |  |  | 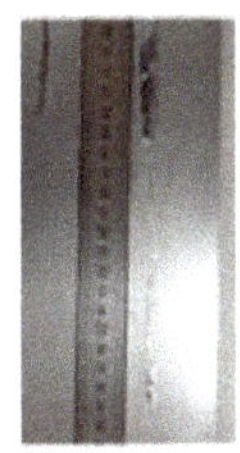 |
| Morphology (SQDS) | 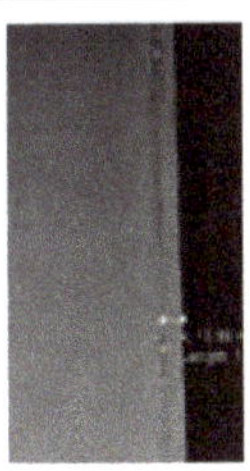 | 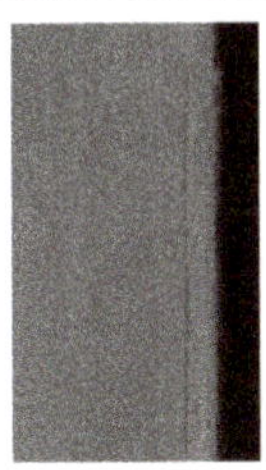 |  |  | 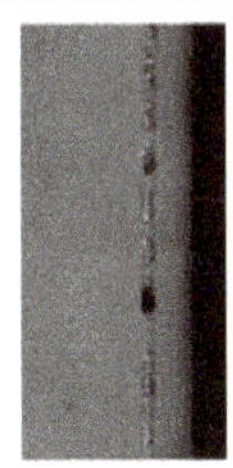 |

*2.2. Convolutional Neural Network Model of Edge Defects*

Traditional machine vision or deep learning intelligent methods can be used to achieve the automatic and high-precision detection of the edge defects of hot rolling strips. They are prone to confusion because of the similarity of edge defect features. If traditional machine vision methods are used to extract, segment, or classify defects' features, obtaining a high recognition accuracy or a strong generalization and perception ability is difficult for the model. As artificial intelligence and deep learning theories are developing, the technology of image detection with a strong similarity is resulting in better performance, such as for face recognition and medical diagnosis [29,30]. However, a CNN, which is a mature deep learning algorithm, has shown excellent performance in many application fields [31–33]. The CNN introduces the convolution linear operation, thereby making it more suitable for processing data similar to a network structure, such as time series and image data. Therefore, according to the defect images taken by an surface quality detection system, this article investigated the intelligent recognition of the edge defects of hot rolling strips based on a CNN.

The structure of the CNN recognition model for hot rolling strip edge defect is shown in Figure 5, which includes a data input layer, multiple sets of convolutional and pooling layers, a fully connected feedforward neural network layer, and an output recognition layer. After the original edge defect image data were subjected to multiple convolutional layers, pooling layers, and a nonlinear activation function mapping operation, the feature information was extracted layer by layer. Finally, the probability of image classification was calculated by the fully connected output layers, and the specific classification of defect images was obtained.

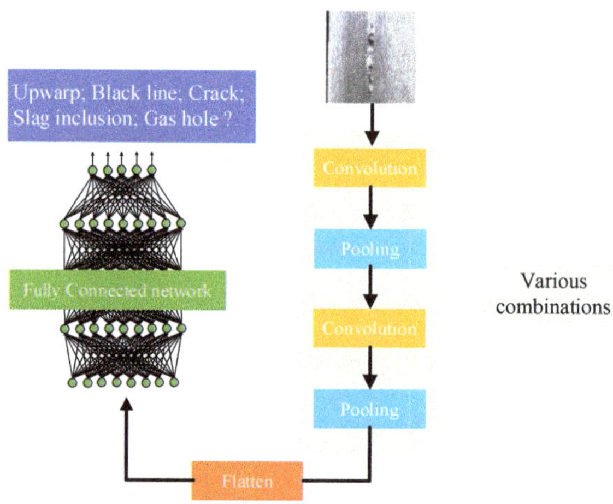


**Figure 5.** Structure of the CNN recognition model for hot rolling strip edge defect.

(1) Input layer

The input layer uses the edge defect images taken by the surface quality detection system of a hot rolling production line. According to the image size, the model numerically characterizes the internal information of defect images, which is used for the subsequent process and training network.

(2) Convolutional layer

The convolutional layer is the core part of the CNN structure. The image features can be extracted through the convolution operation between a group of convolution kernel and input data. Figure 6 shows that during the whole operation process, the convolution kernel slides from left to right for a specified step and implements the convolution operation with the image data of the input layer. When it reaches far right, it returns to the far-left, slides down for a specified step, and continuously slides from left to right until the whole operation is completed. The size of the feature maps obtained by the convolution operation is related to the parameters, such as the original input image size, convolution kernel size, slide step, and padding size. Assuming that the size of the convolution kernel is $m \times m$, the original input image size is $h \times w$, the slide step is $\Delta$, the padding pixel is $p$, and the output size of the feature maps through the convolution operation is $h' \times w'$. The calculation formula is presented in Equation (1).

$$\begin{cases} h' = \left\lfloor \frac{h-m+2p}{\Delta} + 1 \right\rfloor \\ w' = \left\lfloor \frac{w-m+2p}{\Delta} + 1 \right\rfloor \end{cases} \quad (1)$$

where $\lfloor\ \rfloor$ represents the rounding down operation.

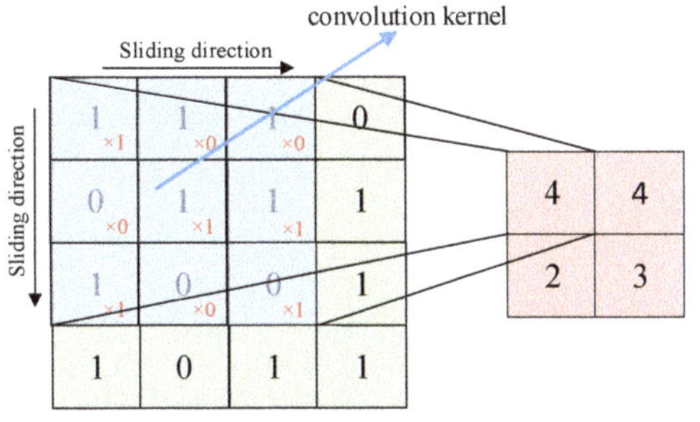


**Figure 6.** Convolution operation process.

The convolution kernel performs the convolution operation with the previous layer through weight sharing to obtain different feature maps. The more convolution kernels, the stronger the ability to extract the features of the input image. The convolution operation formula is described as Equation (2).

$$F_j^l = f[\sum_{i \in U_j} (F_j^{l-1} * \omega_{ij}^l + b_j^l)] \quad (2)$$

where $F_j^l$ is the $j$th output feature map of the $l$th layer, $F_j^{l-1}$ is the input feature map of the $l-1$th layer, $U_j$ is the feature map set of the $l-1$th layer, $\omega_{ij}^l$ is the weight from the $i$th feature map to the $j$th feature map of the $l$th layer, $b_j^l$ is the bias of the $j$th feature map of the $l$th layer, and $f$ is the activation function. To achieve a nonlinear description of the model after the convolution operation, an activation function $f$ is required to implement the nonlinear operation on the linear result, which can enhance the expressive ability of the network model. At present, the commonly used activation functions include: sigmoid, tanh, relu, and prelu. The expression of these activation functions are described in Equations (3)–(6), and their function images are shown in Figure 7.

$$sigmoid(x) = \frac{1}{1+e^x} \quad (3)$$

$$tanh(x) = \frac{e^x + e^{-x}}{e^x + e^{-x}} \quad (4)$$

$$relu(x) = \begin{cases} 0, x < 0 \\ x, x \geq 0 \end{cases} \quad (5)$$

$$prelu(x) = \begin{cases} ax, x < 0 \\ x, x \geq 0 \end{cases} \quad (6)$$

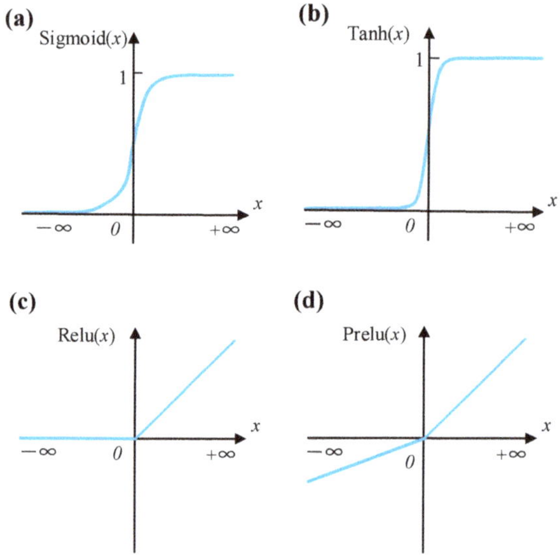


**Figure 7.** Several typical activation function images: (**a**) sigmoid, (**b**) tanh, (**c**) relu, and (**d**) prelu.

(3) Pooling layer

The pooling layer is a down-sampling operation that is usually located after the convolutional layer, and the typical feature information is obtained by down-sampling the original size feature map. Figure 8 shows two commonly used pooling methods, namely max-pooling and average-pooling. The average-pooling takes the average value of the data in the pooling window as the pooling result, and the max-pooling takes the maximum value of the data in pooling window as the pooling result. The max-pooling method is used in most cases. Assuming that the input size of the feature map is $h' \times w'$, the window size of pooling zone is $n \times n$, the slide step is $\Delta'$, and the output size of feature map is $h'' \times w''$. The calculation formula of $h''$ and $w''$ is described as Equation (7).

$$\begin{cases} h'' = \left\lfloor \frac{h'-n}{\Delta'} + 1 \right\rfloor \\ w'' = \left\lfloor \frac{w'-n}{\Delta'} + 1 \right\rfloor \end{cases} \quad (7)$$

where $\lfloor\ \rfloor$ represents the rounding down operation, and the general value $n$ is 2.

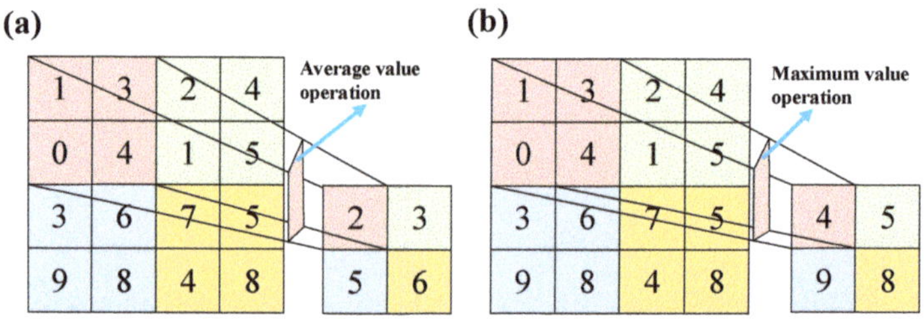


**Figure 8.** Two pooling methods: (**a**) average-pooling and (**b**) max-pooling.

(4) Fully connected layer

The fully connected layer integrates all the feature informations extracted from previous convolutional and pooling layers. Each neuron in the fully connected layer connects all neurons in the previous layer, and the calculation formula is described as Equation (8).

$$y_{w,b}(x) = f(\omega^T x) = f(\sum_{i=1}^{n} \omega_i x_i + b) \quad (8)$$

where $y_{w,b}(x)$ is the output of the fully connected layer, which is a one-dimension vector; $x_i$ is the input of the fully connected layer, which is the feature map values after the convolution and pooling operation; $\omega_i$ is the weight of the network model; $b$ is the bias of the network model; and $f$ is the activation function.

(5) Output layer

The last layer of the model is the output recognition layer. The output of multiple neurons is mapped to (0,1) through the Softmax function; its value represents the probability that the input image belongs to this classification. If the input of Softmax is $y_i(i = 1, 2, \ldots, k)$, then the output probability of the defect classification by Softmax function can be described as Equation (9).

$$Softmax(y_1, y_2, \cdots, y_k) = \begin{cases} e^{y_1} / \sum_{i=1}^{k} e^{y_i} = p_1 \\ e^{y_2} / \sum_{i=1}^{k} e^{y_i} = p_2 \\ \vdots \\ e^{y_k} / \sum_{i=1}^{k} e^{y_i} = p_k \end{cases} \quad (9)$$

Figure 9 shows the basic process of the CNN. We assumed that the input image data was a 10 × 10 matrix, the size of convolution kernel was 3 × 3, the slide step was 1, and the padding pixel was 0. Through Equation (1), the size of the feature map (hidden layer) obtained from the convolution operation was found to be 8 × 8. The size of the pooling zone was set as 2 × 2, and the slide step was 1. Subsequently, through the further operation of Equation (7), the size of the feature map was found to be 4 × 4. After the flatten operation, the size of the fully connected layer became 16 × 1, and the recognition result was outputted through classification. The CNN had various structures through the combination of different convolutional layers, pooling layers, and different numbers of fully connected layers. Different structures of CNNs have different levels of learning ability for different features. For this, the corresponding experiments had to be implemented for different, specific, and practical problems to ensure a better performance for learning and prediction capability.

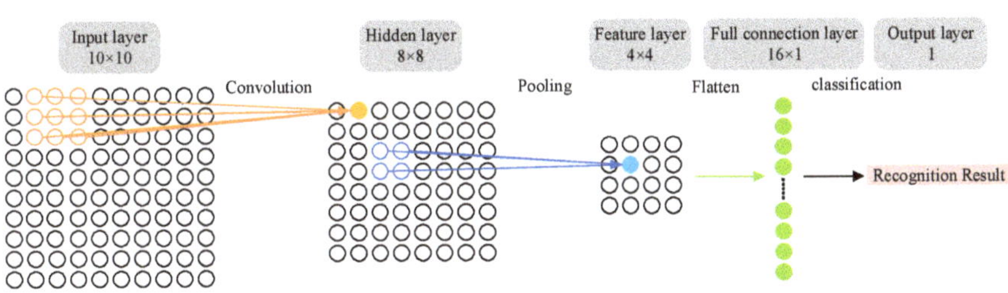


**Figure 9.** Basic process of convolutional neural networks.

## 3. Experiment and Analysis

### *3.1. Edge Defect Dataset*

Taking the five aforementioned types of edge defects as the research object, the edge defect dataset was collected and produced at the hot rolling production line (Figure 10). The total of 2000 edge defect images were found in the dataset, and 400 images were found in each type of defect. After pre-processing, the size of each image in the dataset was unified to 100 × 100. According to the specified proportion, the dataset was divided into three parts, namely training, validation, and test sets. The image distribution of each part is shown in Table 2. The training and validation sets were used for training the model, whereas the test set was used to verify the learning and generalization abilities of the model, and was not used in model training.

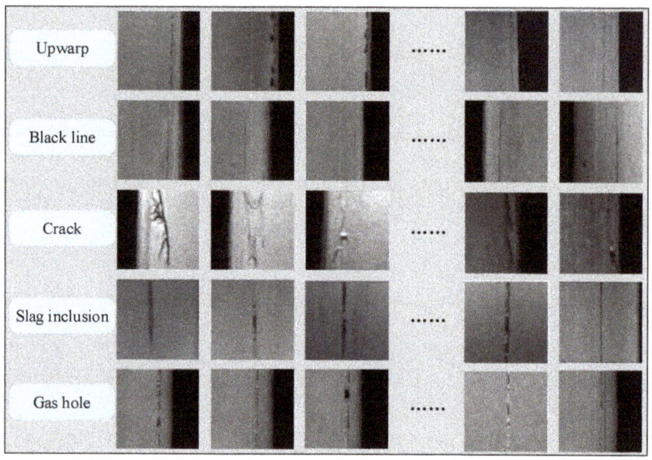


**Figure 10.** Edge defect dataset of a hot rolling strip.

**Table 2.** Image distribution of each part in the training set, the validation set, and the test set.

| Dataset | Upwarp | Black Line | Crack | Slag Inclusion | Gas Hole | Aggregate |
|---|---|---|---|---|---|---|
| Training set | 240 | 240 | 240 | 240 | 240 | 1200 |
| Validation set | 80 | 80 | 80 | 80 | 80 | 400 |
| Test set | 80 | 80 | 80 | 80 | 80 | 400 |
| Aggregate | 400 | 400 | 400 | 400 | 400 | 2000 |

### *3.2. Experimental Process*

At present, when dealing with different practical problems, there is no uniform principle for how to select and determine the structure of a CNN. Therefore, in this paper, three representative CNN structures, namely LeNet-5, AlexNet, and VggNet-16 [34–36], were used to establish the edge defect recognition model for the hot rolling strip. Corresponding training experiments were conducted to analyze the influence of the network structure parameters on the recognition accuracy and operating speed of the edge defects. The LeNet-5 network consisted of an input layer, three convolutional layers, two pooling layers, and two fully connected layers. The AlexNet network consisted of an input layer, five convolutional layers, three pooling layers, and three fully connected layers. The VggNet-16 network consisted of an input layer, thirteen convolutional layers, five pooling layers, and three fully connected layers. The structure and operation process of the three network models are shown in Figure 11. Figure 11 shows that theoretically, as a model's structure increases, the number of parameters will increase correspondingly and the perception of dealing with problems would gradually improve. However, in practical applications, it is not true that the more complex a model structure is, the better the recognition performance. Thus, it was indispensable to conduct a model training experiment. In this paper, the main software and equipment used included the Linux operating system (Ubuntu), Intel E5-2680 V3 CPU (128GB memory), TiTan RTX GPU (24GB video memory), the PyCharm programming environment, and the PyTorch platform. One can also use platforms such as NVIDIA Triton or the Microsoft ONNX server for the model.

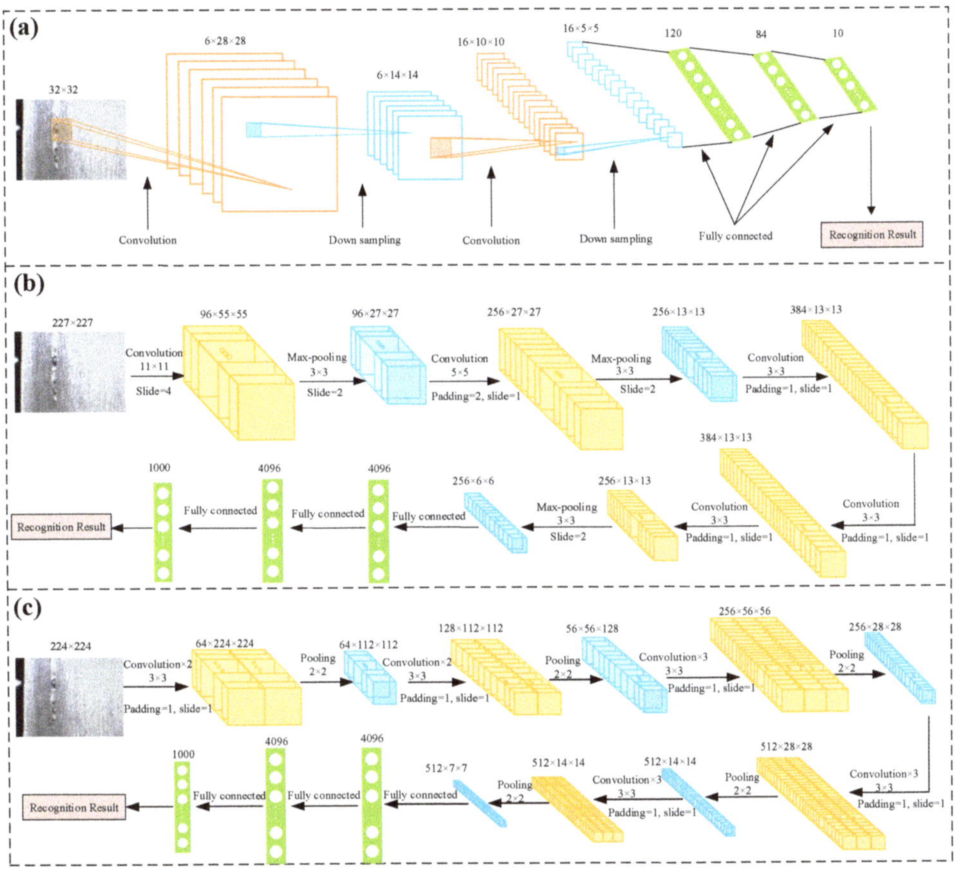


**Figure 11.** Three CNN recognition models for edge defects: (**a**) LeNet-5, (**b**) AlexNet, and (**c**) VggNet-16.

### 3.3. Experimental Results and Discussion

On the basis of three recognition models, experiment results with different learning rates (lr) and sample batches (batch) were recorded. Figure 12 shows the final experiment result of the LeNet-5 model. Figure 12a–c shows the training time and recognition accuracy of the model on the test set under three learning rates (lr = 0.0001, lr = 0.001, and lr = 0.01) and four sample batches (batch = 32, batch = 64, batch = 128, and batch = 256), respectively. The experiment results indicated that the model had the shortest training time of 404 s with lr = 0.001 and batch = 256, but the recognition accuracy of the model on the test set was too low at only 48.5%. When the recognition accuracy of the model reached the highest at 68.5% with lr = 0.01 and batch = 64, its training time was 445 s. In the process of off-line training, if the training time was not much different, so only the recognition accuracy of the test set was regarded as the evaluation standard. The training process of the model with the highest recognition accuracy of 68.5% is shown in Figure 12d,e. The error loss of the training set was slightly lower than the error loss of the validation set during the entire training process, converging to 0.48 and 0.55, respectively. The overall accuracy of the training set was slightly higher than that of the validation set, finally the accuracy reaching 0.75 and 0.68, respectively, thus indicating that the training and learning process of the model was correct and that the model had a certain generalization ability. However, the recognition accuracy of the model could not meet the requirements of practical applications. Further experiments with adjustments of the parameters, such as lr and batch, were carried out, but the final recognition accuracy could not exceed 70%, which indicated that the edge defect recognition model based on LeNet-5 was not effective. To further improve the recognition accuracy, the model structure must be redesigned and adjusted, and a corresponding experiment must be verified. However, ensuring a high recognition accuracy is difficult because of the rather complicated process.

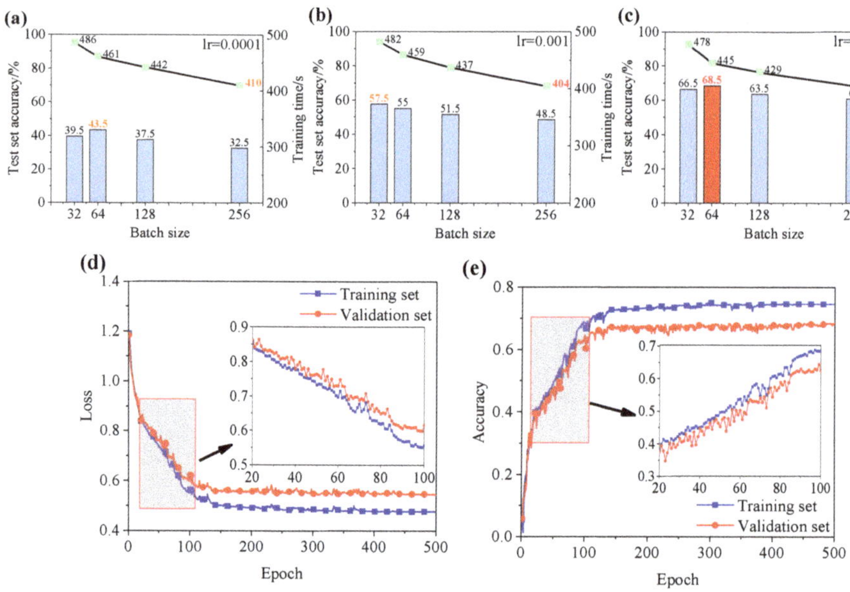


**Figure 12.** Edge defect LeNet-5 CNN model experiment results: (**a**) test set accuracy and training time with lr = 0.0001, (**b**) test set accuracy and training time with lr = 0.001, (**c**) test set accuracy and training time with lr = 0.01, (**d**) error loss of training process with lr = 0.01 and batch = 64, and (**e**) accuracy of training process with lr = 0.01 and batch = 64.

The authors of this paper used the AlexNet convolutional neural network to establish an edge defect recognition model. The experiment results are shown in Figure 13. In Figure 13a–c, two groups with the model's defect recognition accuracy exceeding 90% on the test set can be observed. When lr = 0.001 and batch = 32 and when lr = 0.001 and batch = 64, the accuracy deviation between the two groups was 1% and the training time deviation was 67 s. Similarly, not considering the training time, the training process of the model with the recognition accuracy of 93.5% is shown in Figure 13d,e. The entire training process was relatively stable, and the error loss of the training set and the validation set converged to 0.11 and 0.19, respectively, which were lower than the error loss of the LeNet-5 recognition model. The accuracy of the training set and the validation set reached 0.96 and 0.93, respectively, which were significantly higher than the accuracy of the LeNet-5 recognition model.

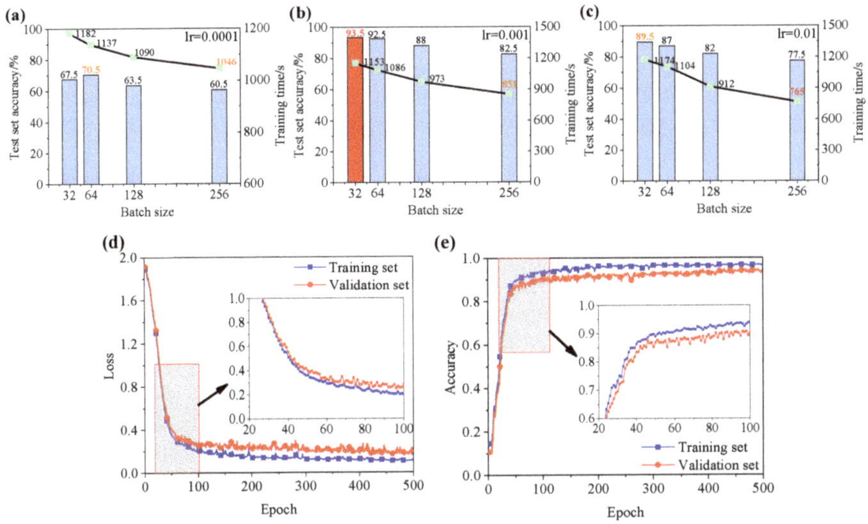

**Figure 13.** Edge defect AlexNet CNN model experiment results: (**a**) test set accuracy and training time with lr = 0.0001, (**b**) test set accuracy and training time with lr = 0.001, (**c**) test set accuracy and training time with lr = 0.01, (**d**) error loss of training process with lr = 0.001 and batch = 32, and (**e**) accuracy of training process with lr = 0.001 and batch = 32.

This paper further used the VggNet-16 convolutional neural network to establish an edge defect recognition model. The experiment results are shown in Figure 14. Figure 14a–c shows that when lr = 0.001 and batch = 128, the defect recognition accuracy of the model on the test set was up to 74% and the training time was 3205 s. Figure 14d,e shows that local oscillations existed during the model training process. Meanwhile, when the number of iterations exceeded 400, the error loss of the validation set had an upward trend, indicating that the model had a certain degree of overfitting.

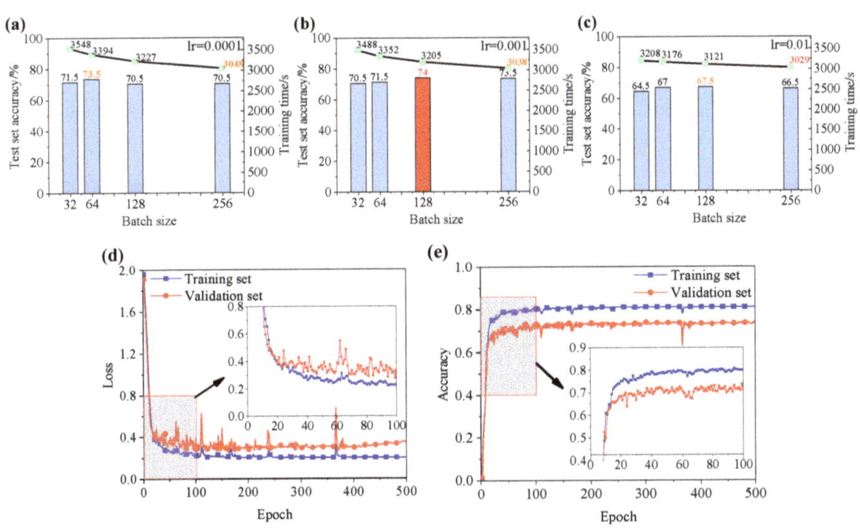

**Figure 14.** Edge defect VggNet-16 CNN model experiment results: (**a**) test set accuracy and training time with lr = 0.0001, (**b**) test set accuracy set and training time with lr = 0.001, (**c**) test set accuracy and training time with lr = 0.01, (**d**) error loss of training process with lr = 0.001 and batch = 128, and (**e**) accuracy of training process with lr = 0.001 and batch = 128.

By comparing Figures 12–14, it can be seen that the training time of the model slightly decreased with the increase in learning rate and greatly decreased with the increase in sample batch, though too large a batch greatly reduced the recognition accuracy of the model. The defect image recognition time is more important to satisfy online applications. In this paper, the recognition experiment of a single edge defect image was conducted for each model with different parameters. The experiment results are shown in Table 3. The results showed that the average single defect recognition times of three types of CNN models were 2.7, 3.5, and 5.4 ms, respectively. The learning rate and sample batch had no obvious influence on the recognition time of a single defect image. In this paper, the AlexNet (lr = 0.001 and batch = 32) model was selected as the hot rolling strip edge defect recognition model because its accuracy and speed of recognition meet the engineering requirements. Figure 15 shows the visualization recognition results of the model on the test set. The recognition results of the defect image were expressed by the probability value. In Figure 15a–e, the recognition result of each image defect classification is expressed by a probability vector (probability 1, probability 2, probability 3, probability 4, and probability 5). The values of probability 1, probability 2, probability 3, probability 4, and probability 5 represented the scores of five defect classifications (upwarp, black line, crack, slag inclusion, and gas hole, respectively), and the sum of the five probability values was 1. In the process of defect image recognition, when the probability value of a certain item in the probability vector exceeded 0.5 (orange boundary in the figure), the classification of the defect of the image was determined. The recognition performance of edge cracks was the best, and no recognition error was recorded. More errors appeared between the upwarp and black line, which also confirmed the conclusion that the two types defect have the same generation mechanism [27,28]. The confusion matrix of the defect recognition results on the test set is shown in Table 4. According to the diagonal value in this matrix, it can be seen that the model had a better overall recognition and classification effect on edge defects. Based on the value distribution on both sides of the diagonal line, it could also be seen that the upwarp and black line easily caused recognition errors, which had a greater impact on the accuracy of the model. Though a small number of recognition errors were observed between slag inclusion and gas hole, the overall recognition accuracy could meet the requirements of practical application. In the future, in order to further improve and optimize the model, the dataset needs to be expanded and processed, especially to expand the defect images of upwarp, black line, and gas hole as much as possible. Meanwhile, an image-enhancing technique can be introduced to optimize the model.

**Table 3.** Recognition time of a single defect image on the test set with different models (ms).

| Batch | LeNet-5 | | | AlexNet | | | VggNet-16 | | |
|---|---|---|---|---|---|---|---|---|---|
| | Lr = 0.0001 | Lr = 0.001 | lr = 0.01 | lr = 0.0001 | lr = 0.001 | lr = 0.01 | lr = 0.0001 | lr = 0.001 | lr = 0.01 |
| 32 | 2.386 | 2.381 | 2.386 | 3.462 | 3.462 | 3.395 | 5.131 | 5.737 | 5.508 |
| 64 | 2.514 | 2.572 | 2.568 | 3.513 | 3.479 | 3.401 | 5.305 | 5.542 | 5.416 |
| 128 | 2.844 | 2.825 | 2.859 | 3.566 | 3.783 | 3.495 | 5.324 | 5.493 | 5.703 |
| 256 | 3.180 | 3.081 | 3.081 | 3.799 | 3.544 | 3.385 | 5.280 | 5.427 | 5.464 |

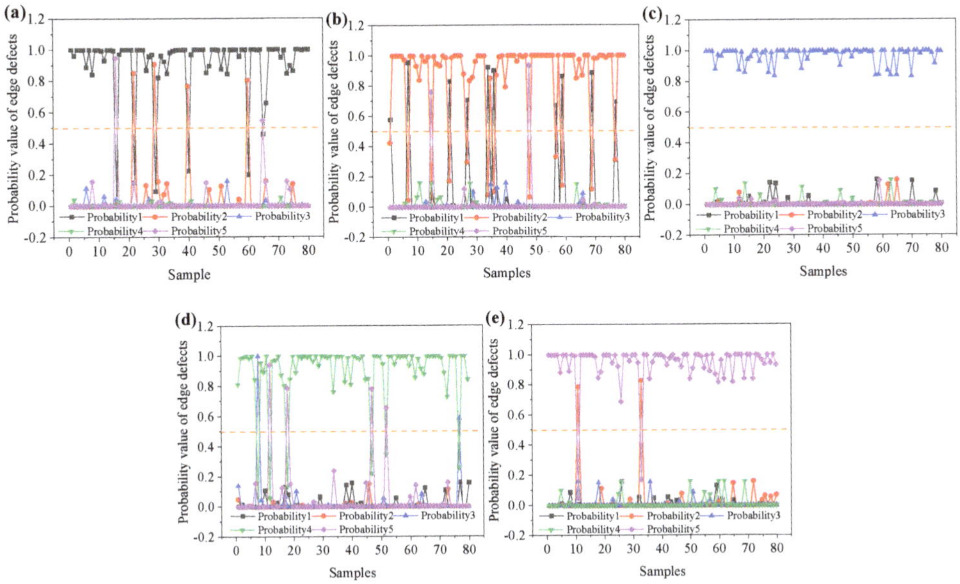


**Figure 15.** Visualization results of five types of edge defect recognition on the test set: (**a**) upwarp, (**b**) black line, (**c**) crack, (**d**) slag inclusion, and (**e**) gas hole.

**Table 4.** Confusion matrix of the edge defect recognition model on the test set.

| Confusion Matrix | | Model Classification | | | | |
|---|---|---|---|---|---|---|
| | | Upwarp | Black Line | Crack | Slag Inclusion | Gas Hole |
| Actual classification | Upwarp | 74 | 4 | 0 | 0 | 2 |
| | Black line | 10 | 68 | 0 | 0 | 2 |
| | Crack | 0 | 0 | 80 | 0 | 0 |
| | Slag inclusion | 0 | 0 | 2 | 74 | 4 |
| | Gas hole | 0 | 2 | 0 | 0 | 78 |

## 4. Conclusions

The edge defects of hot rolling strips have five types: upwarp, black line, crack, slag inclusion, and gas hole. The appearance morphologies of these five types of defects show a certain linear feature. However, the width, length, and color of the lines are not completely consistent with the specific texture features. To improve the detection accuracy of edge defects, edge defect recognition models were established on the basis of LeNet-5, AlexNet, and VggNet-16 by using a convolutional neural network as the core.

The edge defect recognition model based on the LeNet-5 convolutional neural network was found to have the highest accuracy of 68.5% on the test set, and its average recognition time for a single defect image was 2.7 ms. Though the model was found to have a certain generalization ability, its prediction accuracy is a bit low. The edge defect recognition model based on the VggNet-16 convolutional neural network had the highest accuracy of 74% on the test set, and its average recognition time for a single defect image was 5.4 ms. The model was found to have local oscillations and a certain overfitting trend during the training process. The edge defect recognition model based on the AlexNet convolutional neural network had the highest accuracy of 93.5% on the test set, and its average recognition time for a single defect image was 3.5 ms.

Among the three models, the edge defect recognition model based on the AlexNet convolutional neural network was found to have the highest prediction accuracy, a good generalization ability, and the best comprehensiveness. However, the accuracy of the model needs to be further improved, especially because the two defects of upwarp and black line are still easily confused. In future research, we plan to adapt some advanced neural networks (such as EfficientNet, EfficientDet, and RegNet) to further improve model performance (accuracy, training speed, recognition speed, transfer ability, etc.). At the same time, more defect images will be collected for model training and testing.

**Author Contributions:** Conceptualization, D.W.; methodology, D.W. and Y.X.; writing—original draft preparation, D.W. and Y.X.; writing—review and editing, Y.X., B.D., and Y.W.; supervision, H.Y. and M.S.; funding acquisition, D.W. and H.L. All authors have read and agreed to the published version of the manuscript.

**Funding:** This work was supported by the National Natural Science foundation of China (Grant No. 52074242), and the Natural Science Foundation of Hebei Province, China (Grant No. E2016203482).

**Institutional Review Board Statement:** Not applicable.

**Informed Consent Statement:** Not applicable.

**Data Availability Statement:** Restrictions apply to the availability of these data. Data was obtained from Maanshan Iron & Steel Company Limited and are available from the authors with the permission of Maanshan Iron & Steel Company Limited.

**Conflicts of Interest:** The authors declare no conflict of interest.

## References

1. Wang, N.N.; Hou, B.; Song, H.C.; Zhou, L.L.; Liu, Y.X.; Wang, R. Research on the quality online control technology for hot rolled strip. *Steel Roll.* **2017**, *34*, 59.
2. Song, J. Analysis on The Surface Defects of Hot Strips Based on Data Mining Methods. Master's Thesis, Shanghai Jiao Tong University, Shanghai, China, 2008.
3. Wu, P.C.; Lu, T.J.; Wang, Y. Development and perspective of automatic strip surface inspection system. *Iron Steel* **2000**, *35*, 70.
4. Ai, Y.H.; Xu, K. Development and prospect of on-line surface inspection of steel plates. *Met. World* **2010**, *5*, 37–39.
5. Xiao, S.H.; Wu, L.; He, W.; Peng, Y. Application of deep learning in surface quality detection. *Mach. Des. Manuf.* **2020**, *1*, 288.
6. Xu, K.; Yang, C.L.; Zhou, P. Technology of on-line surface inspection for Hot rolling Steel Strips and Its Industrial Application. *J. Mech. Eng.* **2009**, *45*, 111. [CrossRef]
7. Wang, C.M.; Zhou, D.D.; Xu, K.; Zhang, H.N. Review of intelligent manufacturing technology in steel industry. *Chin. Met.* **2018**, *28*, 1.
8. Xu, K.; Wang, L.; Wang, J.Y. Surface defect recognition of hot rolling steel plate based on tetrolet transform. *J. Mech. Eng.* **2016**, *52*, 13. [CrossRef]
9. He, Y.H.; Miao, R.T.; Chen, Y.; Liang, S.; Fei, J.H. Development and application of the on-line surface inspection system for hot rolling strips based on LED light source. *Baosteel Technol.* **2011**, *3*, 1.
10. Gan, S.F.; Sun, L.; Yang, C.; Liu, G.Y. Application of multi-level image classification system in surface defect detection of cold-rolled silicon strip. *Met. Ind. Autom.* **2009**, *33*, 63.
11. Yi, Z.H.A.O. Application of deep learning in defects identification of silicon steel sheet. *Digit. Technol. Appl.* **2017**, *12*, 19.
12. Han, Y.L.; Yan, Y.H. Discernment and classification of banding strip surface defect based on BP neural network. *Chin. J. Sci. Instrum.* **2006**, *27*, 1692.
13. Chu, M.X. Research of Key Technologies for Steel Plate Surface Defects Detection. Ph.D. Thesis, Northeastern University, Shenyang, China, 2014.
14. Xing, J.F. Surface Defect Recognition and System Development of Hot Rolling Strip Based on Convolutional Neural Network. Master's Thesis, Northeastern University, Shenyang, China, 2019.
15. Song, K.C.; Yan, Y.H. A noise robust method based on completed local binary patterns for hot rolling steel strip surface defects. *Appl. Surf. Sci.* **2013**, *285*, 858. [CrossRef]
16. Hu, L.T. Research on Surface Defect Recognition of Steel Plate Based on Machine Vision. Master's Thesis, Wuhan University of Science and Technology, Wuhan, China, 2018.
17. Xie, F. Research on Surface Defect Detection Technology Based on Machine Vision. Master's Thesis, Nanjing University of Aeronautics and Astronautics, Nanjing, China, 2019.
18. Wen, C.; Gao, Y.P.; Gao, L.; Li, X.Y. A new ensemble approach based on deep convolutional neural networks for steel surface defect classification. *Procedia CIRP* **2018**, *72*, 1069.

19. Gao, Y.P.; Gao, L.; Li, X.Y.; Yan, X.G. A semi-supervised convolutional neural network-based method for steel surface defect recognition. *Robot. Comput. Integr. Manuf.* **2020**, *61*, 101825. [CrossRef]
20. Saiz, F.A.; Serrano, I.; Barandiáran, I.; Sánchez, J.R. A robust and fast deep learning-based method for defect classification in steel surfaces. In Proceedings of the 2018 International Conference on Intelligent Systems (IS), Funchal-Madeira, Portugal, 25–27 September 2018; pp. 455–460.
21. Lee, S.Y.; Tama, B.A.; Moon, S.J.; Lee, S. Steel surface defect diagnostics using deep convolutional neural network and class activation map. *Appl. Sci.* **2019**, *9*, 5449. [CrossRef]
22. He, Y.; Song, K.C.; Dong, H.W.; Yan, Y.H. Semi-supervised defect classification of steel surface based on multi-training and generative adversarial network. *Opt. Lasers Eng.* **2019**, *122*, 294. [CrossRef]
23. He, Y.; Song, K.C.; Meng, Q.G.; Yan, Y.H. An end-to-end steel surface defect detection approach via fusing multiple hierarchical features. *IEEE Trans. Instrum. Meas.* **2020**, *69*, 1493–1504. [CrossRef]
24. Dong, H.W.; Song, K.C.; He, Y.; Xu, J.; Yan, Y.H.; Meng, Q.G. PGA-net: Pyramid feature fusion and global context attention network for automated surface defect detection. *IEEE Trans. Ind. Inform.* **2019**, *16*, 7448–7458. [CrossRef]
25. Guan, S.Q.; Lei, M.; Lu, H. A Steel Surface Defect Recognition algorithm based on improved deep learning network model using feature visualization and quality evaluation. *IEEE Access* **2020**, *8*, 49885. [CrossRef]
26. Fu, G.Z.; Sun, P.Z.; Zhu, W.B.; Yang, J.X.; Gao, Y.L.; Yang, M.Y.; Gao, Y.P. A deep-learning-based approach for fast and robust steel surface defects classification. *Opt. Lasers Eng.* **2019**, *121*, 397. [CrossRef]
27. You, H.C.; Wang, D.C.; Gao, T.; Wang, H.B.; Zhang, W. Formulation mechanism of edge seam defects for hot rolling IF steel. *Heavy Mach.* **2019**, *6*, 16.
28. Hu, X.W.; Wang, D.C.; Gao, T.; You, H.C.; Zhao, H. Treatment measures of edge seam defects for hot rolling IF steel. *Heavy Mach.* **2020**, *1*, 53.
29. Chun, P.J.; Yamane, T.; Izumi, S.; Kameda, T. Evaluation of tensile performance of steel members by analysis of corroded steel surface using deep learning. *Metals* **2019**, *9*, 1259. [CrossRef]
30. Kim, M.; Han, J.C.; Hyun, S.H.; Janssens, O.; Van Hoecke, S.; Kee, C.; De Neve, W. Medinoid: Computer-aided diagnosis and localization of glaucoma using deep learning. *Appl. Sci.* **2019**, *9*, 3064. [CrossRef]
31. Huang, J.; Zhang, G. Survey of object detection algorithms for deep convolutional neural networks. *Comput. Eng. Appl.* **2020**, *56*, 12.
32. Javaheri, E.; Kumala, V.; Javaheri, A.; Rawassizadeh, R.; Lubritz, J.; Graf, B.; Rethmeier, M. Quantifying mechanical properties of automotive steels with deep learning based computer vision algorithms. *Metals* **2020**, *10*, 163. [CrossRef]
33. Kesse, M.A.; Buah, E.; Handroos, H.; Ayetor, G.K. Development of an artificial intelligence powered TIG welding algorithm for the prediction of bead geometry for TIG welding processes using hybrid deep learning. *Metals* **2020**, *10*, 451. [CrossRef]
34. Lecun, Y.; Bottou, L.; Bengio, Y.; Haffner, P. Gradient-based learning applied to document recognition. *Proc. IEEE* **1998**, *86*, 2278. [CrossRef]
35. Krizhevsky, A.; Sutskever, I.; Hinton, G.E. ImageNet Classification with deep convolutional neural networks. *Commun. ACM* **2012**, *1*, 84–90. [CrossRef]
36. Simonyan, K.; Zisserman, A. Very deep convolutional networks for large-scale image recognition. *arXiv* **2014**, arXiv:1409.1556.

*metals*

MDPI

*Article*

# A Buckling Instability Prediction Model for the Reliable Design of Sheet Metal Panels Based on an Artificial Intelligent Self-Learning Algorithm

**Seungro Lee [1,†], Luca Quagliato [2,*,†], Donghwi Park [1], Guido A. Berti [3] and Naksoo Kim [1,*]**

[1] Department of Mechanical Engineering, Sogang University, Seoul 04107, Korea; lsr1310@sogang.ac.kr (S.L.); pdhwi93@sogang.ac.kr (D.P.)
[2] Department of Mechanical and Biomedical Engineering, Ewha Womans University, Seoul 03760, Korea
[3] Department of Management and Engineering, University of Padua, 36100 Vicenza, Italy; guido.berti@unipd.it
* Correspondence: lucaq@ewha.ac.kr (L.Q.); nskim@sogang.ac.kr (N.K.)
† These authors contributed equally to this work.

**Abstract:** Sheets' buckling instability, also known as oil canning, is an issue that characterizes the resistance to denting in thin metal panels. The oil canning phenomenon is characterized by a depression in the metal sheet, caused by a local buckling, which is a critical design issue for aesthetic parts, such as automotive outer panels. Predicting the buckling instability during the design stage is not straightforward since the shape of the component might change several times before the part is sent to production and can actually be tested. To overcome this issue, this research presents a robust prediction model based on the convolutional neural network (CNN) to estimate the buckling instability of automotive sheet metal panels, based on the major, minor, and Gaussian surface curvatures. The training dataset for the CNN model was generated by implementing finite element analysis (FEA) of the outer panels of various commercial vehicles, for a total of twenty panels, and by considering different indentation locations on each panel. From the implemented simulation models the load-stroke curves were exported and utilized to determine the presence, or absence, of buckling instability and to determine its magnitude. Moreover, from the computer aided design (CAD) files of the relevant panels, the three considered curvatures on the tested indentation points were acquired as well. All the positions considered in the FEA analyses were backed up by industrial experiments on the relevant panels in their assembled position, allowing to validate their reliability. The combined correlation of curvatures and load-displacement curves allowed correlating the geometrical features that create the conditions for buckling instability to arise and was utilized to train the CNN algorithm, defined considering 13 convolution layers and 5 pooling layers. The trained CNN model was applied to another automotive frame, not used in the training process, and the prediction results were compared with experimental indentation tests. The overall accuracy of the CNN model was calculated to be 90.1%, representing the reliability of the proposed algorithm of predicting the severity of the buckling instability for automotive sheet metal panels.

**Keywords:** sheet metal; buckling instability; oil canning; artificial intelligence; convolution neural network

**Citation:** Lee, S.; Quagliato, L.; Park, D.; Berti, G.A.; Kim, N. A Buckling Instability Prediction Model for the Reliable Design of Sheet Metal Panels Based on an Artificial Intelligent Self-Learning Algorithm. *Metals* **2021**, *11*, 1533. https://doi.org/10.3390/met11101533

Academic Editor: Pedro Prates

Received: 22 August 2021
Accepted: 22 September 2021
Published: 26 September 2021



## 1. Introduction

Denting resistance is an important issue in the design and manufacturing of sheet metal panels and structures and it is strongly influenced by the panel thickness and the manufacturing process conditions [1]. In addition to that, thanks to recent developments in computer-aided design and advanced manufacturing techniques, thinner and lighter complex surface topologies are being designed, even for commercial and mass-production vehicles [2]. Panel thickness, residual stresses induced by the manufacturing process as well as the surface curvature of panels [3] are all influencing factors for the denting resistance and, due to the strong interest of the automotive industry in this issue, several scholars focused their effort on studying this issue from the experimental as well the numerical points of view, as hereafter summarized.

From the experimental point of view, denting resistance can be assessed through indentation tests carried on sheet metal specimens, made of the material that is aimed to be utilized, or directly on the final product. Johnson and Schaffnit [4] investigated the denting resistance of cold-rolled low-carbon steels, used for the manufacturing of automotive outer panels, employing laboratory experiments where the dent depth at specific impact energy was measured. From the results of their analysis, they discovered that the denting resistance is proportional to the material yield strength and the square of the thickness, thus thinner plates require materials with higher yield strength to prevent denting.

In more recent work, Lu et al. [5] investigated the static and dynamic dent resistance of automotive body panels by means of laboratory experiments linking it to several designs and process variables, such as the material properties, the residual stress distribution after the forming process and the shape of the panel. One of the most interesting results of their analysis is that denting resistance is not only influenced by the material properties of the metal sheet but also by its surface curvature. As concerns the influence of the material properties and hardening behavior on the denting resistance, Shih and Horvath [6] employed dome test experiments and shown how a single denting loading, rather than an incremental ones, better replicate the denting conditions in real applications. In addition to that, in an interesting work by Holmberg and Thilderkvist [7], the influence of the original metal sheet properties and the stamping process parameters on the denting resistance were investigated on double-curved panels. From the results of this contribution, it can be inferred that the denting resistance is, as it can be imagined, strongly influenced by the original material's properties of the metal sheet but it is also influenced by the blank holding force applied during the stamping process due to its influences on the residual stresses in the final product. Finally, Ekstrand and Asnafi [8] and Asnafi [9] defined a testing method to assess the denting resistance of automotive body panels and, in their analysis considered both major and minor curvatures of the panel as the main influencing parameters and that the boundary conditions, namely, the panel fixing jig's geometry, have a strong influence on the results of the test.

However, the denting resistance of the original material and that of the manufactured panel are not equal due to the influences of the panel geometry and the manufacturing process. In addition to that, the application of strong fixing devices during the test, such as bolting or bead holders, may influence the results of the rest which may differ from the denting resistance of the final panel. For these reasons, in the research presented in this paper, denting experiments were carried out on an automotive panel already installed on a vehicle, and by considering different locations characterized by different geometrical features.

As concerns the development of numerical models for the prediction of the denting resistance of sheet metal panels, Holmberg and Nejabat [10] developed and validated a numerical model based on an automotive side-door, showing good accordance between numerical and experimental results in the low strain region, whereas the deviation increases for high strains. In [11], as also carried out in the research presented in this paper, the numerical model was developed neglecting the strain history in the panel caused by the manufacturing process. Similarly, concerning the finite element model development for the assessment of the denting resistance, Shen et al. [12] coupled different hardening models with the Baushinger effect to consider their influence on the denting resistance, showing its great influence. However, both the experiments and the numerical model were defined on test plates and not applied to real automotive panels, thus its reliability in real applications cannot be assessed. Finally, Park et al. [13] developed a numerical model for the estimation of the surface deflection of automotive body panels based on the surface curvature that allowed for the plotting of regions of the model where buckling instability is more likely to arise.

In the research presented in this paper, 20 numerical models relevant for 5 different panels belonging to 4 different vehicles were defined in ABAQUS 6.14/Standard (Dessault Systemes). The parts are defined as fender, front and back doors, hood, and roof and, for the sake of confidentiality, their relevant model is identified as M#1, M#2, M#3 and M#4, respectively. Numerical indentation analyses were carried out on different locations on these twenty parts, with 3 different thicknesses, resulting in a total of 1733 load stroke curves that were analyzed to investigate whether, in the considered load range of 0–20 kgf, buckling instability occurred. Afterward, through image analysis, the major, minor and Gaussian curvatures on the indentation points were extracted and correlated to the buckling instability [3,14]. Finally, the data set correlating the buckling instability to the considered curvatures was fed to a convolutional neural network (CNN) algorithm composed of 13 convolution layers and 5 pooling layers for training. Of the 1733 cases, 1386 (80%) were utilized for the training, whereas the remaining 347 (20%) were used for the first testing phase. Self-learning models have shown the capability of improving the performances of regression models [15] since their response can be continuously trained and updated, such as in the cause of machine learning [15,16]. For the case of pattern recognition, the application of deep learning in neural networks [17] allows correlating actions and effects with high accuracy, thus it was applied in this research. In recent years, artificial intelligence models were applied to various engineering topics, such as for the optimization of the friction welding parameters [18], the estimation of the radial-axial ring rolling process [19] and detection of cracking defects on stainless steels [20].

To validate the finite element model implementation procedure, experiments were carried out on some indentation positions of the above-mentioned panels and an additional part (door), completely neglected during the training and the first validation process of the CNN algorithm. By comparing the experimental and FEA load-stroke curves relevant for the same indentation points the average deviation was estimated to be 9.82%, defined as the ratio between experimental and simulated load-stroke curve area integrals. In addition to that, the proposed CNN-based model was utilized for the prediction of the buckling instability tendency on the above-mentioned additional door part, showing that the proposed algorithm can predict the higher or lower risk of instability buckling with an accuracy equal to 90.1%.

Thanks to the defined correlation between the denting resistance to the panel geometry, considered in terms of the major, minor and Gaussian curvatures, the proposed procedure can already be applied in the early stages of the design process and can be utilized by process engineers for the optimizations of the denting resistance with small changes in the aesthetical design. Moreover, the database can be extended by either adding more numerical, experimental or both, results, widening its application capabilities as well as its reliability.

Finally, to demonstrate the improvement in the prediction capability of the buckling instability phenomenon granted by the utilization of the proposed methodology, in comparison to previously published algorithms, the models proposed by Jung [1] and by Kim et al. [21] were applied to the experimental results cases carried out by the authors. From the results of this last cross-validation, it was possible to conclude that the developed methodology is also capable of predicting the buckling instability when its effect is not pronounced or, in other words, it has a higher prediction sensitivity.

In order to provide a clear insight into the main capabilities of the CNN-based model presented in this paper, the main findings of the research are highlighted as follows:

- The buckling instability of sheet metal panels can be estimated in the early design stages based on the modeling curvatures, avoiding expensive and time-consuming redesigns of the forming die;
- Both experimental and finite element results can be included in the training and validation data sets, allowing extension of the range of validity and applicability of the developed algorithm;

- The implemented image-based CNN methodology proved that machine learning algorithms can also already be utilized for optimization during the early stages of the design process. Moreover, although the methodology proposed in this paper was applied to sheet metal panels for the prediction of the buckling instability, it can be extended to different processes by accounting for the desired target function by applying the same implementation procedure presented in this paper.

## 2. Numerical Model Implementation

To train the CNN model, presented in Section 5, indentation finite element simulation models were implemented in ABAQUS/Standard considering the fender, front and back doors, hood and roof geometries. For each part, four different vehicle models, named M#1, M#2, M#3 and M#4, were considered for a total of 20 geometries. In Figure 1, the panels' geometries, used for the training of the CNN model, are reported along with the M#5-Door, utilized only for validation purposes. For the sake of confidentiality, the dimensions of the parts, as well as specific details of the geometries, cannot be disclosed. Even so, since the proposed methodology is based on the local curvatures and not on the overall panel's geometry, this latter information is not strictly related to the research. On the other hand, the local curvatures of the experimentally tested panels, also used for the finite element model validation, are summarized in the Appendix B of the paper.

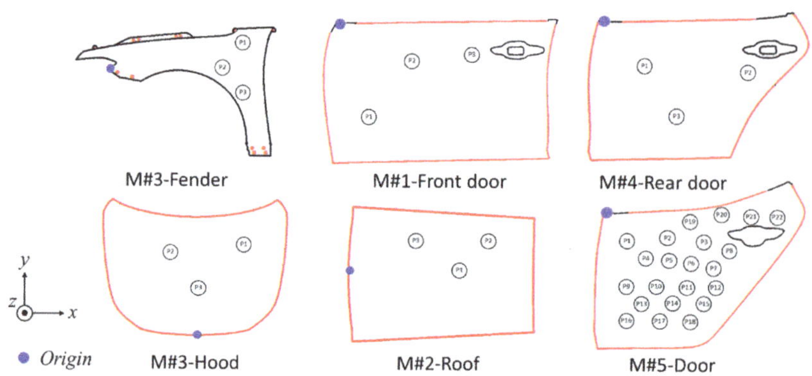


**Figure 1.** Automotive outer panels utilized for the CNN model training and panel utilized for the model validation, with relevant boundary conditions to the FEA models.

All the panels are modeled considering the three initial thicknesses of the steel sheet, equal to 0.5, 0.6 and 0.7 mm, as normally employed in vehicle manufacturing. Moreover, as previously mentioned, since the proposed approach is aimed to be used in the early stages of the design process, where no details of the manufacturing process are available, the thickness variation related to the stamping process is not included in the analysis.

In the numerical simulations, all models were meshed with S4R elements, a 4-node general-purpose shell, reduced integration with hourglass control, for the smooth curvature portions of the model, and with S3 elements, a 3-node triangular general-purpose shell, for the remaining portions of the model. The need for utilizing two different types of elements is given by the complex geometries of the aesthetical details, where the utilization of S4R mesh may result in bad element aspect ratios. The average element side lengths are equal to 3.95 mm for the fender, 5.32 mm for the front door, 5.11 mm for the rear door, 8.07 mm for the hood and 10.0 mm for roof panels. The details of the mesh sensitivity analysis, for one of the considered roof panels, are described in Appendix A.

As concerns the boundary conditions applied to the model, the vehicle full-closed configuration was considered, thus fixed-boundary conditions were applied to the nodes belonging to the contours of doors, hood and roof, whereas the constraints introduced by the assembly bolts were considered for the fender, as shown in Figure 1 (red lines). During the indentation, both in the FEA model as well as in the experiments presented in Section 6, the indenter has a cylindrical shape with a radius equal to 12.5 mm and a fillet radius equal to 0.5 mm. In all the finite element models, the indenter was made to contact the panel surface with an orthogonal vector to the panel surface in the considered spot. Friction was modeled considering a Coulomb friction model with a friction coefficient equal to 0.2 in all the simulation models.

Since all the considered panels are manufactured with AISI-1008 galvanized steel, the same elastic-plastic material properties were utilized as reported in Equation (1), Swift-type hardening model, and shown in Figure 2, where experimental and numerical hardening curves are reported. As concerns the elastic mechanical properties, Young's modulus of 206 GPa and Poisson's ratio of 0.33 were considered. Elastic and plastic mechanical properties data were acquired from the results of tensile tests carried out on ASTM-E8 plate-type specimens carved from the undeformed AISI-1008 plate. Since the considered plates are manufactured through a rolling process, a slight anisotropy arises during the manufacturing operation, causing the rolling direction to have a higher yield and hardening than the transversal one. However, as previously mentioned, since the manufacturing process is not considered in this research, the anisotropy cannot be taken into account and, for this reason, the intermediate curve, relevant for the 45° direction, was considered for the estimation of the Swift model constants, Equation (1), as reported in Figure 2.

$$\sigma_f = 486.7(0.008 + \bar{\varepsilon})^{0.202} \quad \text{(MPa)} \tag{1}$$

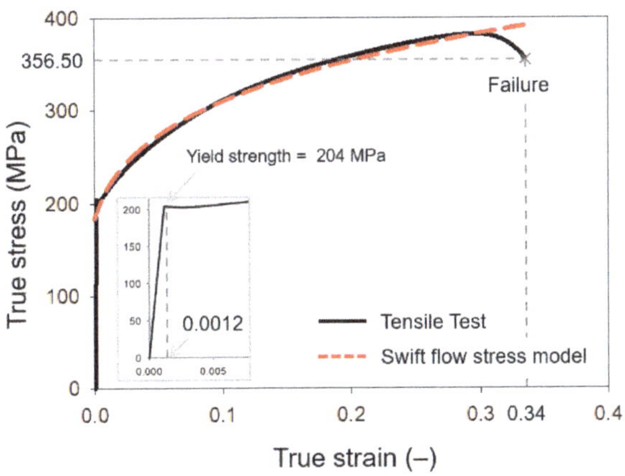


**Figure 2.** Experimental true stress–strain curve and flow stress model for the AISI-1008 material.

## 3. Curvature Calculation Procedure

The algorithm proposed in the paper is aimed to be utilized during the early stages of the design process where no detailed information relevant for the manufacturing is normally available. For this reason, the authors decided to link the buckling instability to the major, minor and Gaussian curvatures, and to the initial sheet thickness; those can all be estimated from CAD files or, as in the case of this research, from the FEA model mesh files. The curvatures and thicknesses extracted from the panels' mesh files were utilized as training input for the developed CNN algorithm, as presented in Section 5. The procedure presented in this section of the paper details the approach adopted for the extrapolation of the nodal curvature for the whole model based on its original mesh and is made necessary since the curvatures cannot be directly determined from the mesh file.

### 3.1. Quadrilateral Mesh into Triangular Mesh Conversion

The numerical models of panels are composed of the quadrilateral mesh (S4R) and the triangular mesh (S3) elements. To calculate the curvatures in the triangular mesh, the quadrilateral mesh is converted to the triangular mesh. The connected four nodes (1, 2, 3, 4) in the element are extracted from the mesh file and the 4-node elements are subdivided into two triangular meshes, (1, 2, 3) and (1, 3, 4) shown in Figure 3a. If the element has multi point constraint (MPC) conditions, the mesh is divided according to the number of MPCs, as shown in Figure 3b.

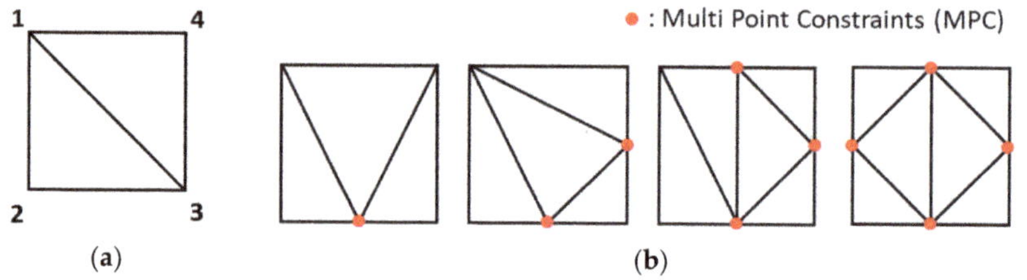


**Figure 3.** Quadrilateral mesh to triangular mesh converting strategy (**a**) without MPC and (**b**) with MPC.

### 3.2. Vertex Normal Vector Calculation

After converting the quadrilateral mesh into the triangular mesh, the connectivity between the three nodes and the corresponding triangular element is checked and the list with each element and the corresponding three vertices is created. Afterward, for each one of the mesh elements (triangular) the element-wise normal vector ($N_i$), Figure 4a, is computed and is used for the calculation of the normal vector of the central vertex among the elements ($N_P$) in its surroundings according to Equation (2), as schematized in Figure 4b [22].

$$N_P = \frac{cN}{|cN|} \quad where\ cN = \sum_{i=0}^{n-1} \frac{N_i \sin\alpha_i}{|V_i||V_{i+1}|} = \sum_{i=0}^{n-1} \frac{V_i \times V_{i+1}}{|V_i|^2|V_{i+1}|^2} \tag{2}$$

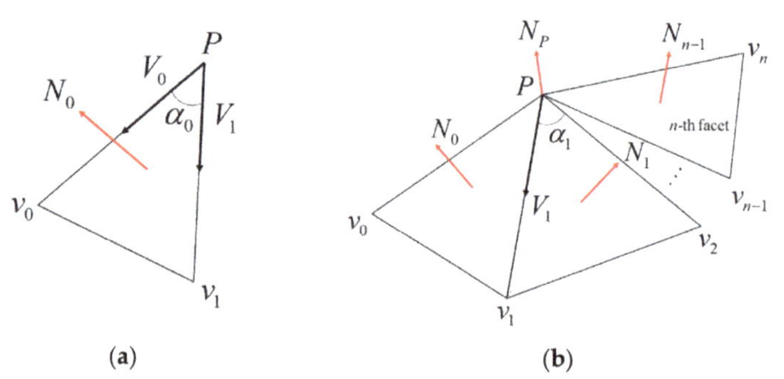


**Figure 4.** (**a**) Elementwise and (**b**) vertexwise normal vector calculation.

### 3.3. Nodal Curvature Calculation

According to the estimation of the normal vector for each vertex (node) of the triangular mesh, the relevant vertex curvature is estimated according to Equation (3) for the unit-length vector ($s$ $t$) in the plane with an orthogonal coordinate of ($u$ $v$) [23]. The unit-length vector is used to normalize the considered vector length to a $-1 \leq s \leq 1$ and $-1 \leq t \leq 1$.

$$\begin{aligned} \kappa_n &= (s\ \ t)\prod\begin{pmatrix} s \\ t \end{pmatrix} = (s'\ \ t')\begin{bmatrix} \kappa_1 & 0 \\ 0 & \kappa_2 \end{bmatrix}\begin{pmatrix} s' \\ t' \end{pmatrix} &= \kappa_1 (s')^2 + \kappa_2 (t')^2 \\ \prod &= (D_u n\ \ D_v n) = \begin{pmatrix} \frac{\partial n}{\partial u} u & \frac{\partial n}{\partial v} u \\ \frac{\partial n}{\partial u} v & \frac{\partial n}{\partial v} v \end{pmatrix} & \end{aligned} \tag{3}$$

From the vertex normal vector information, as calculated in Section 3.2, the $(u\ \ v)$ coordinates system is constructed and the second fundamental tensor $\prod$ is computed.

The principal major ($\kappa_1$) and minor ($\kappa_2$) curvatures are estimated by calculating the eigenvalues of the second fundamental tensor whereas the Gaussian curvature ($\kappa_G$) is determined by the multiplication of the major and minor curvatures ($\kappa_1 \cdot \kappa_2$).

## 4. Buckling Instability Definition and Calculation

The buckling instability phenomenon is characterized by a change from positive to negative derivate in the load-displacement curve, as highlighted in Figure 5. In order to quantify the magnitude of the buckling, the integral of the area beneath the buckling region is divided by the area integral of the load-displacement reference curve, provided by the panels' manufacturer, as shown in Equation (4).

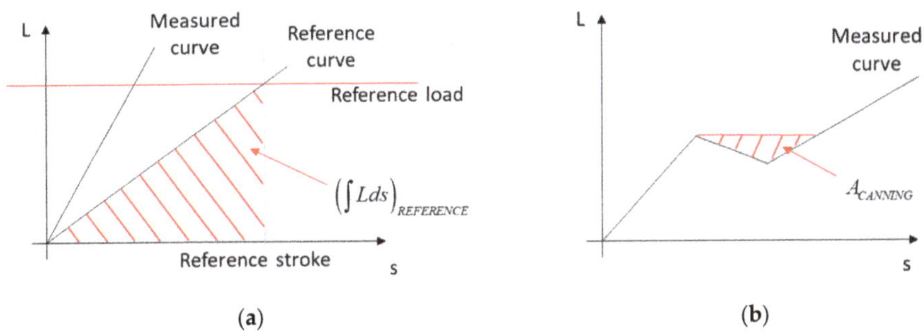


**Figure 5.** Buckling instability definition based on area integral ratio considering (**a**) reference curve area and (**b**) canning area.

The reference curve, as shown in Figure 5, defines the maximum displacement that the panel should show when a 20 kgf load is applied on its surface. In order words, the reference curve defined the minimum stiffness requirement for the panel and can be adjusted according to the requirements. Based on the ratio reported in Equation (4), the three classes of responses are defined as no canning ($AR = 0$), soft canning ($0 < AR < 1$) and hard canning ($AR \geq 1$), respectively.

$$AR = A_{CANNING} / \left(\int Lds\right)_{REFERENCE} \tag{4}$$

## 5. CNN Deep-Learning Algorithm Development and Training

This section of the paper is organized following the same flow utilized during the research for the development of the proposed neural network (NN)-based algorithm, thus, after an introduction of the theoretical backgrounds of the model, the focus will be placed on the methodology for the calculation of the considered three curvatures, utilized as input for the CNN algorithm. Afterward, the definition of the buckling instability, calculated from the results of the load-displacement curve, is provided along with the explanation of the pre-processing operations utilized for the construction of the training and validation data set for the CNN algorithm. Finally, the developed CNN model structure and validation procedure are presented, allowing to have an overall picture of the model background, implementation training and validation activities.

### *5.1. Neural Network (NN) Model Development*

Neural network models consist of: (i) input layers, (ii) hidden layers and (iii) output layers, as can be seen in Figure 6a. Input layers are connected to the hidden layers by the weight functions ($\theta_{i,j}$), which are calculated during the training of the NN algorithm. For the case of a 1D problem, for each one of the nodes, the inputs coming from the previous layer are defined as $x_i$, are multiplied by the weight functions ($\theta_{i,j}$) and summed out to the bias values ($b_i$). The outputs of the layer are derived through the activation function ($f$), as shown in Equation (5) and Figure 6b.

$$\hat{y}_i = f\left(\sum \theta_{i,j} \cdot x_i + b_i\right) \quad (5)$$

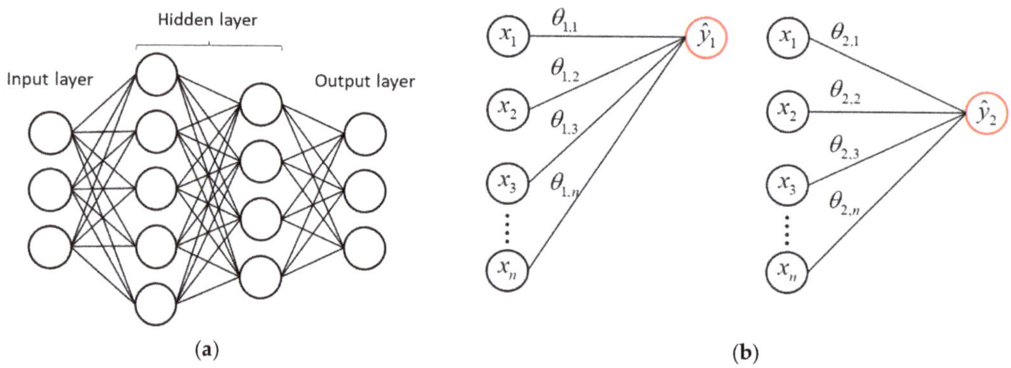


**Figure 6.** (**a**) Neural network model and (**b**) connection between input nodes and an output node.

The logic for the calculation of the weights functions is based on Equation (6), where $E$ is the loss function and $\eta$ the learning rate. The standard NN model computes the weights to minimize the loss function, $E$, by applying the gradient descent and the backpropagation method [24,25]. However, to achieve a quicker convergence in the determination of the components of the filter matrix, $\theta_{i,j}$, the ADAM optimizer, presented in Equation (7), was utilized [26] in this research. The ADAM optimizer utilized a two-moment vector, $\beta_1$ and $\beta_2$, for the update of the filter matrix elements and it grants a higher convergence in comparison to the gradient descent function, which is based on a single moment vector.

Finally, in the developed algorithm, a categorical cross-entropy loss function is defined as in Equation (8), where $y_i$ is the true value and $\hat{y}_i$ the predicted one, defined in Equation (5). The loss function is utilized to evaluate the quality of the prediction carried out by the trained NN algorithm during the training procedure.

$$\theta_{i+1} = \theta_i - \eta \frac{dE_i}{d\theta_i} \quad (6)$$

$$\begin{gathered} \theta_{i+1} = \theta_i - \eta \frac{\hat{m}_{1,i+1}}{\sqrt{\hat{m}_{2,i+1}+\varepsilon}} \\ where\ \ m_{n,i+1} = \beta_n m_{n,i} + (1-\beta_n)\frac{dE}{d\theta_i}\ \ \&\ \ \hat{m}_{n,i+1} = m_{n,i}/\left(1-\beta_n^{timestep}\right)\ ,\ n = 1,2 \end{gathered} \quad (7)$$

$$E(y_i, \hat{y}_i) = -\sum_{i=1}^{n} y_i \log(\hat{y}_i) \quad (8)$$

To avoid 'vanishing gradient' problems, arising if $dE/d\theta_i \leq 0$, a ReLU (rectified linear unit) function $f$ was employed, Equation (9), and allows always obtaining a positive (>0) output from each one of the nodes of the hidden layer. If $x_i \leq 0$, then the 'answer' of the relevant node is not considered. The reasoning behind this choice is given by the fact that, in the input layers, the above-mentioned three curvatures are always inputted as positive values and a negative or null number does not have any physical meaning and should be therefore neglected.

$$f(x) = \begin{cases} 0 & for\ x \leq 0 \\ x & for\ x > 0 \end{cases} \quad (9)$$

For the output layer, the softmax activation function, defined in Equation (10), was applied to normalize the output values from 0 to 1, thus considering the highest number as the correct predicted answer. This strategy allows having a single answer from the trained NN algorithm which is subdivided into three classes: non-buckling, soft buckling and hard buckling.

$$f(x_i) = e^{x_i} / \sum_{i=1}^{3} e^{x_i}\ ,\ i = 1,2,3 \quad (10)$$

In this research, the input layers for the NN algorithm are defined as the 2D image arrays of local major, minor and Gaussian curvatures, whereas the three output layers are set to predict non-buckling, soft buckling and hard buckling. The detailed explanation of the physical interpretation for the non-buckling, soft buckling and hard buckling outputs were provided in Section 4 of the paper.

The CNN methodology was utilized to convert the curvature distribution images into the input layers for the NN algorithm. The CNN algorithm is utilized for image recognition [27] and allows transforming an image into an array, whose features are then extracted by applying a convolution filter. The employed convolution filter has a 3 × 3 size and, as shown in the example reported in Figure 7, the filter was defined to extract only the principal diagonal values of the input image array. The three diagonal values, defined as $\theta_{i,j} \cdot x_{i,j}$ when $i = j$, are fed into Equation (5), where the subscript $j$ is added to the input value $x$ to account for the 2D nature of the inputted image, not considered in the 1D representation of Figure 6. Accordingly, Equation (5) becomes Equation (11), where the applied convolutional filter is shown as well.

$$\hat{y}_{i+k,j+k} = f\left(\sum_{i=1,j=1}^{3} \theta_{i,j} \cdot x_{i,j} + b_i\right) \quad (11)$$

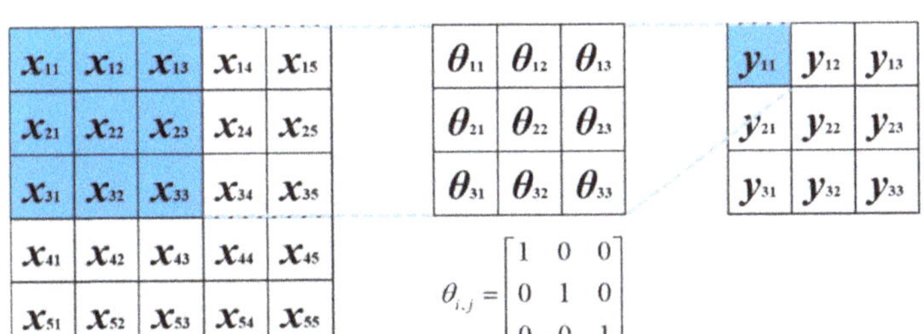

**Figure 7.** Convolutional layer with a stride of 1 and a 3 × 3 size filter.

According to the model defined in Equation (11), the weights, the cross-entropy function of Equations (7) and (8), and the activation function of Equations (9) and (10), a CNN model composed of an input layer made of 3 channels, 22 hidden layers, and 3 channels for the output layer were implemented in Python 3.7.4 and Keras 2.3.1 with TensorFlow backend framework.

The details relevant for the pre-processing of the curvature images extracted from the mesh file, as presented in Section 3, are reported in Section 5.2, whereas the structure, training and cross-validation of the developed CNN algorithm are reported in Section 5.3, respectively.

### 5.2. Pre-Processing

The curvature distribution images of the major, minor and Gaussian curvatures are regarded as grayscale images in the graphical GUI where white is considered as 255 whereas black as 0. This conversion allows inputting the curvature distribution as a number included in a predefined range. The contour intervals are considered separately along with each curvature distribution, where the major curvature has 45 discrete sections of 0.09 to −0.009, the minor curvature is sorted out in 45 sections of 0.009 to −0.09 and the Gaussian curvatures has 36 intervals between 0.009 and −0.009.

These ranges were chosen according to the results of the three considered curvatures, all of which fall inside the above-mentioned ranges. In all images, a black background is added outside the original picture contour to avoid false readings during the conversion from image to channel input values. By considering the boundary projected area (top view) of the considered panel, the curvature distributions are converted into grayscale BMP images and a pixel-to-mm conversion ratio is calculated. This conversion ratio is not constant but depends on the size of the panel in the x–y plane (Figure 8).

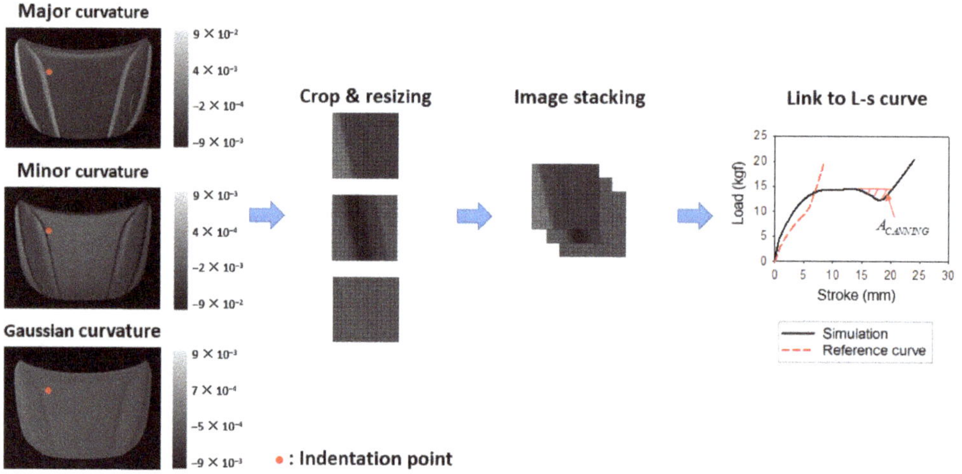


**Figure 8.** Training data set preparation procedure.

In order to correlate the above-mentioned curvatures with the buckling instability data, a series of indentation points were chosen on the panel surface and the surrounding 50 mm × 50 mm indentation area, identified according to the indenter dimensions. Since the number of pixels in the indentation area is not constant, it varies from panel to panel, the original number of pixels is upgraded to 256 pixels by 256 pixels image. By following this procedure, all the images have the same number of pixels along x- and y-directions and can be used for a uniform and reasonable training of the NN algorithm. On the selected indentation points, finite element indentation analyses, as summarized in Section 2, were carried out and, according to the procedure detailed in Sections 3 and 4, allowed linking the local curvatures distributions with the three classes of responses, namely, no canning, soft canning and hard canning.

By following this procedure, the training set, made of 1733 local curvatures distributions and relevant class assignment (569 no canning, 412 soft canning and 752 hard canning), along with the relevant thickness, was constructed and utilized for the training of the proposed CNN algorithm, as detailed in Section 5.3 of the paper. Furthermore, for the case of the validation panel, the same procedure was utilized but the results were not utilized for the training of the CNN algorithm.

### 5.3. CNN Model Structure, Training, and Cross-Validation

To predict the buckling instability phenomenon arising in thin metal panels according to the surface curvatures, as summarized so far, a combination of a pre-trained weight matrix in the VGG-16 model [28] and a self-developed classification layer was utilized according to the architecture shown in Figure 9. The architecture consists of a features extraction part and a classification part.

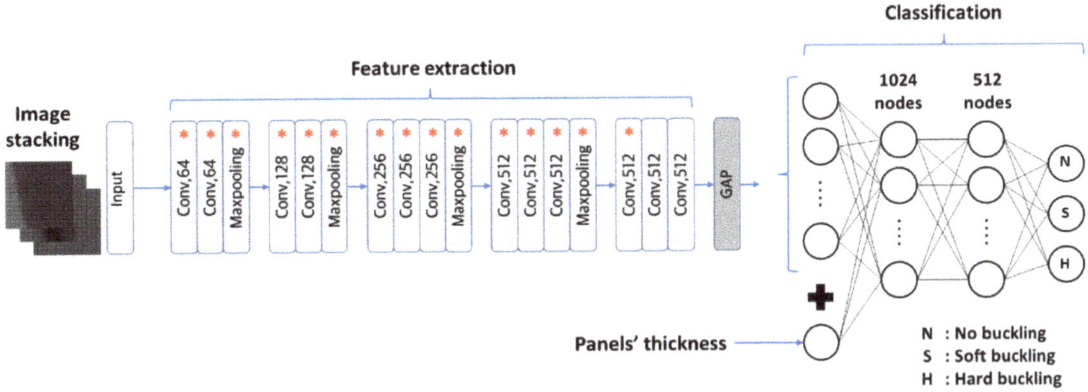


**Figure 9.** The developed CNN model architecture.

In the feature extraction part, the model consists of 13 convolutional layers and 4 max-pooling layers. The self-developed classification part has a global average pooling layer and dense layers, where the last dense layer has 3 nodes that are designed to provide a non-buckling, soft buckling and hard buckling classification. The drop-out method in the classification section is utilized after a dense layer to overcome possible overfitting problems.

The layers highlighted with the symbol '*' in Figure 9 possess pre-trained weight matrices calculated with the ImageNet dataset [27] and are not updated during the model training. These non-trainable weight matrices are based on a huge classification of datasets and help the trained model to recognize the images by pre-trained weights matrix in the early stages, improving the classification results even if the images are different from that of the ImageNet [27,29,30]. In the convolutional layers, a progressively increasing number of filters is applied to recognize ever-smaller features in the images. Hence, by following this procedure, the number following the word 'Conv' in Figure 9 represents the number of filters in the relevant layer, the higher the number of extracted features the lower the size of the relevant feature. This results in the progressive zoom-in of the images, as schematized in Figure 10, where the results are relevant for the output of the first convolution of each convolutional block, from Conv,64 to Conv,512, for the case of the filter matrix reported in the detail of Figure 10. In Figure 10, the extracted features from each of the images are highlighted in yellow and green.

After the convolution layers, the featured images are resized in half by the max-pooling layer to amplify the features and reduce the size of the computed data. After the feature extraction, the 2D image arrays are flattened to an array of 1 row by the global average pooling (GAP in Figure 9) layer allowing matching with the averaged values of a 2D image into a single value for one node.

After exiting from the Conv,512 layer, and after the GAP application, the features are inputted in the Dense, 512 layer and are reduced to 256 features in the dropout layer with a 0.5 dropout ratio [31]. Because the dropout layer selects the 256 nodes randomly at each step during the training process, the model does not depend on specific nodes, allowing overcoming of the overfitting problem. In this step, the selection of the 256 features to be inputted into the Dense, 256 layer is carried out and the randomly selected 128 nodes are linked to 3 nodes in the last layer, thus allowing obtaining of the prediction of no canning, soft canning or hard canning. The hyperparameters are tuned as 0.00005 for the learning rate, with $10^{-8}$ for $\varepsilon$, 0.9 for $\beta_1$ and 0.999 for $\beta_2$ in Equation (7), 128 batch size and 300 learning steps according to computing power.

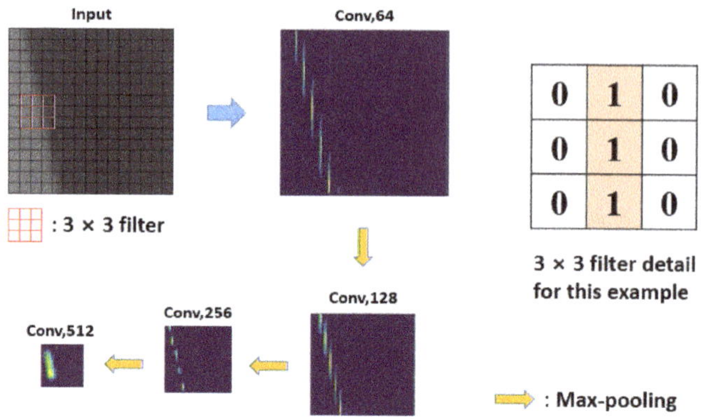


**Figure 10.** Feature recognition for progressive convolutional blocks by 3 × 3 example filter.

The learning procedure is carried out by using the $k$-fold cross-validation process. In order to evaluate the prediction model for the non-trained data, the total dataset is randomly subdivided into $k$ sets. One set is utilized as the test set for estimation of the training results whereas the remaining four sets are used for the training of the CNN algorithm. In this study, the 5-fold cross-validation process was utilized and the best model among the 5 split models was selected to predict the buckling instability of the target panel geometry.

The final results for the test set are evaluated by the average value of classification accuracy and $F_1$ score of the five split model results. The $F_1$ score is calculated by following Equation (12) as a harmonic mean value of precision and sensitivity, as in Equations (13) and (14). In Equations (13) and (14), TP, FP and FN are defined as true positive, false positive and false-negative prediction, respectively.

$$F_1 = 2 \times \frac{Precision \times Sensitivity}{Precision + Sensitivity} \tag{12}$$

$$Precision = \frac{\mathrm{TP}}{\mathrm{TP} + \mathrm{FP}} \tag{13}$$

$$Sensitivity = \frac{\mathrm{TP}}{\mathrm{TP} + \mathrm{FN}} \tag{14}$$

## 6. Denting Resistance Validation Experiments

To validate the finite element model presented in Section 2, indentation experiments were carried out on 5 different panels belonging to the five categories presented in Figure 1. Moreover, 22 indentation tests were carried out on the M#5-Door, only used for the validation of the developed NN algorithm. The six tested parts, with the relevant considered indentation points, are reported in previous Figure 1.

For the measurement of the load-displacement curves during the indentation experiments, a 6-axis robot, Figure 11, having the indenter as the final end-point, was employed. The dimensions of the indenter cylinder are reported in Figure 11 and are the same utilized in the FEA simulations, described in Section 2 and Appendix A.

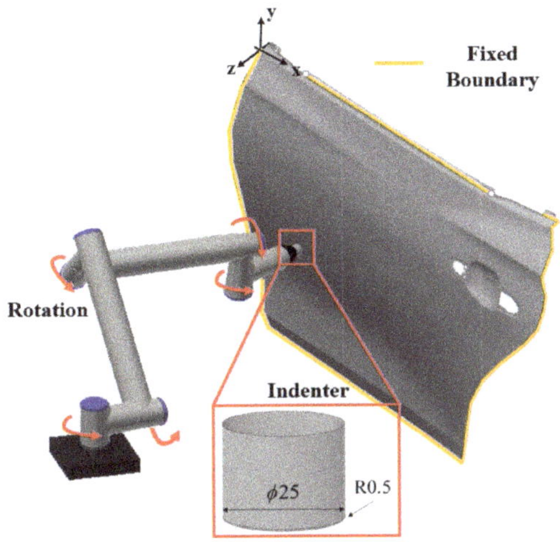


**Figure 11.** Denting resistance experiment measurement system.

The robot is equipped with a load cell, allowing measuring of the load applied during the test, whereas the displacement of the end-point is automatically measured by the robot arm encoder. The indentation points and their relevant coordinates on the panel surface are summarized in Table A1 (Appendix B of the paper) whereas the load-displacement results are reported in the result section where they are compared with the results relevant for the implemented FEA models. In Table A1, in the 'RES' column, the letters 'N', 'S' and 'H' stand for no canning, soft canning and hard canning, respectively.

## 7. Results

To prove the reliability of the implemented FEA model, the results of the indentation experiments, in terms of load-displacement curves, were compared with those of the relevant numerical simulation model. The results of the 18 load-displacement curves are reported in Figure 12 whereas the experimental and numerical area integrals, along with the percentage errors, are reported in Figure 13 and detailed in Table A2 (Appendix B of the paper).

As concerns the developed buckling instability prediction model based on the CNN architecture, its reliability was verified in two different steps. In the former one, the *k*-fold cross-validation process was employed and the relevant accuracy and $F_1$ score were calculated and the results are summarized in Table 1. The calculations were carried out considering a total of 347 points relevant for the panels shown in Figure 1. As previously mentioned, the M#5-Door points, Figure 1, were not used for any of the model training sets since they are aimed to be used only for validation purposes.

**Table 1.** Results of test-set in 5-fold cross validation.

| n-Fold | 1 | 2 | 3 | 4 | 5 | Average |
|---|---|---|---|---|---|---|
| Accuracy | 0.908 | 0.886 | 0.899 | 0.914 | 0.897 | 0.901 |
| $F_1$ score | 0.909 | 0.898 | 0.900 | 0.921 | 0.907 | 0.907 |

For the calculation of both accuracy and $F_1$ score, 80% of the whole data set was used for the CNN algorithm training, whereas the remaining 20% for its validation. All five sets reported in Table 1 are different from each other and were randomly selected to avoid any bias in the validation process.

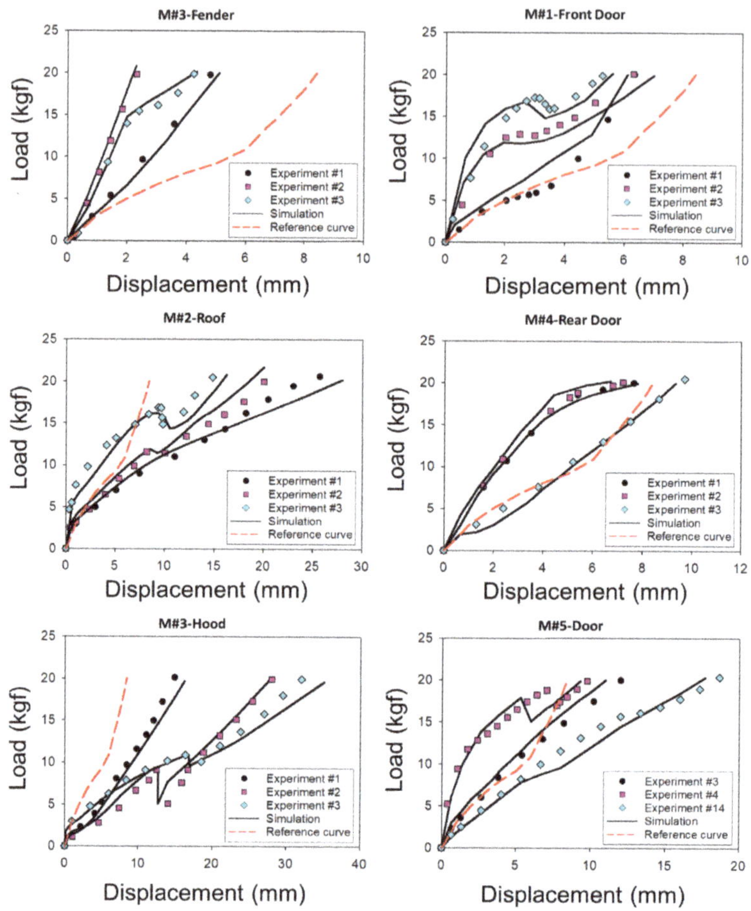


**Figure 12.** Experimental vs. numerical load-displacement curves for the validation points.

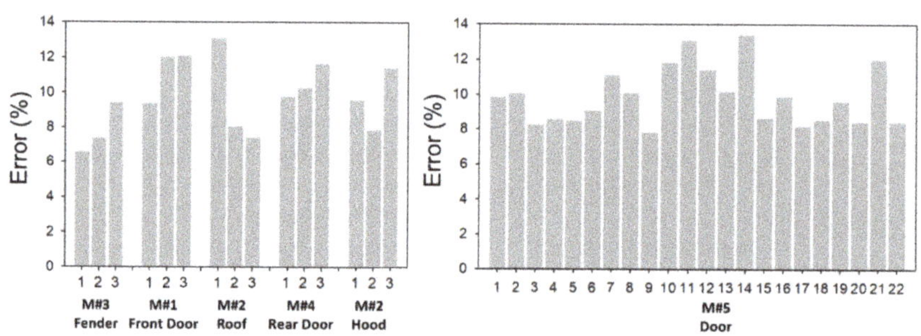


**Figure 13.** Experimental vs. numerical load-displacement curves for the validation points.

The latter validation was carried out considering the prediction for buckling instability on the 22 points of the M#5-Door (Figure 1), where, from the results of the experiments, only point #4 showed hard canning, whereas the remaining 21 showed no canning. Moreover, to verify the reliability of the buckling instability prediction in the considered thickness range of 0.5–0.7 mm, a 0.55 mm thickness M#1-Hood part was numerically tested as well. As concerns the 0.55 mm thickness M#1-Hood, since no experiments are available, the comparison was carried out considering the reference curve criteria for the definition of the relevant canning class, as shown in Section 4. The detailed comparison is reported in Table A3 (Appendix B of the paper).

Before the prediction of the buckling instability of the M#5-Door, the whole 60 panels' data set's (20 different panels and 3 thicknesses) were fed into the CNN algorithm and 3 output channels were identified as red for hard canning, green for soft canning and blue for no canning. The results of the buckling instability for the 22 points of the M#5-Door (Figure 1) and the 18 points for the 0.55 mm thickness M#1-Hood are reported in Figure 14a,b, respectively. The load-displacement curve for point #4 of the M#5-Door, red points in Figure 14a is that of Figure 12.

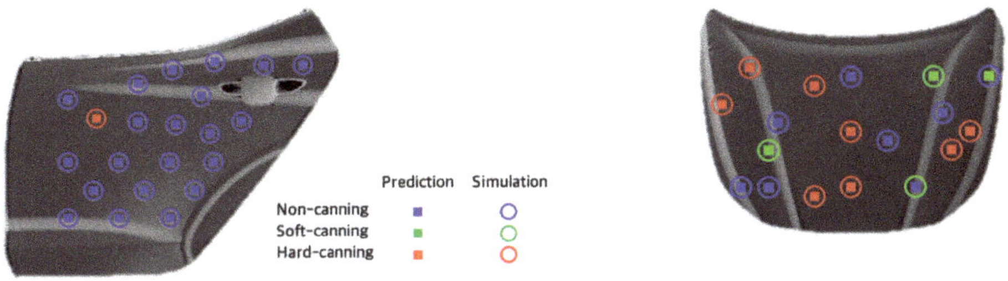


(**a**) (**b**)

**Figure 14.** Buckling instability prediction of indenting points on the (**a**) M#5-Door (0.7 mm thickness) and (**b**) M#1-Hood (0.5 mm thickness).

Finally, to demonstrate the improvement in the prediction of the buckling instability phenomenon in comparison to previously published models, the two contributions of Jung [1] and Kim et al. [21] were considered. The two models detailed in these two contributions were applied to the panels' geometries presented in Figure 1 and, along with the predictions carried out by the developed model, were compared with the experimental results summarized in previous Figure 12. The comparison among these three prediction models, and the relevant experimental results, are reported in Table 2 where N, S, and H stand for no canning, soft canning and hard canning, respectively.

**Table 2.** Buckling instability prediction results comparison among developed and literature models for the prediction of the buckling instability of thin metal sheets.

| Panel Model | P# | Experiments | Proposed Model | Jung [1] | Kim et al. [30] | Panel Model | P# | Experiments | Proposed Model | Jung [1] | Kim et al. [30] |
|---|---|---|---|---|---|---|---|---|---|---|---|
| M#3 Fender | #1 | N | N | N | N | M#5 Door | #5 | N | N | N | N |
| | #2 | N | N | N | N | | #6 | N | N | N | N |
| | #3 | N | N | N | N | | #7 | N | N | N | N |
| M#1 Front Door | #1 | N | N | N | N | | #8 | N | N | N | N |
| | #2 | S | S | N | N | | #9 | N | N | N | N |
| | #3 | H | H | N | N | | #10 | N | N | N | N |
| M#2 Roof | #1 | N | N | N | N | | #11 | N | N | N | N |
| | #2 | S | S | N | N | | #12 | N | N | N | N |
| | #3 | H | H | N | N | | #13 | N | N | N | N |
| M#4 Rear Door | #1 | N | N | N | N | | #14 | N | N | N | N |
| | #2 | N | N | N | N | | #15 | N | N | N | N |
| | #3 | N | N | N | N | | #16 | N | N | N | N |
| M#3 Hood | #1 | N | N | N | N | | #17 | N | N | N | N |
| | #2 | H | H | H | H | | #18 | N | N | N | N |
| | #3 | H | H | N | H | | #19 | N | N | N | N |
| M#5 Door | #1 | N | N | N | N | | #20 | N | N | N | N |
| | #2 | N | N | N | N | | #21 | N | N | N | N |
| | #3 | N | N | N | N | | #22 | N | N | N | N |
| | #4 | H | H | H | H | | | | | | |

The overall accuracy of the two considered literature models is good but they fail to recognize the soft-canning case, where the canning area, defined in Section 4, Figure 5, is not pronounced. This is clear for the case of the M#1-Front Door point #2 and M#2-Roof point #2, where the two literature models predict no canning whereas both experiments and developed CNN model predict experienced and predicted soft canning. Moreover, even if the considered canning area leads to a hard canning evaluation if the result is close to the soft canning, prediction inaccuracies have also been observed, such as the case of the M#1-Front Door point #3, M#2-Roof point #3, M#3-Hood point#3.

Through this additional validation, it is possible to conclude that the proposed methodology, based on the CNN method, is not only capable of accurately predicting the buckling instability phenomenon but it is also more sensitive to the transition between soft and hard canning, further enhancing its reliability.

## 8. Discussion

Based on the comparison between the experimental and FEA results concerning the load-stroke curves reported in Figure 12, some differences are present between the two curves. These differences are mainly caused by the two simplifications introduced in the FEA model, as hereafter summarized. On one hand, for the case of the real panels, the forming process causes a non-uniform thickness distribution, which also slightly alters the local denting resistance. On the other hand, the forming process also causes the arising and relief of residual stresses, which influence the local mechanical properties, thus also the denting resistance. However, since the proposed approach is aimed to be utilized during the early stages of the design process, only the conceptual model might be available, thus the consideration of the forming process, normally defined in a later stage, might not be feasible. For this reason, in the proposed approach, the thickness was modeled as constant considering three different levels, a representative for the sheet thicknesses mostly used in the automotive industry. In addition to that, in order to avoid any possible bias, isotropic mechanical properties were considered as well.

Although these two simplifications introduce some errors into the calculation, the comparison between experimental and FEA load-stroke curves results in the ratio between the integral areas equal to 9.82% being reliable. For the sake of conciseness, not all the results relevant for the tested panels can be included in the paper but the above-mentioned average deviation between experimental and numerical load-stroke curves was calculated on the whole result set and not only on the results included in the paper.

As concerns the conversion strategy from mesh file to grayscale images, as shown in Figure 15, the higher profiles, protruding from the panel surface, are considered to have negative major and minor curvatures, thus a positive Gaussian curvature. For this reason, in Figure 15 for the case of point #2 (minor curvature), the light grey at the bottom of the picture represents a higher z-coordinate (orthogonal to the panel surface) in comparison to the top of the picture (black). In the region in the middle between the light grey and the black, the change in curvature is visible as a progressive grey shading.

Similarly for point #2, the consideration of the Gaussian curvature allows avoiding loss of information when the images are supplied to the CNN algorithm, overcoming the difficulty in identifying surface features in cases of high-darkness images. By employing this approach, the training and validation sets, as presented in Section 5.3, were constructed and the high accuracy and $F_1$ scores allow concluding that the considered dataset construction procedure is balanced and not biased. This important aspect should be considered if the proposed algorithm is aimed to be employed to different components or in case the implemented dataset is aimed to be expanded by considering additional FEA or experimental results.

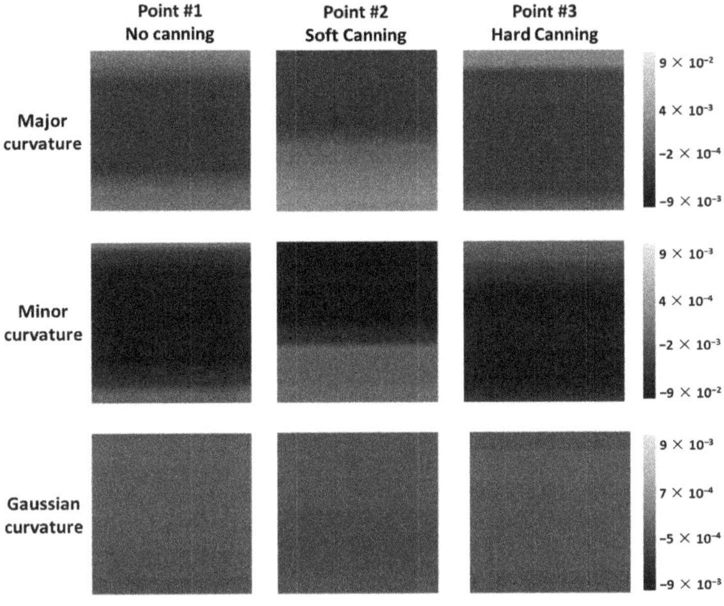


**Figure 15.** Major, minor and Gaussian curvatures for the cases of no canning (point #1), soft canning (point #2) and hard canning (point #3) (indenting points on the M#2-Roof).

## 9. Conclusions

In the research presented in this paper, a methodology for the estimation of the buckling instability in sheet metal panels based on the CNN theory was defined and validated on a wide data set and shown to be accurate and reliable even though some simplifications were introduced in the development of the FEA model utilized for the construction of the training and validation data sets. From a global perspective, the developed image conversion approach and CNN algorithm for the prediction of the buckling instability of panels was shown to be a reliable method which, thanks to its reasonable simplifications, can also be utilized in the early stages of the design process, allowing to correct the panels' geometries if and where required. The overall accuracy in the estimation of the buckling instability, calculated to be 90.1%, was estimated by comparing the predictions carried out by the developed model with the observations relevant for 1733 cases, proving the strong correlation between major, minor and Gaussian curvatures with the panel's denting resistance. In addition to that, the proposed model can also be extended to different panels' geometries and utilized in a more advanced manner by considering a combination between sheet metal forming and structural simulations, allowing to account for the local thickness distribution and residual stresses.

**Author Contributions:** Conceptualization, S.L. and L.Q.; methodology, S.L.; software, S.L.; validation, S.L. and L.Q.; formal analysis, S.L. and D.P.; investigation, S.L.; resources, S.L.; data curation, S.L.; writing—original draft preparation, S.L. and L.Q.; writing—review and editing, L.Q.; visualization, S.L. and L.Q.; supervision, G.A.B. and N.K.; project administration, N.K.; funding acquisition, L.Q. and N.K. All authors have read and agreed to the published version of the manuscript.

**Funding:** This work was supported by the Basic Science Research Program through the National Research Foundation of Korea (NRF) funded by the Ministry of Education (grant number: 2019R1I1A1A01062323) and by the National Research Foundation of Korea (NRF) grant funded by Korea government (MSIT) (grant number: 2019R1F1A1060567).

**Institutional Review Board Statement:** Not applicable.

**Informed Consent Statement:** Not applicable.

**Data Availability Statement:** All the results are available on request to the corresponding author.

**Acknowledgments:** This research was carried out with the help of the 'HPC Support' Project, supported by the Ministry of Science, ICT and NIPA of Korea. This support is gratefully acknowledged.

**Conflicts of Interest:** The authors declare no conflict of interest.

## Appendix A. The Mesh Sensitivity for Roof Panels (h-Convergence)

In order to determine the element size which allows obtaining accurate results, four different element sizes were considered. The procedure hereafter reported is relevant for the M#2-roof panel (Figure A1a), but it has been applied to all the geometries reported in Figure 1. Regardless of the panel size, the same approach for the optimization of the mesh size was utilized and the same contact conditions between indenter and panel were adopted.

The procedure adopted for the mesh sensitivity analysis is hereafter reported considering the M#2-roof, but the same approach was also utilized for all the panels reported in Figure 1. For this analysis, four different mesh sizes were considered, namely, 15 mm, 10 mm, 5 mm and 2.5 mm size elements.

First of all, the results of the buckling area were considered as evaluation criteria and as reported in Figure A1b, no differences were identified between the 2.5 mm and 5 mm mesh size element configurations. Afterward, by comparing the detailed load-stroke curve for the 5 mm and 10 mm meshes, Figure A1c, it was possible to conclude that the differences are almost negligible, making the 10 mm (average element length mesh) the one utilized in the analyses for all the roof panels.

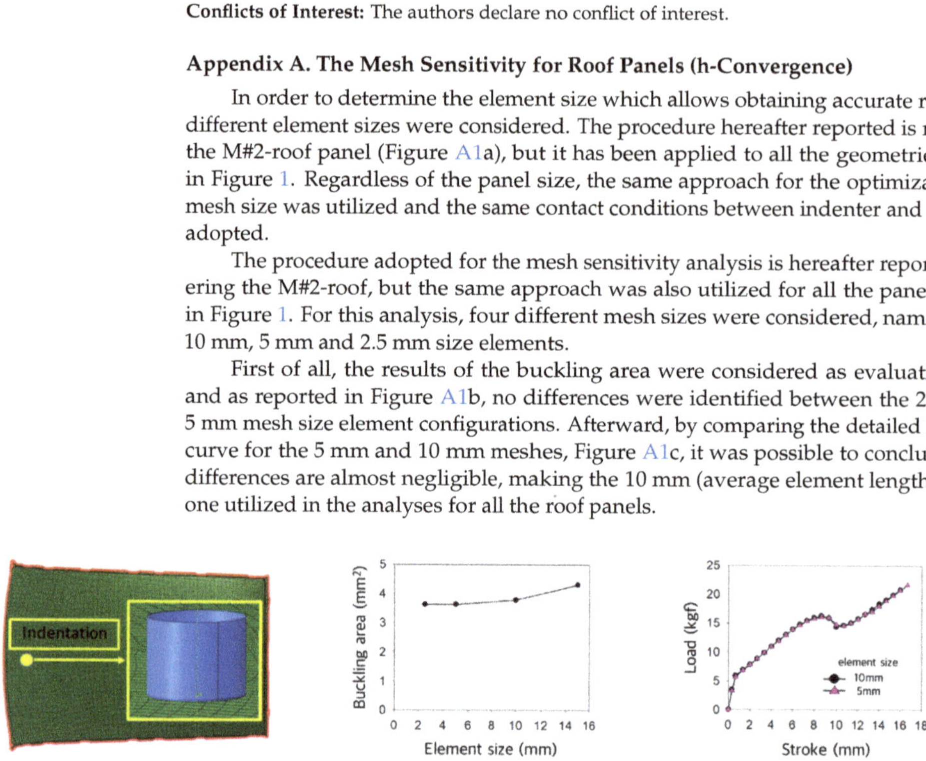


(a) (b) (c)

**Figure A1.** (**a**) M#2-roof FEM model with detail of the indenter and indentation region. (**b**) Mesh sensitivity analysis considering the buckling area as the evaluation criteria and (**c**) comparison between load-stroke FEM results for 5 mm and 10 mm average element lengths for the M#2-roof showing the almost overlapping of the results.

## Appendix B. The Detailed Dimensions and Canning Results for Considered Panels

**Table A1.** Numbering and coordinates of the validation point experiments for 0.7 mm thickness (dimensions in mm).

| Model | P# | x | y | z | Maj.Curv. | Min.Curv. | Gauss.Curv. | RES |
|---|---|---|---|---|---|---|---|---|
| M#3 Fender | #1 | 650 | 100 | 30 | $-1.80 \times 10^{-4}$ | $1.23 \times 10^{-3}$ | $-2.23 \times 10^{-7}$ | N |
| | #2 | 550 | 0 | 68 | $1.76 \times 10^{-4}$ | $1.66 \times 10^{-3}$ | $2.91 \times 10^{-7}$ | N |
| | #3 | 650 | −50 | 70 | $1.37 \times 10^{-4}$ | $1.20 \times 10^{-3}$ | $1.65 \times 10^{-7}$ | N |
| M#1 Front Door | #1 | 370 | −265 | 135 | $-1.19 \times 10^{-3}$ | $-1.37 \times 10^{-5}$ | $1.64 \times 10^{-8}$ | N |
| | #2 | 485 | −97 | 105 | $-2.56 \times 10^{-5}$ | $1.65 \times 10^{-3}$ | $-4.21 \times 10^{-8}$ | S |
| | #3 | 795 | −95 | 105 | $-3.59 \times 10^{-5}$ | $7.26 \times 10^{-4}$ | $-2.60 \times 10^{-8}$ | H |
| M#2 Roof | #1 | 500 | 0 | 112 | $-9.08 \times 10^{-3}$ | $-2.90 \times 10^{-4}$ | $2.63 \times 10^{-6}$ | N |
| | #2 | 465 | 320 | 108 | $-9.65 \times 10^{-3}$ | $-5.03 \times 10^{-5}$ | $4.85 \times 10^{-7}$ | S |
| | #3 | 600 | 320 | 119 | $-8.60 \times 10^{-3}$ | $-5.38 \times 10^{-5}$ | $4.62 \times 10^{-7}$ | H |

**Table A1.** *Cont.*

| Model | P# | x | y | z | Maj.Curv. | Min.Curv. | Gauss.Curv. | RES |
|---|---|---|---|---|---|---|---|---|
| M#4 Rear Door | #1 | 240 | −200 | 105 | $-1.81 \times 10^{-3}$ | $-1.62 \times 10^{-5}$ | $2.92 \times 10^{-8}$ | N |
| | #2 | 860 | −230 | 100 | $-9.16 \times 10^{-4}$ | $2.56 \times 10^{-5}$ | $-2.35 \times 10^{-8}$ | N |
| | #3 | 450 | −450 | 103 | $-5.69 \times 10^{-4}$ | $1.86 \times 10^{-4}$ | $-1.06 \times 10^{-7}$ | N |
| M#3 Hood | #1 | 450 | 750 | 245 | $-2.49 \times 10^{-4}$ | $-8.62 \times 10^{-5}$ | $2.15 \times 10^{-8}$ | N |
| | #2 | −200 | 700 | 251 | $-1.90 \times 10^{-4}$ | $-1.14 \times 10^{-4}$ | $2.17 \times 10^{-8}$ | H |
| | #3 | 0 | 450 | 210 | $-2.29 \times 10^{-4}$ | $-1.96 \times 10^{-4}$ | $4.50 \times 10^{-8}$ | H |
| M#5 Door | #1 | 130 | −130 | 88 | $-1.78 \times 10^{-3}$ | $-3.64 \times 10^{-5}$ | $6.46 \times 10^{-8}$ | N |
| | #2 | 350 | −80 | 74 | $-4.43 \times 10^{-2}$ | $4.67 \times 10^{-5}$ | $-2.07 \times 10^{-6}$ | N |
| | #3 | 550 | −120 | 80 | $1.61 \times 10^{-5}$ | $8.40 \times 10^{-3}$ | $1.35 \times 10^{-7}$ | N |
| | #4 | 220 | −190 | 100 | $-3.87 \times 10^{-3}$ | $-2.48 \times 10^{-5}$ | $9.60 \times 10^{-8}$ | H |
| | #5 | 350 | −200 | 100 | $-3.05 \times 10^{-3}$ | $4.51 \times 10^{-5}$ | $-1.38 \times 10^{-7}$ | N |
| | #6 | 470 | −210 | 100 | $-1.51 \times 10^{-3}$ | $1.20 \times 10^{-4}$ | $-1.80 \times 10^{-7}$ | N |
| | #7 | 580 | −240 | 104 | $-1.61 \times 10^{-3}$ | $1.56 \times 10^{-4}$ | $-2.51 \times 10^{-7}$ | N |
| | #8 | 680 | −200 | 99 | $-1.81 \times 10^{-3}$ | $1.05 \times 10^{-4}$ | $-1.90 \times 10^{-7}$ | N |
| | #9 | 130 | −330 | 104 | $-9.75 \times 10^{-4}$ | $2.23 \times 10^{-5}$ | $-2.17 \times 10^{-8}$ | N |
| | #10 | 290 | −330 | 103 | $-9.41 \times 10^{-4}$ | $1.20 \times 10^{-4}$ | $-1.13 \times 10^{-7}$ | N |
| | #11 | 450 | −330 | 105 | $-7.78 \times 10^{-4}$ | $2.56 \times 10^{-4}$ | $-2.00 \times 10^{-7}$ | N |
| | #12 | 590 | −330 | 110 | $-6.26 \times 10^{-4}$ | $2.87 \times 10^{-4}$ | $-1.80 \times 10^{-7}$ | N |
| | #13 | 210 | −420 | 100 | $-6.80 \times 10^{-4}$ | $1.44 \times 10^{-4}$ | $-9.82 \times 10^{-8}$ | N |
| | #14 | 370 | −420 | 100 | $-5.59 \times 10^{-4}$ | $4.75 \times 10^{-4}$ | $-2.65 \times 10^{-7}$ | N |
| | #15 | 530 | −420 | 107 | $-4.29 \times 10^{-4}$ | $5.25 \times 10^{-4}$ | $-2.25 \times 10^{-7}$ | N |
| | #16 | 130 | −510 | 88 | $1.30 \times 10^{-4}$ | $1.65 \times 10^{-3}$ | $2.13 \times 10^{-7}$ | N |
| | #17 | 290 | −510 | 88 | $-3.96 \times 10^{-4}$ | $6.17 \times 10^{-4}$ | $-2.44 \times 10^{-7}$ | N |
| | #18 | 450 | −510 | 98 | $-6.66 \times 10^{-4}$ | $8.57 \times 10^{-4}$ | $-5.71 \times 10^{-7}$ | N |
| | #19 | 460 | −35 | 55 | $-3.16 \times 10^{-3}$ | $5.37 \times 10^{-5}$ | $-1.70 \times 10^{-7}$ | N |
| | #20 | 595 | −10 | 42 | $-2.26 \times 10^{-5}$ | $8.93 \times 10^{-3}$ | $-2.02 \times 10^{-7}$ | N |
| | #21 | 750 | −20 | 25 | $-3.57 \times 10^{-4}$ | $9.15 \times 10^{-3}$ | $-3.26 \times 10^{-6}$ | N |
| | #22 | 875 | −20 | 54 | $-6.11 \times 10^{-3}$ | $-2.66 \times 10^{-4}$ | $1.63 \times 10^{-6}$ | N |

**Table A2.** Experimental vs. numerical load-displacement curve's area integrals and errors.

| Model | P# | EXP. Area ($mm^2$) | FEA Area ($mm^2$) | Error (%) | AR | Model | P# | EXP. Area ($mm^2$) | FEA Area ($mm^2$) | Error (%) | AR |
|---|---|---|---|---|---|---|---|---|---|---|---|
| M#3 Fender | #1 | 44.45 | 47.38 | 6.59 | 0 | M#5 Door | #5 | 158.65 | 172.12 | 8.49 | 0 |
| | #2 | 21.26 | 22.83 | 7.40 | 0 | | #6 | 131.62 | 143.57 | 9.08 | 0 |
| | #3 | 50.01 | 54.72 | 9.43 | 0 | | #7 | 84.09 | 93.45 | 11.13 | 0 |
| M#1 Front Door | #1 | 49.95 | 54.65 | 9.40 | 0 | | #8 | 80.75 | 72.60 | 10.09 | 0 |
| | #2 | 79.94 | 89.57 | 12.05 | 0.07 | | #9 | 96.79 | 89.22 | 7.82 | 0 |
| | #3 | 72.49 | 81.27 | 12.11 | 2.59 | | #10 | 77.05 | 86.17 | 11.84 | 0 |
| M#2 Roof | #1 | 313.01 | 354.07 | 13.12 | 0 | | #11 | 101.89 | 115.26 | 13.12 | 0 |
| | #2 | 229.83 | 248.36 | 8.06 | 0.54 | | #12 | 121.12 | 107.23 | 11.47 | 0 |
| | #3 | 204.45 | 219.63 | 7.43 | 2.52 | | #13 | 76.76 | 68.94 | 10.18 | 0 |
| M#4 Rear Door | #1 | 94.43 | 103.66 | 9.77 | 0 | | #14 | 222.01 | 192.16 | 13.44 | 0 |
| | #2 | 100.60 | 90.23 | 10.31 | 0 | | #15 | 326.91 | 355.21 | 8.66 | 0 |
| | #3 | 97.51 | 86.13 | 11.67 | 0 | | #16 | 99.85 | 109.74 | 9.91 | 0 |
| M#3 Hood | #1 | 132.75 | 145.46 | 9.58 | 0 | | #17 | 265.83 | 244.01 | 8.21 | 0 |
| | #2 | 244.89 | 264.16 | 7.87 | 15.95 | | #18 | 376.46 | 344.23 | 8.56 | 0 |
| | #3 | 340.96 | 379.95 | 11.44 | 2.89 | | #19 | 328.76 | 360.36 | 9.61 | 0 |
| M#5 Door | #1 | 72.19 | 79.31 | 9.87 | 0 | | #20 | 370.89 | 402.18 | 8.44 | 0 |
| | #2 | 177.16 | 159.32 | 10.07 | 0 | | #21 | 268.97 | 301.27 | 12.01 | 0 |
| | #3 | 135.57 | 124.34 | 8.29 | 0 | | #22 | 312.28 | 285.94 | 8.43 | 0 |
| | #4 | 144.81 | 132.39 | 8.58 | 3.09 | | | | | | |

**Table A3.** Coordinates and results of the validation points for 0.55 mm thickness M#1-Hood (dimensions in mm).

| Model | P# | x | y | z | Maj. Curv. | Min. Curv. | Gauss. Curv. | Pred. Res. | FEA Res. |
|---|---|---|---|---|---|---|---|---|---|
| M#1 Hood | #1 | 750 | 450 | 177 | $-9.90 \times 10^{-3}$ | $-2.16 \times 10^{-5}$ | $2.14 \times 10^{-7}$ | S | S |
| | #2 | 700 | −200 | 187 | $-2.17 \times 10^{-4}$ | $-1.09 \times 10^{-4}$ | $2.36 \times 10^{-8}$ | H | H |
| | #3 | 450 | 0 | 466 | $-2.60 \times 10^{-4}$ | $-2.54 \times 10^{-4}$ | $6.62 \times 10^{-8}$ | H | H |
| | #4 | 250 | −600 | 43 | $-5.60 \times 10^{-4}$ | $3.79 \times 10^{-3}$ | $-2.12 \times 10^{-6}$ | N | N |
| | #5 | 550 | 650 | 121 | $-2.38 \times 10^{-3}$ | $-3.16 \times 10^{-4}$ | $7.53 \times 10^{-7}$ | H | H |
| | #6 | 700 | −700 | 137 | $-1.00 \times 10^{-2}$ | $-2.63 \times 10^{-4}$ | $2.64 \times 10^{-6}$ | H | H |
| | #7 | 850 | 750 | 137 | $-3.76 \times 10^{-4}$ | $4.88 \times 10^{-3}$ | $-1.84 \times 10^{-6}$ | S | N |
| | #8 | 250 | −450 | 68 | $-1.03 \times 10^{-3}$ | $-3.32 \times 10^{-4}$ | $3.43 \times 10^{-7}$ | N | N |
| | #9 | 450 | 550 | 112 | $-1.09 \times 10^{-3}$ | $-3.53 \times 10^{-4}$ | $3.84 \times 10^{-7}$ | H | H |
| | #10 | 450 | −450 | 114 | $-7.45 \times 10^{-4}$ | $-2.72 \times 10^{-4}$ | $2.02 \times 10^{-7}$ | S | S |
| | #11 | 650 | 500 | 146 | $-6.96 \times 10^{-4}$ | $-2.25 \times 10^{-4}$ | $1.56 \times 10^{-7}$ | N | N |
| | #12 | 900 | −550 | 175 | $-1.02 \times 10^{-4}$ | $5.83 \times 10^{-4}$ | $-5.96 \times 10^{-8}$ | H | H |
| | #13 | 250 | 350 | 73 | $-4.68 \times 10^{-4}$ | $8.77 \times 10^{-3}$ | $-4.10 \times 10^{-6}$ | N | S |
| | #14 | 600 | −400 | 148 | $-1.43 \times 10^{-2}$ | $-2.29 \times 10^{-4}$ | $3.27 \times 10^{-6}$ | N | N |
| | #15 | 200 | −200 | 82 | $-6.59 \times 10^{-4}$ | $-3.66 \times 10^{-4}$ | $2.41 \times 10^{-7}$ | H | H |
| | #16 | 500 | 200 | 147 | $-3.60 \times 10^{-4}$ | $-1.95 \times 10^{-4}$ | $7.02 \times 10^{-8}$ | N | N |
| | #17 | 250 | 0 | 102 | $-5.83 \times 10^{-4}$ | $-3.80 \times 10^{-4}$ | $2.22 \times 10^{-7}$ | H | H |
| | #18 | 850 | 0 | 466 | $-1.52 \times 10^{-4}$ | $-1.35 \times 10^{-4}$ | $2.05 \times 10^{-8}$ | N | N |

## References


1. Jung, D.W. A parametric study of sheet metal denting using a simplified design approach. *KSME Int. J.* **2002**, *16*, 1673–1686. [CrossRef]
2. Dicello, J.A.; George, R.A. Design criteria for the dent resistance of auto body panels. *SAE Tech. Pap.* **1974**, 389–397. [CrossRef]
3. Guo, M.; Hu, Y.; Sanghera, R. *Finite Element Analyses and Correlations on Oil Canning of a Door Outer Panel*; SAE Technical Paper; SAE International: Warrendale, PA, USA, 2009; pp. 1–6. [CrossRef]
4. Johnson, T.E.; Schaffnit, W.O. *Dent Resistance of Cold-Rolled Low-Carbon Steel Sheet*; SAE Technical Paper; SAE International: Warrendale, PA, USA, 1973; Volume 82, pp. 1719–1730. [CrossRef]
5. Lu, H.; Ma, M.; You, J.; Li, Z. Dent resistance for automobile body panels. *Chin. J. Mech. Eng.* **2009**, *22*, 903–911. [CrossRef]
6. Shih, H.C.; Horvath, C.D. *Effects of Material Bending and Hardening on Dynamic Dent Resistance*; SAE Technical Paper; SAE International: Warrendale, PA, USA, 2005. [CrossRef]
7. Holmberg, S.; Thilderkvist, P. Influence of material properties and stamping conditions on the stiffness and static dent resistance of automotive panels. *Mater. Des.* **2002**, *23*, 681–691. [CrossRef]
8. Ekstrand, G.; Asnafi, N. On testing of the stiffness and the dent resistance of autobody panels. *Mater. Des.* **1998**, *19*, 145–156. [CrossRef]
9. Asnafi, N. On strength, stiffness and dent resistance of car body panels. *J. Mater. Process. Tech.* **1995**, *49*, 13–31. [CrossRef]
10. Holmberg, S.; Nejabat, B. Numerical assessment of stiffness and dent properties of automotive exterior panels. *Mater. Des.* **2004**, *25*, 361–368. [CrossRef]
11. Shen, H.; Li, S.; Chen, G. Numerical analysis of panels' dent resistance considering the Bauschinger effect. *Mater. Des.* **2010**, *31*, 870–876. [CrossRef]
12. Shen, H.; Li, S.; Chen, G. Quantitative analysis of surface deflections in the automobile exterior panel based on a curvature-deviation method. *J. Mater. Process. Technol.* **2012**, *212*, 1548–1556. [CrossRef]
13. Park, C.D.; Chung, W.J.; Kim, B.M. A numerical and experimental study of surface deflections in automobile exterior panels. *J. Mater. Process. Technol.* **2007**, *187–188*, 99–102. [CrossRef]
14. Soltoggio, A.; Stanley, K.O.; Risi, S. Born to learn: The inspiration, progress, and future of evolved plastic artificial neural networks. *Neural Netw.* **2018**, *108*, 48–67. [CrossRef]
15. Patel, H.V.; Panda, A.; Kuipers, J.A.M.; Peters, E.A.J.F. Computing interface curvature from volume fractions: A machine learning approach. *Comput. Fluids* **2019**, *193*, 104263. [CrossRef]
16. Yusri, I.M.; Abdul Majeed, A.P.P.; Mamat, R.; Ghazali, M.F.; Awad, O.I.; Azmi, W.H. A review on the application of response surface method and artificial neural network in engine performance and exhaust emissions characteristics in alternative fuel. *Renew. Sustain. Energy Rev.* **2018**, *90*, 665–686. [CrossRef]
17. Schmidhuber, J. Deep Learning in neural networks: An overview. *Neural Netw.* **2015**, *61*, 85–117. [CrossRef] [PubMed]
18. Winiczenko, R. Effect of friction welding parameters on the tensile strength and microstructural properties of dissimilar AISI 1020-ASTM A536 joints. *Int. J. Adv. Manuf. Technol.* **2016**. [CrossRef]
19. Mirandola, I.; Berti, G.A.; Caracciolo, R.; Lee, S.; Kim, N.; Quagliato, L. Machine learning-based models for the estimation of the energy consumption in metal forming processes. *Metals* **2021**, *11*, 833. [CrossRef]

20. Feng, S.; Zhou, H.; Dong, H. Using deep neural network with small dataset to predict material defects. *Mater. Des.* **2019**, *162*, 300–310. [CrossRef]
21. Kim, H.; Al-Saeedi, S.; Jang, C.; Quagliato, L.; Kim, N. Development of an index model for oil canning of steel sheet metal forming products. *Int. J. Adv. Manuf. Technol.* **2018**. [CrossRef]
22. Max, N. Weights for Computing Vertex Normals from Facet Normals. *Graph. Tools—Jgt Ed. Choice* **2005**, 75–81. [CrossRef]
23. Rusinkiewicz, S. Estimating curvatures and their derivatives on triangle meshes. In Proceedings of the Second International Symposium on 3D Data Processing, Visualization, and Transmission: 3DPVT 2004, Thessaloniki, Greece, 6–9 September 2004; pp. 486–493. [CrossRef]
24. Rumelhart, D.E.; Hinton, G.E.; Williams, R.J. Learning representations by back-propagating errors. *Nature* **1986**, *323*, 533–536. [CrossRef]
25. Cherry, J.M.; Adler, C.; Ball, C.; Chervitz, S.A.; Dwight, S.S.; Hester, E.T.; Jia, Y.; Juvik, G.; Roe, T.; Schroeder, M.; et al. SGD: Saccharomyces genome database. *Nucleic Acids Res.* **1998**, *26*, 73–79. [CrossRef]
26. Kingma, D.P.; Ba, J.L. Adam: A method for stochastic optimization. In Proceedings of the 3rd International Conference on Learning Representations, ICLR 2015, San Diego, CA, USA, 7–9 May 2015; pp. 1–15.
27. Krizhevsky, A.; Sutskever, I.; Hinton, G.E. ImageNet classification with deep convolutional neural networks. *Adv. Neural Inf. Process. Syst.* **2012**, *2*, 1097–1105. [CrossRef]
28. Simonyan, K.; Zisserman, A. Very deep convolutional networks for large-scale image recognition. In Proceedings of the 3rd International Conference on Learning Representations, ICLR 2015, San Diego, CA, USA, 7–9 May 2015; pp. 1–14.
29. Yosinski, J.; Clune, J.; Bengio, Y.; Lipson, H. How transferable are features in deep neural networks? *Adv. Neural Inf. Process. Syst.* **2014**, *4*, 3320–3328.
30. Vogado, L.H.S.; Veras, R.M.S.; Araujo, F.H.D.; Silva, R.R.V.; Aires, K.R.T. Leukemia diagnosis in blood slides using transfer learning in CNNs and SVM for classification. *Eng. Appl. Artif. Intell.* **2018**, *72*, 415–422. [CrossRef]
31. Srivastava, N.; Hinton, G.; Krizhevsky, A.; Sutskever, I.; Salakhutdinov, R. Dropout: A simple way to prevent neural networks from overfitting. *J. Mach. Learn. Res.* **2014**, *15*, 1929–1958.

metals

MDPI

*Article*

# Novel Prediction Model for Steel Mechanical Properties with MSVR Based on MIC and Complex Network Clustering

**Yuchun Wu, Yifan Yan and Zhimin Lv ***

Collaborative Innovation Center of Steel Technology, University of Science and Technology Beijing, Beijing 100083, China; s20191303@xs.ustb.edu.cn (Y.W.); b20170502@xs.ustb.edu.cn (Y.Y.)
* Correspondence: lvzhimin@nercar.ustb.edu.cn

**Abstract:** Traditional mechanical properties prediction models are mostly based on experience and mechanism, which neglect the linear and nonlinear relationships between process parameters. Aiming at the high-dimensional data collected in the complex industrial process of steel production, a new prediction model is proposed. The multidimensional support vector regression (MSVR)-based model is combined with the feature selection method, which involves maximum information coefficient (MIC) correlation characterization and complex network clustering. Firstly, MIC is used to measure the correlation between process parameters and mechanical properties, based on which a complex network is constructed and hierarchical clustering is performed. Secondly, we evaluate all parameters and select a representative one for each partition as the input of the subsequent model based on the centrality and influence indicators. Finally, an actual steel production case is used to train the MSVR prediction model. The prediction results show that our proposed framework can capture effective features from the full parameters in terms of higher prediction accuracy and is less time-consuming compared with the Pearson-based subset, full-parameter subset, and empirical subset input. The feature selection method based on MIC can dig out some nonlinear relationships which cannot be found by Pearson coefficient.

**Keywords:** mechanical properties prediction; high-dimensional data; feature selection; maximum information coefficient; complex network clustering

**Citation:** Wu, Y.; Yan, Y.; Lv, Z. Novel Prediction Model for Steel Mechanical Properties with MSVR Based on MIC and Complex Network Clustering. *Metals* **2021**, *11*, 747. https://doi.org/10.3390/met11050747

Academic Editor: Pedro Prates

Received: 6 April 2021
Accepted: 28 April 2021
Published: 1 May 2021



## 1. Introduction

The level of steel industry is an important indicator to measure the industrialization of the country. At present, all walks of life have more and more stringent requirements for iron and steel products. The mechanical properties of steel can often mean the difference between a long, efficient life in the most abrasive and wear-intensive applications, and frequent or even catastrophic failure. Understanding these properties is absolutely important because all production activities are ultimately to satisfy the actual quality requirements. To maintain and improve the product quality, energy efficiency, and economic profits, the quality prediction and control based on some mechanical properties are essential and have been investigated quite extensively in recent years [1]. Among numerous indicators, tensile strength, yield strength, and elongation are the most commonly used measurements for product's mechanical property, which are affected by a variety of comprehensive factors [2]. However, the production process of steel products contains complex physical and chemical changes with intricate technological processes, which means that property prediction and control have always been a difficult problem in the metallurgical industry. In the traditional practice, property prediction depends on the experience and destructive test, which are costly, time-consuming, and laborious. If the prediction could consider the relevant process parameters, and accordingly optimize the metal composition and process technology, it can greatly reduce the testing time and improve the production efficiency of iron and steel enterprises. Based on this idea, two main methods for that are the empirical and statistical models respectively, while the prediction accuracy can still be improved. This is primarily because these methods mostly depend on the experience and mechanism and ignore the value of data. The parameters selected are too few to represent the actual situation.

By contrast, data-driven prediction methods do not require deep understanding of mechanism but depend on the collected process data only [3]. With the application of big data platform in the steel industry, the real-time data of the throughout production process can be easily obtained, and the number and dimension of samples increase explosively. High-dimensional data have high value theoretically, but they will greatly increase the complexity of modeling and bring the curse of dimensionality [4]. Owing to the process complexity and intricate variable interactions, the major problem is that the nonlinearity and coupling between variables restrict the choice of prediction models and methods. Therefore, how to extract knowledge from the throughout process data, select effective features from the full-parameter set and ultimately establish a more accurate performance prediction model is our main job.

As a dimensionality reduction method, feature selection aims to select the most representative feature subset from the original data set [5], which mainly involves two steps: feature subset selection and feature subset evaluation. The prevailing approaches to feature selection fall into three categories: (a) filters, (b) wrapper methods, and (c) embedded methods. Filter is a single feature selection process, which is independent of subsequent learners. It usually ranks features in the parameter space to obtain subsets [6]. The performance evaluation of the learners works as the evaluation criteria of the wrapper methods. As a representative, Las Vegas Wrapper (LVW) method uses random strategy to search feature subset, and takes the error of the final classifier as the subset evaluation standard [7,8]. Embedded methods integrate feature selection and training learners, which automatically select features during learner training [9]. Moreover, some dimension reduction methods such as principal component analysis (PCA), singular value decomposition (SVD), linear discriminant analysis (LDA), and ISOMAP algorithm can also be regarded as feature selection methods for special basic data [10–13]. However, such methods do not consider the correlation and redundancy between attributes before and after dimensionality reduction, and the results are lack of interpretability.

In fact, when analyzing the relationship between the high-dimensional variables, a variety of distance and similarity indicators can be used to measure the correlation and redundancy between attributes, such as distance, information gain, mutual information, dependency, consistency, etc. Meanwhile, the higher the correlation between attributes, the stronger the necessity and operability of feature selection. Narayana designed an artificial neural network (ANN) model to correlate the complex relations among composition, temperature, and mechanical properties of steels. The ANN predictions are more accurate with experimental results as compared with the calculated properties of the existing model [14,15]. Some studies improve the performance of feature selection by choosing effective measurement indicators [16,17]. Nevertheless, many indicators such as Pearson coefficient, maximum information compression coefficient, and least squares regression error can only measure the linear relationship between features but not the nonlinear relationship. On the basis of information theory, Reshef proposed the concept of maximum information coefficient (MIC), which can widely measure the linear and nonlinear correlation between features and capture many functional or nonfunctional relationships [18]. Moreover, it is confirmed that MIC can accurately measure the correlation between attributes in large data sets. In addition, an intelligent MIC is presented to quickly approach the optimal value [19].

Clustering methods can be used for feature selection, which divide all the nodes in the network into several discrete subgroups based on the correlation metrics [20]. In the theory of complex network, the reason why an actor has power is because of its relationship with other actors. Therefore, it can be considered that the nodes in one cluster have similar "power" or "importance", and nodes with most centrality can be selected as the representation of each partition. If we regard all process parameters as a whole node set, the process of feature selection can be implemented as follows: (a) clustering all the process parameters and (b) selecting representative ones for each group. Some researchers have explored the centrality and influence indicators in complex networks to reflect the importance of nodes in the network [21]. The patterns among nodes, including the differences and connections, can also be studied to find the key network participants [22]. However, the key parameters selected based on experience virtually ignore the parameter interactions such as the similarity between them and their importance in the network. Moreover, many feature extraction methods transform the original data set to another by recombining existing features into new features, which may destroy the original physical structure of data and cause the new features to lose their physical meaning. Therefore, based on the characteristics of the steel product data set, all variables can be clustered according to the correlation coefficient, and the relationships between them can be measured by the centrality and influence indicators, so as to complete the feature selection and obtain the input parameters for the subsequent learners.

With the continuous development of data mining technology, artificial intelligence methods such as neural network [23], fuzzy control [24], and expert system [25] have become more and more popular. Among them, support vector machine (SVM) is an efficient learning machine based on statistical learning theory and structural risk minimization principle proposed by Vapnik. It can deal with problems with multiple input and single output. However, problems in the steel production process often have multiple outputs which are not mutually independent. If multiple support vector machine regression (SVR) algorithms are used to estimate multiple output functions, each sample point cannot be treated equally, so the accuracy is poor. Therefore, in order to improve the accuracy of estimation and reduce the computational workload of multidimensional regression problems, multi-output support vector machine regression (MSVR) can be used for performance prediction for the steel products [26].

Motivated by the above considerations, we propose a novel prediction model for steel mechanical properties, with MSVR based on MIC and complex network clustering. In our model, we measure the correlation between features with MIC, employ hierarchical clustering analysis based on the complex network theory, quantitatively evaluate each feature by centrality and influence indicators, then choose a feature subset as a parameter input which could represent a large amount of information. The MSVR is used to predict the mechanical properties and its accuracy can verify our proposed framework. By the case analysis of the practical steel production data in a steel company in Central China, we compare our method with the full-parameter subset input, empirical subset input, and Pearson-based subset input. It turns out that our scheme has the lowest computational complexity and the highest prediction accuracy.

The remaining sections of this article are organized as follows: preliminaries about the correlation evaluation index, theory of complex network, and the performance prediction model are briefly introduced in Section 2; in Section 3, the detailed development of the proposed novel prediction model with MSVR based on MIC and complex network clustering is presented; in Section 4, an actual case of steel production is studied and the comparison analyses of prediction results are provided; and Section 5 gives conclusions.

## 2. Preliminaries

### 2.1. Correlation Analysis Methods

Correlation analysis is a basic issue in statistics that aims to quantify the association between two variables from limited data, which can be divided into linear and nonlinear. Linear correlation refers to the case that the output and input are in positive proportion or inverse proportion. When two variables share a linear relationship, the Pearson correlation is the standard measure of dependence, while it is not applicable when relationships are highly nonlinear. The nonlinear correlation is more complex and may be formed by the superposition of a variety of complex functional relationships. Therefore, it is natural to ask how to measure statistical correlation in a way that treats relationships of different types equally.

As is well known, mutual information (MI) is already widely employed to quantify associations no matter what relationship types [27]. Even though it was proposed in the communications systems, MI has been repeatedly proved to be applicable in various statistical problems. In units known as "bits", MI strictly determines how much information one variable reveals about another. The MI between two random variables $X$ and $Y$ is defined in terms of their joint probability distribution $p(X,Y)$ as

$$I(X,Y) = \iint p(x,y) log_2 \frac{p(x,y)}{p(x)p(y)} dxdy \quad (1)$$

On the basis of MI, Reshef et al. proposed the concept of maximal information coefficient (MIC), a statistic measure other than a dependence one [18]. Compared with MI, *MIC* captures a wider range of associations both functional and not. In principle, *MIC* is based on the idea that if there is a certain relationship between two variables, a grid can be drawn on the scatter diagram of the two variables and the data can be partitioned to encapsulate this relationship. Indeed, to calculate the *MIC* of two variables, explore all grids at the maximum resolution and calculate the largest possible mutual information. Therefore, the heart of *MIC* is a naive mutual information estimate $I(x,y)$ computed using a data-dependent grid scheme. Let $x$ and $y$ respectively denote the number of bins imposed on the $x$ and $y$ axes. The *MIC* grid scheme is chosen so that (i) the total number of bins $xy$ does not exceed some user-specified value $B$ and (ii) the value of the ratio where $Z = log_2(min(x,y))$ is maximized.

The ratio computed using this data-dependent grid scheme is how MIC is defined.

$$MIC(X,Y) = max\left\{\frac{I(x,y)}{Z}\right\} = \frac{I(x,y)}{log_2(min(x,y))}, x \times y < n^6 \quad (2)$$

Note that $B = n^6$. $MIC(X,Y)$ is always nonnegative and $MIC(X,Y) = 0$ only when $X$ and $Y$ are mutually independent. Besides, *MIC* values will be greater than zero when $X$ and $Y$ show any correlations, regardless of how nonlinear that relationship is. Moreover, the stronger the correlation is, the larger the value of $MIC(X,Y)$.

### 2.2. Complex Network Theory

A network consists of nodes that represent individual entities and links between each other. Actually, whether you realize it or not, we are surrounded by all kinds of networks, including transportation networks, social networks, and manufacturing networks; building networks are a good way of modeling. Based on the findings that a scale-free network has the outstanding features of strong connectivity and survivability, Barabâsi and Albert have further developed for network science a tool called complex network theory to study the topology for networks [28]. We have noticed that increasing network sizes and nontrivial topological structures concur with the increasing richness and variety of attribute information associated with the nodes in network.

Complex network is a kind of abstract model which maps the real complex system. It abstracts the entities in the complex system into nodes and the relationships between entities into lines. It can be divided into weighted network and unweighted network. The former has a binary nature where the edges between nodes are either present or not, while the latter displays a large heterogeneity in the capacity and the intensity of the connections. The adjacency matrix is a binary square matrix with the same row and column label, which is commonly used to represent the actual relationships and construct a complex network. Complex network theory is widely used to study the characteristics of various networks and further improve the network performance. The relationship between nodes in the network can be quantitatively studied by centrality analysis, binary relationship research, block-modeling analysis, and cohesive subgroup analysis, etc. [29,30].

### 2.2.1. Complex Network Clustering

Clustering, also known as transitivity, is a typical property of complex networks, where two nodes associated with a common node are likely to be similar. White et al. (1976) proposed the block-modeling theory [31], which can simplify the complex network according to the degree of associations between nodes. Specifically, the nodes are rearranged into blocks by clustering, and the basic characteristics of the whole network can be reflected by each block. Recently, some scholars combined the stochastic block model with clustering to define the relationship between nodes and find subgroups [32,33].

In particular, the first step of block-modeling is to partition the actors, that is, to divide them into different groups based on methods of clustering and scaling. In particular, the Convergent Correlation (CONCOR) procedure is a method of hierarchical clustering for relational data which begins by forming a new square matrix of product–moment correlations between the columns (or rows) of the original data and is found to give results that are highly compatible with analyses and interpretations of the same data using the block-modeling approach [34]. CONCOR is an iterative convergence algorithm, which measures the network structure by repeatedly calculating the correlation matrix. Each iteration of CONCOR contains a hierarchical clustering to achieve partition. According to the correlation matrix between nodes, the data set is divided into different levels and can obtain the tree clustering structure. CONCOR is an iterative convergence algorithm, which measures the network structure by repeatedly calculating the correlation matrix. Each iteration of CONCOR contains a hierarchical clustering to achieve partitions.

The purpose of complex network clustering is to find the subgroups existing in the whole network. According to the correlation, the nodes with high degree of similarity are automatically clustered into one group. Selecting the representative nodes for each group based on the importance and power indicators and eventually forming a representative node set will be better than picking up typical nodes in the whole network. The partition process of the block model is shown in Figure 1, where we can see that several scattered nodes are divided into 16 clusters according to their similarity. The similarity between nodes in one cluster is high, and the importance of each node can be evaluated.

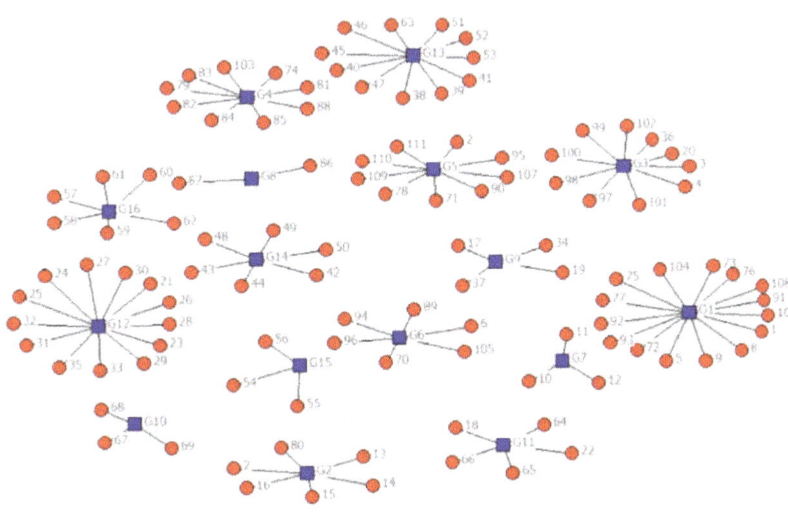

**Figure 1.** Partition example of block model.

### 2.2.2. Centrality Evaluation of Nodes in the Complex Network

In the complex network, how to judge the power and importance of each node mainly depends on its centrality and influence. Based on the actual relationship data, we measure the "power and status" of nodes by the following four commonly used indicators, namely degree, closeness, betweenness, and katz.

Degree Centrality

Degree centrality is defined as the number of links incident upon a node. If the network is directed, then two separate measures of degree centrality are defined, namely, in-degree and out-degree. In-degree is a count of the number of ties directed to the node and out-degree is the number of ties that the node directs to others. In many cases, the degree is the sum of in-degree and out-degree. This index reflects the "power" of a node in the network and nodes with high degree are more likely to be the center of the network.

Betweenness Centrality

Betweenness centrality is a way of detecting the amount of influence a node has over the flow of information in a graph. It is often used to find nodes that serve as a bridge from one part of a graph to another. For every pair of vertices in a connected graph, there exists at least one shortest path between the vertices such that either the number of edges that the path passes through (for unweighted graphs) or the sum of the weights of the edges (for weighted graphs) is minimized. The betweenness centrality for each vertex is the number of these shortest paths that pass through the vertex. This index measures the ability of controlling the resources of each actor. If an actor is on the shortest path of many other actor-pairs, its degree is generally low, but it may play an intermediary role so as to be the center of the network.

Closeness Centrality

Closeness centrality is a way of detecting nodes that are able to spread information very efficiently through a graph. The closeness centrality of a node measures its average farness to all other nodes. Nodes with a high closeness score have the shortest distances to all other nodes. This index reflects the inverse distance of nodes to other points. If an actor is closer to other actors, it is easier to transmit information; therefore, it is more likely to be the center of the network.

Katz Centrality

In graph theory, the Katz centrality is used to measure the relative degree of influence of an actor within a social network. Unlike typical centrality measures which consider only the shortest path between a pair of actors, Katz centrality measures influence by taking into account the total number of walks between a pair of actors. Katz centrality computes the relative influence of a node within a network by measuring the number of the immediate neighbors and also all other nodes in the network that connect to the node under consideration through these immediate neighbors. This index considers the direct and indirect relationship between node and other nodes. The shorter the distance between node $i$ and node $j$, the greater the impact of node $i$ on node $j$.

### *2.3. Mechanical Property Prediction Model*

#### 2.3.1. Support Vector Regression

In contrast to simple linear regression, SVR gives us the flexibility to define how much error is acceptable in our model and will find an appropriate line (or hyperplane in higher dimensions) to fit the data. Specifically, set a threshold $\alpha$ and calculate the loss of data points when $|f(x) - y| > \alpha$, supposing that the data points within the threshold are predicted accurately. One of the main advantages of SVR is that its computational complexity does not depend on the dimensionality of the input space. Additionally, it has excellent generalization capability, with high prediction accuracy.

The objective function of SVR is to minimize the coefficients—more specifically, the error term, which is instead handled in the constraints, where we set the absolute error less than or equal to a specified margin, called the maximum error, ϵ (epsilon). We can tune epsilon to gain the desired accuracy of our model.

Suppose that $x \in \mathbb{R}^d$, $y_i \in \mathbb{R}$, $y_i$ is the output of $x_i$, $d$ is the dimension, and $l$ is the number of samples. Given the training set $\{(x_i, y_i)\}_{i=1}^{l}$, the goal of SVR is to find an optimal equation $f$ from the set of hypothesis equations by minimizing the error term. The optimal equation $f$ is as follows, where $w$ is the weight vector and $b$ is the threshold.

$$\left\{f \middle| f(x) = w^T x + b, w \in \mathbb{R}^d, b \in \mathbb{R}\right\} \quad (3)$$

2.3.2. Multidimensional Support Vector Regression

Assume $Y = \{y_1, y_2, y_3 \ldots\}$ is the quality index set of steel products and $X = \{X^B, X^C, X^R\}$ is the process parameter set from three stages: smelting, continuous casting, and rolling. Each stage consists of many specific process parameters, such as $X^B = \{x_1^B, x_2^B, \ldots, x_a^B\}$, which means the number of variables in the steelmaking stage is $a$. The mean absolute percentage error (MAPE) can be set as the algorithm evaluation index, and the quality modeling considering the effect of process parameters on quality index can be abstracted as:

$$Y^T = \{y_1, y_2, y_3 \ldots\}^T \approx f\left(X^B, X^C, X^R\right) \quad (4)$$

## 3. Mechanical Prediction Model with MSVR Based on MIC and Complex Network Clustering

### 3.1. Problem Description

The data collected from the intricate process of steel production are high-dimensional and coupled with each other. There are complex linear or nonlinear relationships between them; meanwhile, their impact on product quality is hereditary. If we use the full-parameter data to model, not only is the calculation complex, but also the modeling is often inefficient and cannot well reflect the real problem because of the redundant features. If the important and representative features can be selected from the high-dimensional process data to simplify the complex problem, the subsequent modeling will be simpler and the effect will be more obvious. The emphasis of this paper is how to select the representative feature subset from the full-feature set, and then predict the performance of steel products more accurately.

Taking the throughout process of steel production as an example, after cleaning and deduplication of the original data set O, which means removing the parameters that are completely irrelevant to the mechanical properties, namely the MIC between them being less than 0.05. The left process parameters from three typical stages, namely, steelmaking, continuous casting, and rolling are defined as $F = \{X^B, X^C, X^R\} = \{x_1, x_2, x_3 \ldots, x_m\}$ and the number set of each stage is {a, b, c}, which means that the number of total parameters is $(a + b + c = m)$ [35]. Define the mechanical property set $Y = \{y_1, y_2, y_3\}$ which contains three indicators: tensile strength, yield strength, and elongation. The purpose of this study is to use a certain feature selection method to obtain representative and low-dimensional feature subset $X = \{x_1, x_2, x_3 \ldots, x_t\}, t \ll a + b + c$ from high-dimensional variable set $\{X^B, X^C, X^R\}$, and perform the subsequent MSVR performance prediction modeling $Y^T = f(x_1, x_2, x_3 \ldots, x_t)$, which could effectively simplify the calculation and improve prediction accuracy at the same time.

### 3.2. Model and Algorithm

Based on the relevant basic theories in Section 2 and the requirements in Section 3.1, we propose an algorithm that firstly uses MIC to measure the linear and nonlinear correlation relationships between high-dimensional parameters. Secondly, we construct a complex network, and quantitatively evaluate each feature by CONCOR clustering method and centrality and influence analysis. Eventually, we could obtain the feature subset that could represent the full parameters efficiently, which could be used as the input of MSVR and to predict the mechanical properties. In order to verify the effectiveness and feasibility of the algorithm, the full-parameter set, empirical subset, and the best feature subsets selected based on MIC and Pearson coefficients are used as input for MSVR respectively, and the method with the least error and the optimal feature subset could be obtained.

The model and algorithm of this paper can be divided into two parts. One is the prediction model based on MSVR, the other is the feature selection algorithm based on correlation measurement and complex network, as shown in Figure 2.

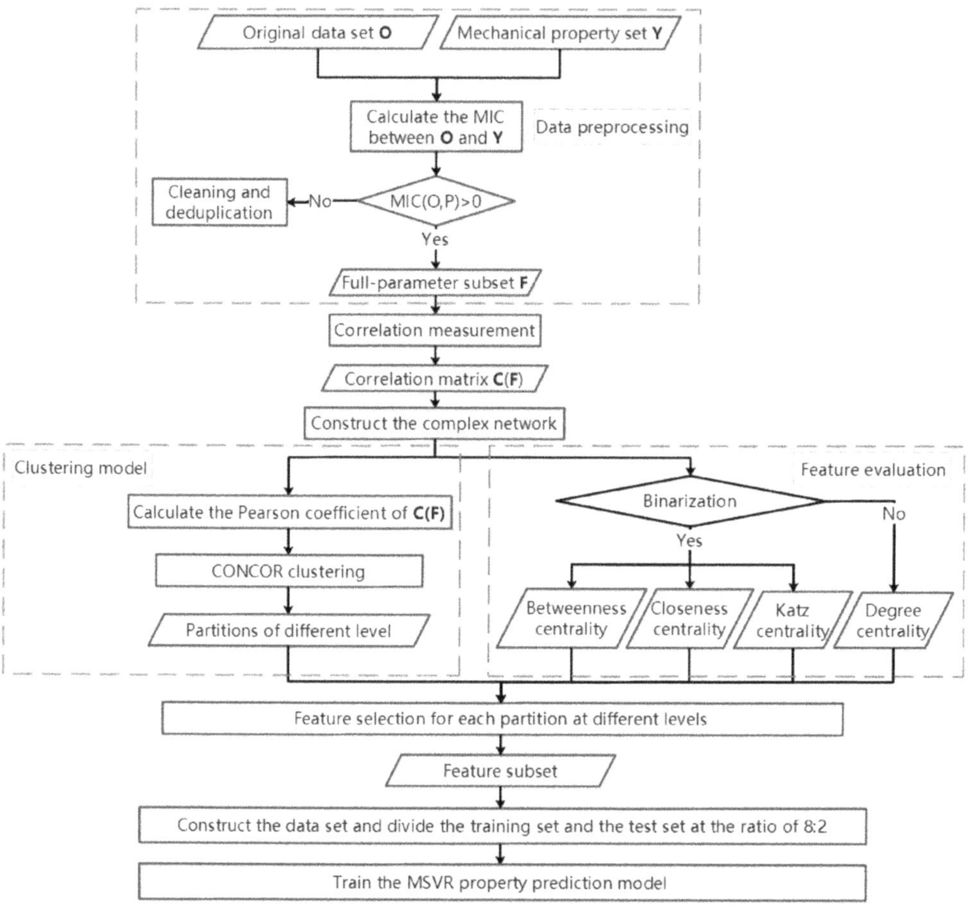


**Figure 2.** The prediction model for steel mechanical properties with multidimensional support vector regression (MSVR) based on maximum information coefficient (MIC) and complex network clustering.

### 3.2.1. Correlation Measurement

Suppose that $C(x_i, x_j)$ is the correlation coefficient between $x_i$ and $x_j$. In this paper, MIC is used to measure the linear and nonlinear correlation between attributes. In order to verify the representation effect of MIC, the Pearson coefficient between attributes is also calculated for modeling.

Create the correlation matrix C by the correlation coefficient between features, and construct the complex network that characterizes the correlation between features. This matrix is a symmetric matrix with diagonal 1.

$$C = \begin{bmatrix} 1 & C(x_1,\ x_2) & C(x_1,\ x_3) & \cdots & C(x_1,\ x_m) \\ C(x_2,\ x_1) & 1 & \cdots & \cdots & C(x_2,\ x_m) \\ C(x_3,\ x_1) & \vdots & \ddots & \cdots & \vdots \\ \cdots & \cdots & \cdots & 1 & \cdots \\ C(x_m,\ x_1) & C(x_m,\ x_2) & \cdots & \cdots & 1 \end{bmatrix} \tag{5}$$

3.2.2. The Clustering Model Based on the Complex Network and CONCOR Algorithm

A complex network is constructed based on the correlation matrix C and the CONCOR algorithm is employed to build the block model. The CONCOR algorithm calculates the Pearson correlation coefficient of the correlation matrix iteratively and carries out the hierarchical clustering, starting from the initial correlation matrix. The flow of the algorithm is shown in Algorithm 1. After the CONCOR, the partition of features is realized. Define the subgroup as $G = \{g_1, g_2, \ldots, g_t\}$ where $t$ is the number of subgroups, and $g_i = \{x_1, x_2, \ldots, x_j\}, i \leq t, j \ll a + b + c$, where $j$ is the number of features in the subgroup $g_i$.

**Algorithm 1.** The clustering model based on CONCOR.

Input : correlation matric C1 and the partition level at which any pair of actors is aggregated.
Output : C2, which denotes the correlation coefficient matrix of C1 and blocks represented in terms of a clustering dendrogram clustering graph under different levels.
Step 1 : Calculate C2 which is the Pearson correlation coefficient of C1.
Step 2 : The blocks are given for each level at which any pair of actors is aggregated. Carry out the hierarchical clustering from the max level, and combine two features with the highest similarity. The similarities of partitions from the same level should all reach one corresponding value, and one feature can only exist in one group.
Step 3: Reduce level by 1, which means reducing the corresponding similarity value of clusters, and look for the features with highest similarity to the clustered partitions from the unclustered features, which could cluster by themselves, or be added into the existing partition.
Step 4: Iterate Step 3 until level = 1 when all features enter the same group.

3.2.3. Feature Evaluation

Given the different partitions of different levels and the similarity complex network, we comprehensively evaluate the nodes in each subgroup for feature selection with four centrality and influence indicators that we mentioned before. What should be pointed out is that the measure of degree centrality is based on the weighted matrix which is the initial correlation matrix, while the measures of betweenness, closeness, and Katz centrality are based on the unweighted matrix which is the binarization of initial correlation matrix. Suppose that $n$ denotes the number of nodes in the network.

Degree Centrality

The absolute degree $C_{AD}(x)$ is the sum of the weights between node $x$ and all other nodes, and the relative degree $C_{RD}(x_i)$ is the absolute centrality divided by the maximum possible degree $(n-1)$.

$$C_{RD}(x_i) = \frac{\sum_j C(x_i,\ x_j)}{n-1} \tag{6}$$

Betweenness Centrality

Define $g_{jk}$ as the geodesic between node $j$ and $k$, and $g_{jk}(x_i)$ as the number of geodesics that go through the node $x_i$. The absolute betweenness centrality $C_{AB}(x_i)$ is the sum of the probabilities that node $x_i$ is on the shortest path between all pairs of points. The relative betweenness $C_{RB}(x_i)$ is the absolute betweenness divided by the maximum possible betweenness $(n^2 - 3n + 2)/2$.

$$C_{AB}(x_i) = 2\sum \frac{g_{jk}(x_i)}{g_{jk}} / \left(n^2 - 3n + 2\right) \quad (7)$$

Closeness Centrality

Define $Farnessx_i$ as the sum of the geodesic distances between node $x_i$ and all other nodes, $d_{ij}$ as the geodesic distances between node $x_i$ and $x_j$, and the absolute closeness centrality $C_{APi}$ is the reciprocal of $Farnessx_i$. The relative closeness centrality $C_{RP}(x_i)$ is $C_{AP}(x_i)$ divided by the maximum possible closeness $1/(n-1)$.

$$C_{RP}(x_i) = \frac{\frac{1}{Farness(x_i)}}{\frac{1}{n-1}} = \frac{n-1}{Farness(x_i)} = \frac{n-1}{\sum_{j=1}^{n} d_{ij}} \quad (8)$$

Katz Centrality

Katz centrality measures the influence by considering the direct and indirect support or attention between nodes. Define $S$ as a matrix consisting of 0 and 1 that reflects the direct-connection relationships between actors when the path length is 1, and $S_{ij} = 1$ denotes that the actor $j$ connects to actor $i$ directly and the length is 1. The sum of $j$-column represents the total number of times that actor $j$ connects to other actors by 1; define ${S_{ij}}^2$ as the number of paths that connect the actor $i$ and $j$ by length 2 and ${S_{ij}}^3$ by length 3, and so on. Considering that the higher the power of the matrix ${S_{ij}}^*$, the lower the effect of the influence, so an attenuation factor $\alpha$ is introduced to characterize this performance. The value of $\alpha$ depends on the situation and $1/a \in (b, 2b)$. When $\alpha$ = 0, it decays completely and when $\alpha$ = 1, it does not. For a matrix where the elements are nonnegative, $a$ simple upper limit of the maximum eigenvalue $b$ is the maximum sum of rows.

Define $P$ = [Degree, Betweenness, Closeness, Katz]. In order to eliminate the influence of dimension, we sort the four indicator values and get four ranking values to measure their comprehensive centrality and influence. Define $R = [R_D, R_B, R_C, R_K, R_T]$ where $R_D, R_B, R_C, R_K, R_T$ represent the ranking values of four centrality indicators and the total ranking respectively.

$$R_T = R_D + R_B + R_C + R_K \quad (9)$$

3.2.4. Feature Selection

Suppose that $R_T^i = \left\{{R_T^i}^1, {R_T^i}^2 \ldots {R_T^i}^p\right\}$ is the total ranking matrix of the features in the subgroup $g_i$, where $p$ denotes the feature number of $g_i$. Select the feature with the top total ranking as the subgroup representation, namely ${R_T^i}^q = minR_T^i$. In this way, explore all subgroups and obtain the feature set $\{x_1, x_2, x_3 \ldots, x_t\}$, where t is the number of subgroups.

3.2.5. Mechanical Property Prediction Based on MSVR

The above work can obtain the feature selection results respectively based on the MIC and Pearson correlation characterization. Moreover, in order to verify the effect of our proposed method, the empirical subset and full-parameter subset are used for comparative experiments. Applying the above four feature sets to construct the data set for MSVR modeling, we divide the training set and the test set, and perform cross validation test to verify the error. It should be pointed out that even though we are using the same correlation characterization, different partition levels get different feature selection results, corresponding to different MSVR prediction results.

## 4. Case Study and Discussion

In order to test the feasibility and efficiency of the proposed prediction model, we collected a total of 1607 data samples of the whole production process from a steel company in Central China and verified our model. The product is the cold-rolled strip and the steel grades selected in our experiment include DR01, DR02, DR04, DR06, DX51, DX52, DX53, SPCC, SPCD, SPCE, SPCF, SPCG, etc. The data come from four main processes: smelting, continuous casting, hot rolling, and cold rolling. The original parameters influence each other and contain a lot of linear and nonlinear relationships, of which the number is 211. The deduplication process is described as follows: calculate the MIC values between the original parameters and three mechanical properties, and remove the ones completely irrelevant to properties, namely the MIC between them is less than 0.05. Finally, a total of 111 process parameters were obtained as the full-parameter subset. The number of parameters in each process stage is shown in Table 1. The exactly defined chemical compositions include C, Si, Mn, P, S, Ni, CR, Cu, ALS, ALT, AS, B, MO, N, NB, PB, SN, TI, which are all collected in the smelting stage. Among the 1607 data samples, the average, maximum value, and variance of the chemical composition contents are listed in Table 2.

**Table 1.** Number of parameters in each process stage.

| Process Stage | Number of Process Parameters |
|---|---|
| Smelting | 18 |
| Continuous casting | 32 |
| Hot rolling | 42 |
| Cold rolling | 19 |
| Sum | 111 |

**Table 2.** The average and variance of the chemical composition contents.

| Composition | Maximum | Average | Variance | Composition | Maximum | Average | Variance |
|---|---|---|---|---|---|---|---|
| C | 0.203 | 0.011914 | 0.020479 | ALT | 0.245 | 0.052757 | 0.028082 |
| Si | 0.1299 | 0.008354 | 0.006990 | AS | 0.0245 | 0.005758 | 0.002594 |
| Mn | 0.8415 | 0.170039 | 0.081668 | B | 0.0009 | 0.000213 | 0.000183 |
| P | 0.071 | 0.015293 | 0.006540 | MO | 0.0089 | 0.001468 | 0.000834 |
| S | 0.0335 | 0.008336 | 0.004066 | N | 0.0074 | 0.001969 | 0.001137 |
| Ni | 0.038 | 0.00974 | 0.004460 | NB | 0.0374 | 0.000866 | 0.001014 |
| CR | 0.0684 | 0.020961 | 0.010005 | PB | 0.0025 | 0.000771 | 0.000296 |
| Cu | 0.084 | 0.019762 | 0.010511 | SN | 0.0192 | 0.00617 | 0.002472 |
| ALS | 0.2235 | 0.050082 | 0.026820 | TI | 0.3064 | 0.042642 | 0.042190 |

### *4.1. Correlation Calculation and Partition Results*

The distribution of MIC values among the 111 process parameters is shown in Figure 3. It can be seen that nearly 50% of MIC values are greater than 0.43 and 34 values are more than 0.8, which indicates that there are indispensable correlation relationships between these features. It is necessary to mine these relationships and remove redundant features, so as to clarify the nature of the relationships between features and simplify the input data set of subsequent modeling.

We construct a complex network based on the MIC matrix, and carry out the CONCOR to build a block model. Set the initial clustering level as 4, and Figure 4 shows the number of partitions under different clustering levels. It can be seen that the number of partitions gradually increases with the rise of clustering level, and the clustering stops when the clustering level is 9, meanwhile the number of partitions is the maximum, 71.

Combined with the partition results shown in Figure 5 of which the clustering level is 4 to 9 respectively, it can be seen that the higher the level, the more partitions. This is because the next level of clustering is based on the previous level, which means expanding the feature numbers within a partition by reducing the similarity of the group, so the number of partitions will decrease. The first clustering level is 9, then the next clustering is based on level 9 which expands the members of each group and reduces the partition number. When the clustering level is 1, all features are in the same partition.

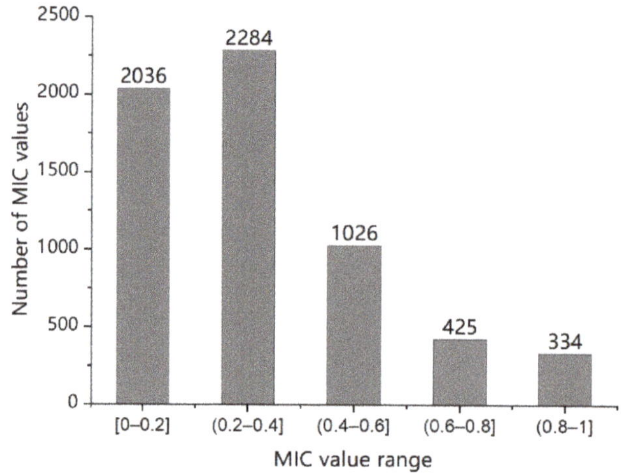


**Figure 3.** MIC value distribution among full process parameters.

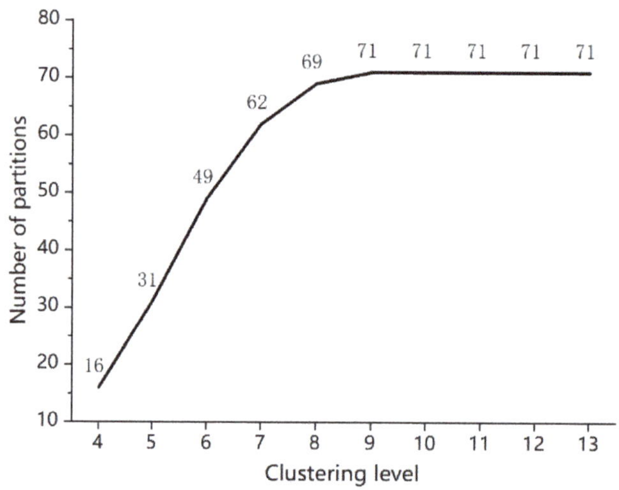


**Figure 4.** The corresponding relation between the clustering level and the number of partitions.

A partition at level 6 is used to show the clustering process. Define $\{a : b\}$ or $\{a\}$ as the process parameter information where $a$ denotes the serial number and $b$ denotes the name. Starting from level 9, {72: RF_IN_TT, 73: RF_EX_TT} and {8: COIL_THK_MAX, 9: COIL_THK_MIN} are assigned to a respective partition firstly because the MIC values between them are 0.9685 and 0.9759 separately, which are the highest. When the level is 7, {1: THK_ACT} joins the partition {8: COIL_THK_MAX, 9: COIL_THK_MIN} because the MIC values between them are both 0.998706, which is the highest. In the same way, when the level is 6, {1: THK_ACT, 8: COIL_THK_MAX, 9: COIL_THK_MIN} and {72: RF_IN_TT, 73: RF_EX_TT} are clustered into a larger partition. The dendrogram is shown in Figure 6, and the MIC values between 5 parameters are shown in Table 3.

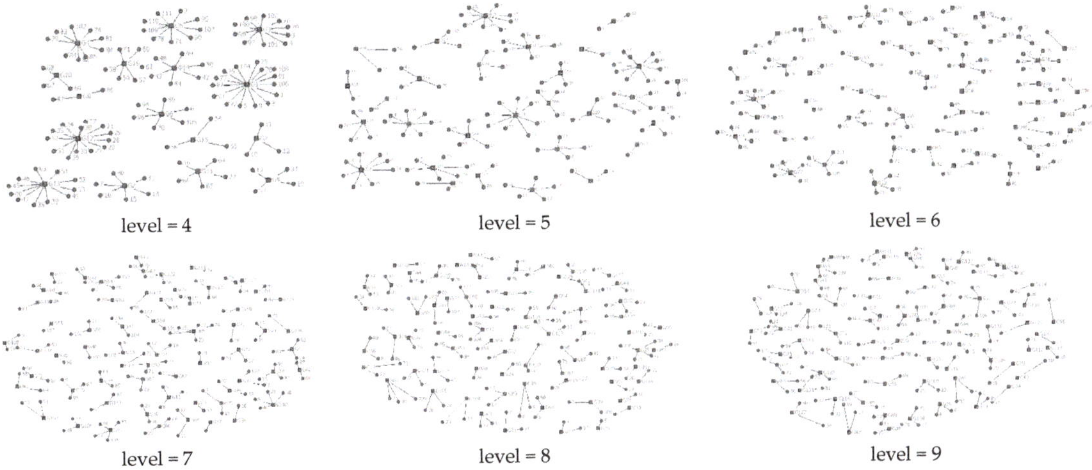


**Figure 5.** Partition results under different clustering levels.

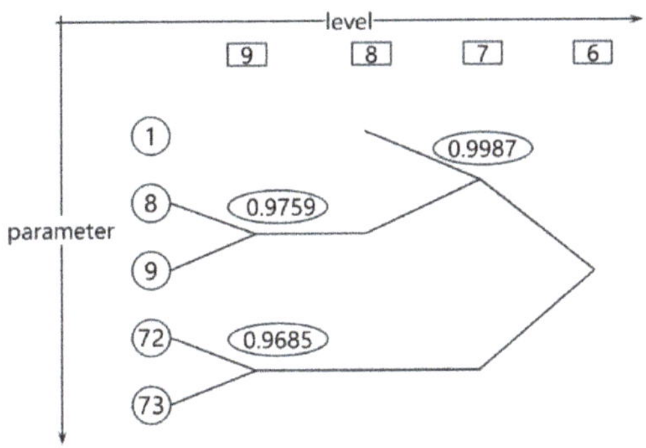


**Figure 6.** Clustering process of feature 1, 8, 9, 72, 73 (the number in the ellipse is the maximum information coefficient (MIC) value between related parameters).

**Table 3.** Maximum information coefficient (MIC) values between features 1, 8, 9, 72, 73.

| Feature Name | Serial Number | 1 | 8 | 9 | 72 | 73 |
|---|---|---|---|---|---|---|
| THK_ACT | 1 | 1 | 0.9987 | 0.9987 | 0.8318 | 0.7765 |
| COIL_THK_MAX | 8 | 0.9987 | 1 | 0.9759 | 0.8926 | 0.8547 |
| COIL_THK_MIN | 9 | 0.9987 | 0.9759 | 1 | 0.8714 | 0.7941 |
| RF_IN_TT | 72 | 0.8318 | 0.8926 | 0.8714 | 1 | 0.9685 |
| RF_EX_TT | 73 | 0.7765 | 0.8547 | 0.7941 | 0.9685 | 1 |

### *4.2. Feature Evaluation and Selection*

As mentioned above, four centrality and influence indicators are selected to evaluate the importance of each parameter and we rank them by category. Table 4 shows the top 20 features with the highest total ranking and their respective rankings of the four indicators. Table 5 shows the detailed information of the top 20 features including the feature name and the cluster number using the MIC-based model at level 4. The last column indicates whether the feature is selected in its cluster. It can be found that the five rankings are highly related. The parameters with the high total rankings tend to rank at the top of the four separate indicators. Among them, the process parameter "TI" ranks respectively 1, 9, 1, 1 at degree, betweenness, closeness, and Katz centrality and the total ranking is 12, which means that this feature owns greater power and is the most representative in the partition.

**Table 4.** The top 20 features with the highest total ranking and their respective rankings of the four indicators.

| Feature | nDegree | D-Rank | nBetweenness | B-Rank | nCloseness | C-Rank | Katz | K-Rank | Total Ranking |
|---|---|---|---|---|---|---|---|---|---|
| 36 | 0.51 | 1 | 0.02 | 9 | 0.14 | 1 | 3.32 | 1 | 12 |
| 81 | 0.50 | 2 | 0.03 | 8 | 0.14 | 7 | 3.14 | 4 | 21 |
| 82 | 0.48 | 5 | 0.03 | 3 | 0.14 | 10 | 3.07 | 9 | 27 |
| 64 | 0.45 | 11 | 0.05 | 1 | 0.14 | 2 | 3.01 | 14 | 28 |
| 6 | 0.47 | 7 | 0.01 | 16 | 0.14 | 8 | 3.12 | 6 | 37 |
| 89 | 0.47 | 6 | 0.01 | 17 | 0.14 | 11 | 3.12 | 5 | 39 |
| 70 | 0.43 | 24 | 0.02 | 11 | 0.14 | 3 | 3.20 | 3 | 41 |
| 105 | 0.43 | 23 | 0.02 | 12 | 0.14 | 4 | 3.20 | 2 | 41 |
| 65 | 0.44 | 14 | 0.03 | 4 | 0.14 | 5 | 2.95 | 20 | 43 |
| 22 | 0.46 | 8 | 0.02 | 10 | 0.14 | 9 | 2.95 | 19 | 46 |
| 66 | 0.43 | 21 | 0.03 | 5 | 0.14 | 6 | 2.86 | 23 | 55 |
| 74 | 0.48 | 4 | 0.01 | 29 | 0.14 | 15 | 3.07 | 8 | 56 |
| 94 | 0.43 | 22 | 0.01 | 18 | 0.14 | 13 | 3.08 | 7 | 60 |
| 79 | 0.48 | 3 | 0.00 | 33 | 0.14 | 17 | 3.04 | 10 | 63 |
| 96 | 0.43 | 26 | 0.01 | 23 | 0.14 | 18 | 3.02 | 13 | 80 |
| 83 | 0.46 | 10 | 0.01 | 30 | 0.14 | 23 | 2.95 | 18 | 81 |
| 2 | 0.40 | 43 | 0.01 | 24 | 0.14 | 14 | 3.02 | 11 | 92 |
| 28 | 0.40 | 39 | 0.03 | 6 | 0.14 | 19 | 2.71 | 32 | 96 |
| 47 | 0.38 | 52 | 0.04 | 2 | 0.14 | 12 | 2.71 | 31 | 97 |
| 90 | 0.40 | 44 | 0.01 | 25 | 0.14 | 16 | 3.02 | 12 | 97 |

**Table 5.** The detailed information of the top 20 features with the highest total ranking.

| Feature Number | Feature Name | Stage | Cluster Number | Selected |
|---|---|---|---|---|
| 36 | TI | Smelting | 3 | 1 |
| 81 | CT_TT | Hot rolling | 4 | 1 |
| 82 | FET_PT | Hot rolling | 4 | 0 |
| 64 | IDN_MIN | Continuous casting | 11 | 1 |
| 6 | COIL_WID | Cold rolling | 6 | 1 |
| 89 | CWC_1 | Hot rolling | 6 | 0 |
| 70 | SLAB_LEN | Hot rolling | 6 | 0 |
| 105 | IM_LEN_1 | Hot rolling | 6 | 0 |
| 65 | IDN_MAX | Continuous casting | 11 | 0 |
| 22 | MN | Smelting | 11 | 0 |
| 66 | IND_AVE | Continuous casting | 11 | 0 |
| 74 | FET_TT | Hot rolling | 4 | 0 |
| 94 | WID_AVE | Hot rolling | 6 | 0 |
| 79 | FDT_TT | Hot rolling | 4 | 0 |
| 96 | WID_MIN | Hot rolling | 6 | 0 |
| 83 | FET_NT | Hot rolling | 4 | 0 |
| 2 | MAT_WID_ACT | Cold rolling | 5 | 1 |
| 28 | ALS | Smelting | 12 | 1 |
| 47 | ST_AVE | Continuous casting | 13 | 1 |
| 90 | CWC | Hot rolling | 5 | 0 |

Finally, the feature selection is based on the partition situation and the feature evaluation results. At different clustering levels, compare the centrality and influence rankings of different features in each partition. The top 1 feature is selected as the representative of the partition, also as a member of the selected feature subset. For example, when the clustering level is 4, 111 features are divided into 16 subgroups. The feature distribution of the first subgroup $g_1$ is shown in Figure 7.

$$R_T^1 = \left\{R_T^{1\,1}, R_T^{1\,2} \ldots R_T^{1\,15}\right\} = \{240, 160, 226, 247, 175, 191, 183, 187, 191, 187, 167, 159, 165, 194, 196\} \quad (10)$$

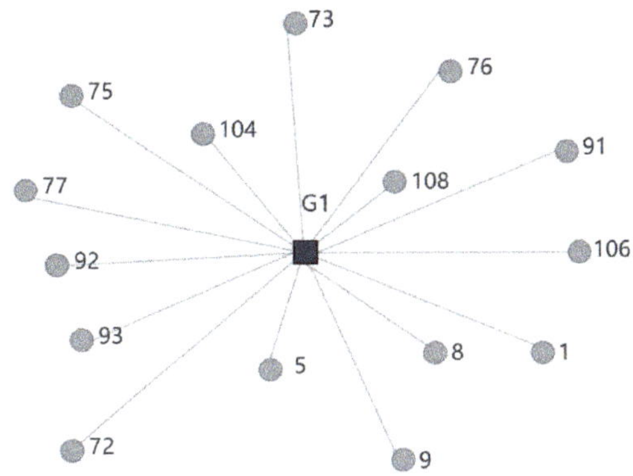


**Figure 7.** The individual network of subgroup $g_1$ when the clustering level is 4.

There are 15 features in subgroup $g_1$, and Figure 8 shows the total ranking scatter diagram of each feature. It can be seen that the ranking distribution within the subgroup is relatively concentrated among [150, 200], which also verifies the rationality of clustering, that is, the rankings of similar features should also be similar. Among them, the top ranking is feature 93 whose total ranking is 159, so select feature 93 as the representative feature of subgroup $g_1$ and add it into the final feature subset.

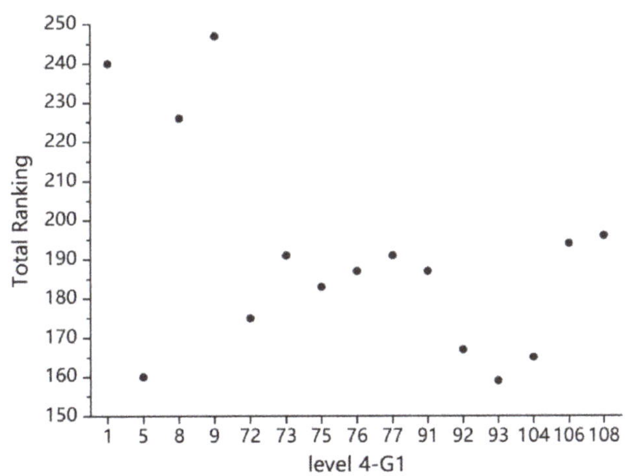


**Figure 8.** The total ranking of each feature in subgroup $g_1$ when the clustering level is 4.

The rest can be deduced by analogy. Select the representative features of all partitions at level 4, and then expand the clustering level. Finally, the feature subsets at level 4–9 are obtained, as shown in Table 6.

Compared with level 8, level 9 has two new representative features, which are {24: S} and {38: PS_MIN}. This is because at level 8, {24} enters the partition {23,25} with the corresponding correlation coefficient {0.5175,0.7232}, 39 enters {38,40} with {0.9021,0.9749}. Both of them are the closest feature to the corresponding partition, and the process conforms to the clustering rules mentioned before.

In addition, we can discover that the clustering at level 9 is most concise with the least feature numbers and the highest correlation in each partition. Therefore, it can be estimated that the representative features selected at this level may have the best prediction effect, which can also be proved in the follow-up article.

**Table 6.** Feature subset selection at cluster level 4–9.

| Level | Feature Number | Added Feature Number | Added Feature Subset |
|---|---|---|---|
| 4 | 16 | — | {93,7,36,81,2,6,10,86,17,67,64,28,47,43,56,59} |
| 5 | 31 | 15 | {104,80,101,79,107,70,12,34,69,18,30,53,48,54,62} |
| 6 | 49 | 18 | {72,75,14,4,103,84,71,111, 94,22,26,31,39,51, 42,48,50,57,61} |
| 7 | 62 | 13 | {8,5,16,3,98,82,95,66,35,21,32,41,46} |
| 8 | 69 | 7 | {1,92,99,74,78,23,63} |
| 9 | 71 | 2 | {24,38} |

### 4.3. MSVR Property Prediction Model

According to the feature selection results, the original sample data are divided into the training set and test set at the ratio of 8:2 to train the MSVR model. Three mechanical properties are selected, which are lower yield strength, tensile strength, and elongation, separately. The mean absolute percentage error (MAPE) is chosen as the evaluation index of the effectiveness of the proposed algorithm. We calculate three MAPE values and the average of them to represent the prediction accuracy. We choose four parameter sets as the input, which are MIC-based subset, Pearson-based subset, full-parameter subset, and empirical subset to perform the comparison experiment.

MSVR Prediction Results at Different Clustering Levels

Figure 9 shows the MAPE comparison between feature selection results based on MIC-based subset, full-parameter subset, and the empirical subset. It can be seen that starting from level 5, four kinds of prediction error (including three MAPE values of three mechanical properties and their average) of our proposed algorithm are all lower than the other two input sets. In addition, as the number of selected features increases from level 4 to level 9, the growth rate slows down and the prediction error decreases. At level 9, the number of features reaches the maximum 71, while the four MAPE values all reach the lowest.

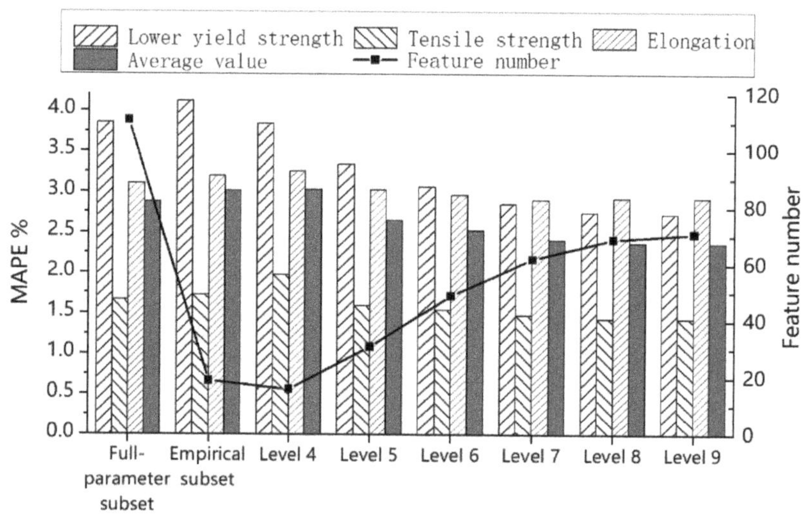


**Figure 9.** Error comparison between MIC-based feature selection and full-parameter, empirical subset modeling.

As shown in Figure 10, when the clustering level is 9, the prediction error of the optimal feature subset is significantly lower than that of the full-parameter and empirical subset. Therefore, it can be concluded that the feature selection method proposed in this paper can select a small number of parameters from the full-parameter set to represent the whole, and the prediction effect is better.

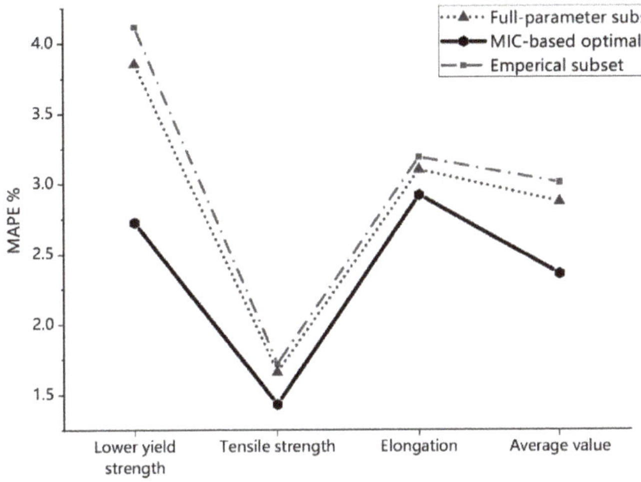


**Figure 10.** Prediction error comparison between the optimal subset of MIC and full-parameter, empirical subset.

In order to verify that the MIC-based feature selection method can characterize the nonlinear correlation relationships between features more reasonably, we use the Pearson coefficient matrix to represent the initial correlation, and compare the prediction error under two correlation measures. The prediction error of "lower yield strength" (left) and the average error of three mechanical properties (right) are shown in Figure 11. It can be seen that the overall error of MIC-based feature selection method is lower than that of Pearson-based method. With the increase of clustering level, the prediction accuracy difference between the two methods gradually becomes smaller.

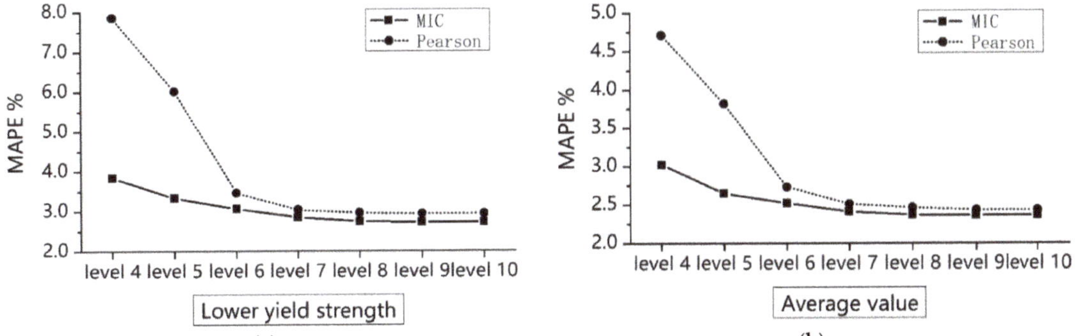


**Figure 11.** Comparison of prediction errors of feature selection methods based on MIC and Pearson coefficient. (**a**) The prediction error of "lower yield strength". (**b**) The average error of three mechanical properties.

When the clustering level is 4, the prediction accuracy of MIC method is 1.69% higher than that of Pearson and only one feature 10 is coincident, which means that the two similarity measurement methods are quite different, among which MIC is better. Apparently, compared with the Pearson coefficient, MIC can widely explore the linear and nonlinear relationship between process parameters.

To sum up, the feature selection method based on MIC and complex network clustering can represent the global situation with fewer features and better prediction effect than the full-parameter subset. At the same time, compared with the empirical subset and the similarity measurement based on Pearson coefficient, our model has higher prediction accuracy. It should be pointed out that no matter whether the feature selection is based on MIC or Pearson coefficient, the prediction accuracy is both higher than that of full-parameter subset, which indicates that there are a lot of linear and nonlinear relationships in the original data set. If it can be well mined and analyzed, the difficulty of subsequent modeling can be greatly reduced.

## 5. Conclusions

Aiming at the complex industrial process of steel production, this paper proposes a property prediction model based on MIC and complex network clustering, which adopts the MSVR on the basis of attribute selection. Compared with full-parameter subset, empirical subset, and feature selection subset based on Pearson coefficient, our scheme has the lowest computational complexity and the highest prediction accuracy.

The innovation and research significance of this paper are as follows:

1. The feature selection method based on MIC and complex network theory can effectively solve the attribute reduction problem of complex process data. The MIC-based subset input has higher prediction accuracy compared with the Pearson-based subset, full-parameter subset, and empirical subset input. Specifically, the average prediction errors of the three mechanical properties with the four different inputs are 2.359%, 2.428%, 2.872%, and 3.010%, respectively;
2. Using MIC to measure the similarities between process parameters can mine many linear and nonlinear relationships, including all kinds of interesting correlations, which are strongly meaningful in constructing the relationship complex network;
3. The centrality and influence theory of complex network can be used to measure the importance of process parameters in the network efficiently;
4. The feature selection method proposed in this paper reduces the calculation complexity and simplifies the actual problem. The selected feature subset still has physical meaning and does not destroy the original feature structure. At the same time, the average error of the MSVR prediction model is also greatly reduced.

**Author Contributions:** Conceptualization, Y.W. and Z.L.; data curation, Y.W. and Z.L.; funding acquisition, Z.L.; methodology, Y.W. and Z.L.; software, Y.W. and Y.Y.; validation, Y.W. and Y.Y.; visualization, Y.W.; writing—original draft, Y.W.; writing—review and editing, Y.W. and Z.L. All authors have read and agreed to the published version of the manuscript.

**Funding:** This work was funded by Fundamental Research Funds for the Central Universities, grant number FRF-MP-20-08.

**Conflicts of Interest:** The authors declare no conflict of interest.

## References

1. Qiao, L.; Wang, Z.; Zhu, J. Application of improved GRNN model to predict interlamellar spacing and mechanical properties of hypereutectoid steel. *Mater. Sci. Eng. A* **2020**, *792*, 139845. [CrossRef]
2. Sui, X.Y.; Lv, Z.M. Prediction of the mechanical properties of hot rolling products by using attribute reduction ELM. *Int. J. Adv. Manuf. Technol.* **2016**, *85*, 1395–1403. [CrossRef]
3. Gao, H.; Xu, Y.; Zhu, Q. Spatial Interpretive Structural Model Identification and AHP-Based Multimodule Fusion for Alarm Root-Cause Diagnosis in Chemical Processes. *Ind. Eng. Chem. Res.* **2016**, *55*, 3641–3658. [CrossRef]
4. Ayesha, S.; Hanif, M.K.; Talib, R. Overview and comparative study of dimensionality reduction techniques for high dimensional data. *Inf. Fusion* **2020**, *59*, 44–58. [CrossRef]

5. Cano, A.; Ventura, S.; Cios, K.J. Multi-objective genetic programming for feature extraction and data visualization. *Soft Comput.* **2017**, *21*, 2069–2089. [CrossRef]
6. Bommert, A.; Sun, X.; Bischl, B.; Rahnenführer, J.; Lang, M. Benchmark for filter methods for feature selection in high-dimensional classification data. *Comput. Stat. Data Anal.* **2020**, *143*, 106839. [CrossRef]
7. González, J.; Ortega, J.; Damas, M.; Martín-Smith, P.; Gan, J.Q. A new multi-objective wrapper method for feature selection—Accuracy and stability analysis for BCI. *Neurocomputing* **2019**, *333*, 407–418. [CrossRef]
8. Karasu, S.; Altan, A.; Bekiros, S.; Ahmad, W. A new forecasting model with wrapper-based feature selection approach using multi-objective optimization technique for chaotic crude oil time series. *Energy* **2020**, *212*, 118750. [CrossRef]
9. Wu, J.-S.; Song, M.-X.; Min, W.; Lai, J.-H.; Zheng, W.-S. Joint Adaptive Manifold and Embedding Learning for Unsupervised Feature Selection. *Pattern Recognit.* **2020**, *112*, 107742. [CrossRef]
10. Najafi, A.; Joudaki, A.; Fatemizadeh, E. Nonlinear Dimensionality Reduction via Path-Based Isometric Mapping. *IEEE Trans. Pattern Anal. Mach. Intell.* **2016**, *38*, 1452–1464. [CrossRef] [PubMed]
11. Wilson, S.R.; Close, M.E.; Abraham, P. Applying linear discriminant analysis to predict groundwater redox conditions conducive to denitrification. *J. Hydrol.* **2018**, *556*, 611–624. [CrossRef]
12. Guillemot, V.; Beaton, D.; Gloaguen, A.; Lofstedt, T.; Levine, B.; Raymond, N.; Tenenhaus, A.; Abdi, H. A constrained singular value decomposition method that integrates sparsity and orthogonality. *PLoS ONE* **2019**, *14*, e0211463. [CrossRef]
13. Kumar, N.; Singh, S.; Kumar, A. Random permutation principal component analysis for cancelable biometric recognition. *Appl. Intell.* **2018**, *48*, 2824–2836. [CrossRef]
14. Narayana, P.L.; Lee, S.W.; Park, C.H.; Yeom, J.-T.; Hong, J.-K.; Maurya, A.K.; Reddy, N.S. Modeling high-temperature mechanical properties of austenitic stainless steels by neural networks. *Comput. Mater. Sci.* **2020**, *179*, 109617. [CrossRef]
15. Narayana, P.L.; Kim, J.H.; Maurya, A.K.; Park, C.H.; Hong, J.-K.; Yeom, J.-T.; Reddy, N.S. Modeling Mechanical Properties of 25Cr-20Ni-0.4C Steels over a Wide Range of Temperatures by Neural Networks. *Metals* **2020**, *10*, 256. [CrossRef]
16. Sun, L.; Yin, T.; Ding, W.; Qian, Y.; Xu, J. Multilabel feature selection using ML-ReliefF and neighborhood mutual information for multilabel neighborhood decision systems. *Inf. Sci.* **2020**, *537*, 401–424. [CrossRef]
17. Zhang, Y.; Zhu, R.; Chen, Z.; Gao, J.; Xia, D. Evaluating and selecting features via information theoretic lower bounds of feature inner correlations for high-dimensional data. *Eur. J. Oper. Res.* **2021**, *290*, 235–247. [CrossRef]
18. Reshef, D.N.; Reshef, Y.A.; Finucane, H.K.; Grossman, S.R.; McVean, G.; Turnbaugh, P.J.; Lander, E.S.; Mitzenmacher, M.; Sabeti, P.C. Detecting Novel Associations in Large Data Sets. *Science* **2011**, *334*, 1518–1524. [CrossRef] [PubMed]
19. Wang, S.; Zhao, Y.; Shu, Y.; Yuan, H.; Geng, J.; Wang, S. Fast search local extremum for maximal information coefficient (MIC). *J. Comput. Appl. Math.* **2018**, *327*, 372–387. [CrossRef]
20. Manbari, Z.; AkhlaghianTab, F.; Salavati, C. Hybrid fast unsupervised feature selection for high-dimensional data. *Expert Syst. Appl.* **2019**, *124*, 97–118. [CrossRef]
21. Chen, W.; Teng, S.-H.; Zhang, H. A graph-theoretical basis of stochastic-cascading network influence: Characterizations of influence-based centrality. *Theor. Comput. Sci.* **2020**, *824*, 92–111. [CrossRef]
22. Long, J.C.; Cunningham, F.C.; Carswell, P.; Braithwaite, J. Patterns of collaboration in complex networks: The example of a translational research network. *BMC Health Serv. Res.* **2014**, *14*, 1–10. [CrossRef]
23. Ahmadi, H.; Rezaei Ashtiani, H.R.; Heidari, M. A comparative study of phenomenological, physically-based and artificial neural network models to predict the Hot flow behavior of API 5CT-L80 steel. *Mater. Today Commun.* **2020**, *25*, 101528. [CrossRef]
24. Mohammadian, M. Modelling, Control and Prediction using Hierarchical Fuzzy Logic Systems: Design and Development. *Int. J. Fuzzy Syst. Appl.* **2017**, *6*, 105–123. [CrossRef]
25. Yousefi, S.; Zohoor, M. Effect of cutting parameters on the dimensional accuracy and surface finish in the hard turning of MDN250 steel with cubic boron nitride tool, for developing a knowledged base expert system. *Int. J. Mech. Mater. Eng.* **2019**, *14*, 1. [CrossRef]
26. Shahri, A.A.; Moud, F.M.; Mirfallah Lialestani, S.P. A hybrid computing model to predict rock strength index properties using support vector regression. *Eng. Comput.* **2020**, 1–16. [CrossRef]
27. Kinney, J.B.; Atwal, G.S. Equitability, mutual information, and the maximal information coefficient. *Proc. Natl. Acad. Sci. USA* **2014**, *111*, 3354–3359. [CrossRef]
28. Albert, R.; Jeong, H.; Barabási, A.-L. Error and attack tolerance of complex networks. *Nature* **2000**, *406*, 378–382. [CrossRef]
29. Kitsak, M.; Gallos, L.K.; Havlin, S.; Liljeros, F.; Muchnik, L.; Stanley, H.E.; Makse, H.A. Identification of influential spreaders in complex networks. *Nat. Phys.* **2010**, *6*, 888–893. [CrossRef]
30. Ahn, Y.-Y.; Bagrow, J.P.; Lehmann, S. Link communities reveal multiscale complexity in networks. *Nature* **2010**, *466*, 761–764. [CrossRef]
31. White, H.C.; Boorman, S.A.; Breiger, R.L. Social Structure from Multiple Networks. *Amer. J. Sociology* **1976**, *81*, 730–780. [CrossRef]
32. Decelle, A.; Krzakala, F.; Moore, C.; Zdeborova, L. Asymptotic analysis of the stochastic block model for modular networks and its algorithmic applications. *Phys. Rev. E* **2011**, *84*. [CrossRef]
33. Lei, J.; Rinaldo, A. Consistency of spectral clustering in stochastic block models. *Ann. Stat.* **2015**, *43*, 215–237. [CrossRef]

34. Breiger, R.L.; Boorman, S.A.; Arabie, P. An algorithm for clustering relational data with applications to social network analysis and comparison with multidimensional scaling. *J. Math. Psychol.* **1975**, *12*, 328–383. [CrossRef]
35. Yan, Y.F.; An, L.D.; Lv, Z.M. Prediction of mechanical properties of cold rolled products based on maximum information coefficient attribute selection. *J. Cent. South Univ. (Sci. Technol.)* **2020**, *051*, 68–75.

metals

MDPI

*Article*

# Topological Optimization of Artificial Neural Networks to Estimate Mechanical Properties in Metal Forming Using Machine Learning

**David Merayo *, Alvaro Rodríguez-Prieto and Ana María Camacho**

Department of Manufacturing Engineering, UNED, Juan del Rosal 12, 28040 Madrid, Spain; alvaro.rodriguez@ind.uned.es (A.R.-P.); amcamacho@ind.uned.es (A.M.C.)
* Correspondence: dmerayo1@alumno.uned.es

**Abstract:** The ability of a metal to be subjected to forming processes depends mainly on its plastic behavior and, thus, the mechanical properties belonging to this region of the stress–strain curve. Forming techniques are among the most widespread metalworking procedures in manufacturing, and aluminum alloys are of great interest in fields as diverse as the aerospace sector or the food industry. A precise characterization of the mechanical properties is crucial to estimate the forming capability of equipment, but also for a robust numerical modeling of metal forming processes. Characterizing a material is a very relevant task in which large amounts of resources are invested, and this paper studies how to optimize a multilayer neural network to be able to make, through machine learning, precise and accurate predictions about the mechanical properties of wrought aluminum alloys. This study focuses on the determination of the ultimate tensile strength, closely related to the strain hardening of a material; more precisely, a methodology is developed that, by randomly partitioning the input dataset, performs training and prediction cycles that allow estimating the average performance of each fully-connected topology. In this way, trends are found in the behavior of the networks, and it is established that, for networks with at least 150 perceptrons in their hidden layers, the average predictive error stabilizes below 4%. Beyond this point, no really significant improvements are found, although there is an increase in computational requirements.

**Keywords:** aluminum alloy; artificial neural network; mechanical property; *UTS*; machine learning; topological optimization; metal forming

**Citation:** Merayo, D.; Rodríguez-Prieto, A.; Camacho, A.M. Topological Optimization of Artificial Neural Networks to Estimate Mechanical Properties in Metal Forming Using Machine Learning. *Metals* **2021**, *11*, 1289. https://doi.org/10.3390/met11081289

Academic Editor: Umberto Prisco

Received: 24 July 2021
Accepted: 13 August 2021
Published: 16 Auguat 2021



## 1. Introduction

Aluminum alloys are among the most widely used materials in the industry, and, although their use is still far from being as widespread as that of steel, they have many advantages that make them a very interesting material whose use is growing regularly [1]. There is a huge number of aluminum alloys, but few of them are typically used in the industrial field [2], sometimes because it is difficult to find new solutions and, sometimes, because they are special materials with optimized properties to fulfill their requirements, according to their application [3].

Aluminum alloys are manufactured by different techniques [4]. Among its many properties of industrial interest, it can be noted that it is a material with a high formability, so it is especially suitable for metal forming processes. Moreover, a precise characterization of the mechanical properties is crucial to estimate the forming capability of equipment but also for a robust numerical modeling of metal forming processes. Among all the mechanical properties, the ultimate tensile strength (*UTS*) plays a key role in the definition of the onset of the plastic instability by tensile tension [3]. This mechanical property is closely related to the strain hardening of the metal and, therefore, to its forming capacity under metal forming processes. In uniaxial tension, the plastic deformation is limited by the value of

$UTS$ since, once it is exceeded, the onset of the plastic instability takes place, and from then on, the behavior of the material is almost unpredictable until fracture [5].

Therefore, it is well known that aluminum alloys present a high ductility [6] because they can withstand a lot of plastic deformation before fracture and, so, they are widely used in metal forming operations [7]; such metals can be manufactured by metal sheet operations and bulk forming processes, such as forging, rolling, extrusion and/or drawing [2]. Ductile materials are able to absorb a great amount of energy before failing, otherwise known as toughness [8].

The design and production of some industrial components rely on the knowledge of the mechanical properties obtained and tensile testing [9]. The $UTS$ may serve to determine the beginning of the plastic instability and provides an insight into the initiation of the fracture or the necking [10]. It also can be used as an input to estimate the forming force in the conventional forming procedures such as stretch-bending or deep-drawing [10].

Metal forming is frequently employed to manufacture components. The microstructure and the mechanical properties of these parts are modified by these processes [2]. For example, the increase in the ultimate stress and hardness observed during the A-6063 extrusion is attributed to the grain size reduction and the temperature increase [11].

In addition to $UTS$, the strain hardening exponent, the yield strength ($YS$), the process-induced residual stresses, and the hardness are also important mechanical characteristics [1,3,10]. These properties offer an idea of the in-service behavior of the formed part. Furthermore, their correlations help to understand the response of the component [12], i.e., knowing the difference between $UTS$ and $YS$ can help the designer to predict how much additional stress a component can withstand before failure [5] because $YS$ defines the onset of plastic deformation, and $UTS$ defines the onset of plastic instability.

Knowing the expected behavior of the materials used in industrial designs is critical; however, obtaining these data frequently requires accessing large amounts of resources, which are normally not accessible [1]. Many tests are required to obtain relevant information, which entail having enough time, personnel and facilities be available [13]. Characterizing a metal comprises many tests that require non-trivial quantities of resources [8].

Although it is a relatively new technology and not as widespread in materials science and manufacturing as in other areas [13], artificial intelligence (AI) and machine learning (ML) techniques have been successfully used to make predictions about the metallurgical properties of some materials [1,3,14–16].

Over recent years, AI and ML have received much attention in the field of materials modeling, due to their outstanding ability to analyze a huge amount of data and expose correlations between complex interrelated properties [17]. ML is, perhaps, the most relevant branch of AI, and is the science of making computers learn and act like humans without being explicitly programmed [18,19]. It is often used to discover hidden patterns in complex systems through a training process in which a great amount of noisy data is furnished as input [20,21]. ML can be classified into supervised learning (the machine learns from labeled data) and unsupervised learning (the machine finds patterns in the data without any external help) [20,22,23].

Among the most challenging topics in this field is the search for the best representation of input variables in ML models, which is commonly called feature engineering and comprises a set of activities, such as feature extraction, feature construction and feature selection [23,24]. Feature engineering research is crucial to the application of the ML.

Within current materials science, the scale and speed of data acquisition, the accuracy of the data and the volatility of the data are additional challenges for researchers [3]. It raises the question of how to use and analyze these data in a useful way that supports the decisions of developers and designers [18,23]. Material data tend to be wide in scope and, often, shallow in depth. Here, depth should be understood as the number of observations of the state of a system. The lack of observations is due not only to the cost and difficulty of acquiring data (especially through experimentation), but also to the nature of the data itself. However, fully employing the data is a key part of advanced design systems [23,25].

In recent years, an incipient trend in materials science research is the combination of existing experimental and numerical modeling methodologies with AI techniques [26–28]. In general, materials science advances thanks to accumulated experience and already established rules [29,30]. New advances in numerical modeling facilitate the methodical acquisition of large amounts of data, while complicating analysis, hypothesis formulation and pattern prediction. The rise of AI techniques makes up for this deficiency to a great extent [31].

Multilayer artificial neural networks (ANN) can be considered the most remarkable methodology of those that are included into the field of AI because they have demonstrated their capabilities in almost all branches of knowledge and because they are currently receiving a lot of attention from investigators [32]. A multilayer network is able to learn a function by training on a labeled dataset that can be used to perform regressions [33]. ANN are made up of perceptrons (neurons) that regroup forming layers (clusters of neurons) that communicate with each other (in general, perceptrons do not communicate with their own layer companions) [13]. For a fully connected multilayer neural network, the time complexity of the backpropagation training is given by Equation (1). So, it is highly recommended to minimize the number of hidden nodes to reduce the training time [1].

$$\mathcal{O}\left(n \cdot m \cdot o \cdot N \cdot \prod_{i=1}^{k} h_i\right), \quad (1)$$

where $n$ is the size of the training dataset, $m$ is the number of features, $o$ is the number of output perceptrons, $N$ is the number of iterations and $k$ is the number of hidden layers (each of them containing $h_i$ nodes).

The main objective of this work is to develop a working methodology that allows optimizing the topology of a multilayer neural network in such a way that it is capable of making predictions about the $UTS$ of wrought aluminum alloys [34], maximizing the precision and accuracy of the estimation and minimizing the computational resources [20]. Although this paper only takes into account the $UTS$, this same approach could be applied to other properties that have already been mentioned, such as the $YS$ or the elongation at break ($A$).

## 2. Methodology

This work is developed following a three-stage scheme, and the data generated in each of them are used as input for the subsequent one [13]. This workflow guarantees that the data that reach each phase are correctly prepared and processed, and are ready to be employed. Therefore, the information resulting from the entire process is a consequence of the initial dataset.

Figure 1 schematically shows the three stages that compose the methodology of this work: in the first stage, an initial input dataset is created; in the second stage, the optimization process is carried out through training-prediction cycles; and in the third phase, all available information is analyzed.

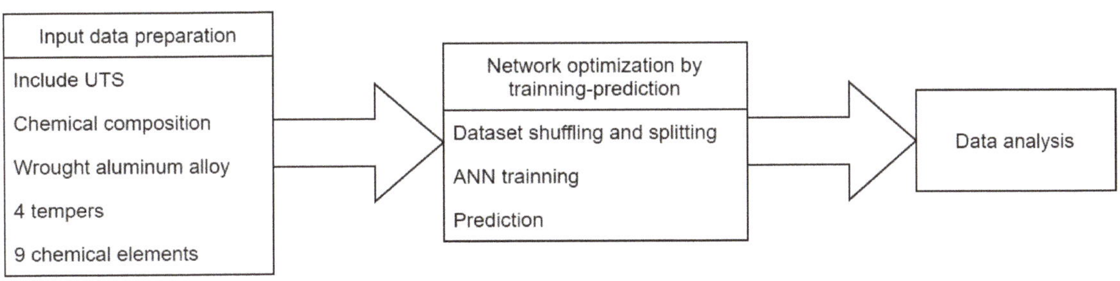


**Figure 1.** Overview of the methodology.

### 2.1. Input Dataset Preparation

All initial data on the properties of the materials were obtained from Matmatch GmbH (Munich, Germany) [35]. It is an online library that contains freely accessible specification sheets about material properties [13]. These include a large number of aluminum alloys with very heterogeneous information. The volume of data initially obtained comprises more than 5000 materials and more than 350 properties [1,3].

After obtaining these data, each record must be read and interpreted in an automated way. Each specification sheet contains much more information than is used during this study, so it is necessary to discard the irrelevant data [35]. The following considerations are taken into account to select the records that are found to be useful:

- Only datasheets that contain the value of the ultimate tensile stress ($UTS$) at 20 °C are considered.
- Only alloys whose chemical composition is defined at more than 95% are taken into account [13] (note that some datasheets are poorly defined).
- Only 9 chemical elements are considered to define the chemical composition of the alloys [29]: Al, Zn, Cu, Si, Fe, Mn, Mg, Ti and Cr. The mass contribution of all other chemical elements is regrouped as "Other".
- Only wrought alloys are considered in this study.
- Only the specification sheets that include the temper of the alloy are considered. This study only considers the following tempers: F (as fabricated), O (annealed), H (strain hardening) and T (thermally treated) [7].

After taking all these considerations into account, only 2671 materials (the obviated records do not meet the aforementioned conditions), 11 input properties and the $UTS$ are considered. These alloys constitute the initial dataset on which the entire study is developed. One of these properties is categorical (temper) and must be mapped as an integer, while the other 11 are numerical ($UTS$ and chemical composition) and must be normalized to avoid bias [20,22]. Normalization is carried out using Equation (2).

$$\tilde{x_i} = \frac{x_i - x_{min}}{x_{MAX} - x_{min}} \tag{2}$$

where $x_i$ is each of the non-normalized input values, $\tilde{x_i}$ is the related normalized value $[0 \ldots 1]$, $x_{min}$ is the minimum value for that parameter and $x_{MAX}$ is the maximum value.

### 2.2. Network Optimization by Training-Prediction

The ANN topology denotes the way in which perceptrons are associated and is an essential characteristic in the performance of the network [19]. Layers are shapeless in the sense that all of its nodes are equally relevant, connected the same way, and lack differentiators [22]. Only the initialization step and the following training make its importance change [25].

In this study, the neural network is defined as a fully connected multilayer feedforward topology [33], which comprises an input layer, two hidden layers and an output layer. In this topology, all the perceptrons in each layer are only connected to all ones in the next layer so that the information only flows in one direction, from the input layer to the output layer [22].

The neural network receives an input vector of 11 elements (chemical composition and temper) and returns a prediction about the value of the expected $UTS$ [3]. Therefore, the input layer is made up of 11 nodes, and the output layer only contains a single node. Additionally, the network topology contains two hidden layers whose number of nodes is to be optimized [19]. Figure 2 shows a schematic representation of the network in which the hidden layers are represented as squares.

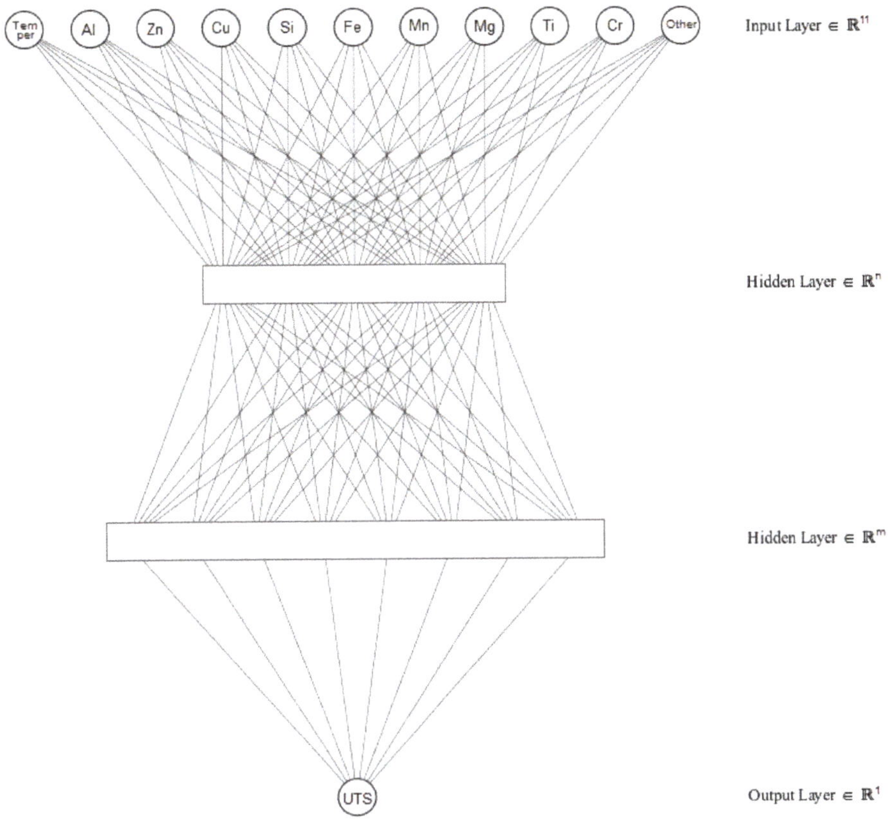


**Figure 2.** Multilayer artificial neural network scheme.

Different topologies are tested to carry out the network optimization: the number of nodes in both hidden layer changes in increments of 10 nodes (from 10 to 200). In this way, 400 topologies are obtained with a number of nodes that varies between 32 and 412. For each of these networks, 10 independent training and prediction iterations are carried out [1,3].

Each iteration consists of four phases:

- Dataset shuffling and splitting to create a training subset (80% of records, so 2137) and a testing subset (remaining 20%, so 534).
- ANN training, using the data contained in the training subset.
- Prediction of the properties for the records in the testing subset.
- Results and data storage for further analysis.

The training is configured as follow [1]:

- Calculation of the learning rate for each parameter using adaptive moment estimation (ADAM) with $\beta_1 = 0.9$, $\beta_2 = 0.999$ (algorithm parameters), $\eta = 0.001$ (step size) and $\epsilon = 10^{-8}$ (stability factor) [36].
- Early stopping after 20 iterations without significant changes.
- Training stops when a training error of less than 0.1 is reached.
- Maximum of 100,000 training epochs to avoid infinite loops.

### 2.3. Data Analysis

The optimization process generates a large amount of data that must be processed to generate information that allows to draw conclusions about the network performance [3,13]. For each of the considered 400 topologies, the predictive performance of the 10 iterations is calculated and stored.

With the information obtained about the tests on these topologies, it is possible to build a performance map of 20 × 20 cells; each of these cells represents a topology described by the number of nodes in the two hidden layers. In each position of this data structure, it is possible to store statistical information that allows making comparisons between the different topologies.

Network topologies with the lowest average predictive error (highest accuracy), lowest standard deviation of predictive error (highest precision), and shortest training time (lowest resource usage) are preferred [20].

## 3. Results and Discussion

Table 1 contains some statistical metrics about the information contained in the input dataset. It is interesting to highlight the wide range of values associated with the *UTS*.

**Table 1.** Statistical information about the input dataset.

| | *UTS* [MPa] | Al [%] | Zn [%] | Cu [%] | Si [%] | Fe [%] | Mn [%] | Mg [%] | Ti [%] | Cr [%] | Other [%] |
|---|---|---|---|---|---|---|---|---|---|---|---|
| Average | 246.3 | 95.1 | 0.6 | 1.0 | 0.5 | 0.5 | 0.5 | 1.4 | 0.2 | 0.2 | 0.2 |
| Std. dev. | 115.1 | 2.6 | 1.3 | 1.6 | 0.6 | 0.2 | 0.4 | 1.4 | 0.2 | 0.1 | 0.3 |
| Median | 230.0 | 95.9 | 0.3 | 0.3 | 0.4 | 0.5 | 0.5 | 0.9 | 0.1 | 0.1 | 0.0 |
| Min. | 40.0 | 83.1 | 0.0 | 0.0 | 0.0 | 0.0 | 0.0 | 0.0 | 0.0 | 0.0 | 0.0 |
| Max. | 700.0 | 99.7 | 7.8 | 6.3 | 12.3 | 1.6 | 1.6 | 5.5 | 0.8 | 0.9 | 2.0 |

As expected, more complex topologies tend to have higher precision than those with fewer perceptrons and, in fact, the lowest error is related to a network with 160 and 200 nodes in its two hidden layers (2.88%). Moreover, the highest error rate (95.28%) occurs for the simplest network of those that are considered (10 and 10 nodes). More detailed information can be found on Table A1.

Figure 3 graphically shows the average predictive error (values above 20% are trimmed to avoid scale-related issues). A region with a very low precision (error greater than 10%) can be seen for topologies with less than 150 perceptrons, while for more complex networks, the error remains lower (less than 10%). It is interesting to note that, as can be seen in the three-dimensional figure, the transition between both regions is quite abrupt, and a step (yellow zone) is formed. This transition zone (yellow) separates an almost flat area from a steep one.

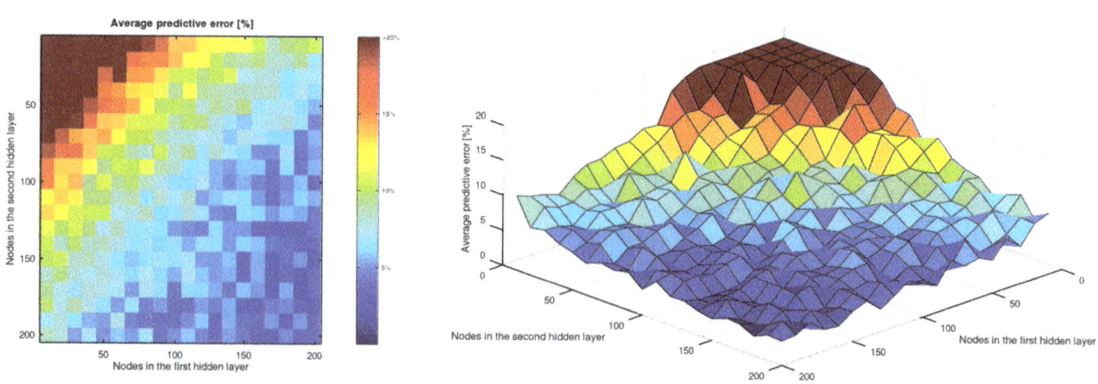


**Figure 3.** Average predictive error [%] (trimmed above 20%).

This transition can be interpreted as a frontier before which the predictions cannot be trusted because the error is excessive and the model is very unstable (small changes in the network produce large differences).

Figure 4 shows the value of the mean predictive error as a function of the number of nodes in the hidden layers. It can be seen that the error asymptotically tends to a value close to 2%. Note that the number of nodes in the hidden layers is the sum of the number of perceptrons in both layers.

It is interesting to highlight that, for neural networks with more than 300 nodes, the average predictive error remains, in all cases, approximately stable at around 4%. It is a very interesting result since it establishes a boundary beyond which no significant improvement can be seen, although there is an increase in computational requirements.

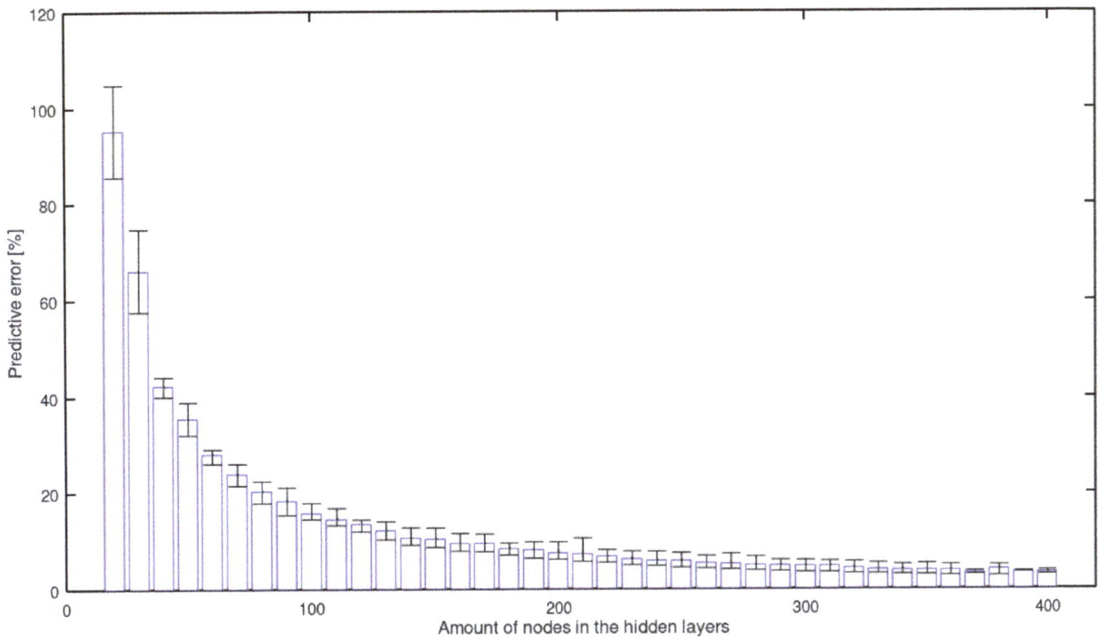


**Figure 4.** Predictive error [%] as a function of the amount of nodes in both hidden layers.

In view of these results, it is clear that complex topologies should be privileged over simpler ones; however, it should also be considered that, as the number of nodes increases, achieving significant improvements becomes very expensive in computational terms and, in fact, it is found that a more complex topology does not always guarantee better accuracy.

On the other hand, it is not only necessary to take into account the accuracy (related to the average predictive error) of the results, but also the precision (related to the predictive error standard deviation).

The standard deviation of the error gives an idea of the repeatability of the estimates, and, together with the average error, allow to identify the confidence range in which a prediction is. In a similar way to what happened in the case of the average error, the more complex networks are more precise (the minimum is reached for a network with 180 and 160 perceptrons, respectively), whereas the simpler ones obtain a higher standard deviation value (the maximum is reached for a network with 30 and 10 nodes, respectively). More detailed information can be found on Table A2.

Figure 5 graphically shows the standard deviation of the predictive error (values above 10% are trimmed to avoid scale-related issues). The distribution of values is much more irregular than in the case of the average error. However, three areas can be seen: for networks with less than 150 perceptrons, the standard deviation is high (mostly above 10%); for topologies with between 150 and 250 nodes, a very irregular transition zone is produced with scattered high values; and, for the more complex networks (more than 250 nodes), the values are mostly kept below 5%.

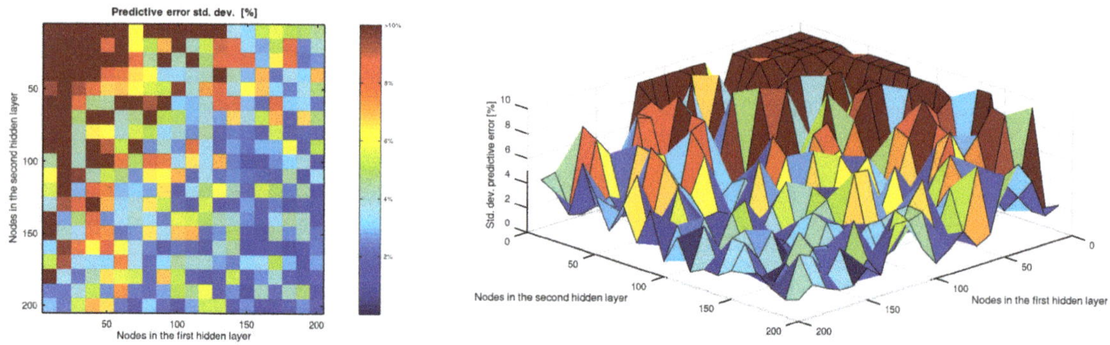


**Figure 5.** Standard deviation of the predictive error [%] (trimmed above 10%).

As can be seen by comparing Figures 3 and 5, although the standard deviation is distributed in a much more irregular way, the trends of both statistical metrics are similar. For networks whose hidden layers contain 150 or more perceptrons, the accuracy and precision stabilize, and there are hardly any significant differences in the performance of these topologies.

## 4. Conclusions and Future Work

This paper studies how to optimize the topology of a multilayer artificial neural network to carry out predictions about mechanical properties of aluminum alloys, such as *UTS*, using machine learning. It is a contribution of great industrial interest since it allows exploring how to obtain sufficiently precise estimates with minimal computational cost and, therefore, using fewer resources. Therefore, the main conclusions of this work are presented as follows:

- An artificial neural network with two hidden layers can predict the *UTS* of wrought aluminum alloys by taking its chemical composition and temper as input. The accuracy of this prediction stabilizes below 4% and even reaches 2.88% in this study.
- The predictive ability of an ANN with two hidden layers to estimate the *UTS* of aluminum alloys is stabilized for topologies that include 150 or more perceptrons in both hidden layers. The precision and accuracy of these networks do not show significant differences that allow us to affirm that one topology is really better than the others.
- A multilayer ANN can be used as a tool to, through machine learning, make predictions about the mechanical behavior of a piece of aluminum alloy subjected to forming processes. In theory, these networks can learn to approximate any nonlinear function if the input data set is large enough and has enough perceptrons [37].

This study presents a methodology that allows optimizing the topology of a neural network whose task is to make predictions about the *UTS* using techniques based on machine learning. In the same way, it would be possible to use this same approach with other properties and even with other materials.

Since this scheme of work is shown to work adequately, a similar method could be applied to test other more complex network architectures. There are a multitude of architectures related to machine learning that allow different problems to be addressed [33].

**Author Contributions:** Conceptualization, D.M., A.R.-P. and A.M.C.; methodology, D.M.; software, D.M.; validation, D.M., A.R.-P. and A.M.C.; formal analysis, D.M.; investigation, D.M.; resources, A.R.-P. and A.M.C.; data curation, D.M.; writing—original draft preparation, D.M.; writing—review and editing, D.M., A.R.-P. and A.M.C.; visualization, D.M.; supervision, A.R.-P. and A.M.C.; project administration, A.R.-P. and A.M.C.; funding acquisition, A.R.-P. and A.M.C. All authors have read and agreed to the published version of the manuscript.

**Funding:** This work was developed within the framework of the Doctorate Program in Industrial Technologies of the UNED and was funded by the Annual Grants Call of the E.T.S.I.I. of the UNED via the projects of reference 2021-ICF07 and 2021-ICF08, as well as by the Innovation Teaching Project of the GID2016-28 focused on "Reliability and Advanced Failure Prognosis in industrial applications applied to the teaching of Materials Technology and Processing".

**Data Availability Statement:** All data can be found in Appendix A.

**Acknowledgments:** We extend our acknowledgments to the Research Group of the UNED "Industrial Production and Manufacturing Engineering (IPME)" and the Industrial Research Group "Advanced Failure Prognosis (AFP) for Engineering Applications". We also thank Matmatch GmbH for freely supplying all the material data employed to accomplish this study.

**Conflicts of Interest:** The authors declare no conflict of interest.

## Abbreviations

The following abbreviations and symbols are used in this manuscript:

| | |
|---|---|
| ADAM | Adaptive moment estimation |
| AI | Artificial intelligence |
| ANN | Artificial neural networks |
| $\beta_n$ | ADAM algorithm parameter |
| $\epsilon$ | ADAM stability factor |
| $\eta$ | ADAM step size |
| $m$ | ADAM first moment estimate |
| ML | Machine learning |
| $UTS$ | Ultimate tensile strength |
| $S_x$ | Standard deviation |
| $v$ | ADAM second moment estimate |
| $\bar{x}$ | Average |
| $YS$ | Yield strength |

## Appendix A. Numerical Results

Table A1 shows the average predictive error after testing each of the 400 topologies through 10 independent training–prediction iterations.

**Table A1.** Average predictive error (%) for each topology as a function of the amount of nodes in the hidden layers.

| | Nodes in the First Hidden Layer | | | | | | | | | | | | | | | | | | | |
|---|---|---|---|---|---|---|---|---|---|---|---|---|---|---|---|---|---|---|---|---|
| Nodes in the second hidden layer | 10 | 20 | 30 | 40 | 50 | 60 | 70 | 80 | 90 | 100 | 110 | 120 | 130 | 140 | 150 | 160 | 170 | 180 | 190 | 200 |
| 10 | 95.28 | 74.79 | 44.11 | 34.89 | 28.74 | 23.07 | 20.56 | 19.26 | 17.96 | 13.79 | 13.98 | 12.86 | 12.55 | 11.70 | 11.60 | 10.48 | 8.68 | 9.42 | 9.86 | 10.62 |
| 20 | 57.57 | 39.99 | 36.42 | 29.08 | 22.67 | 21.50 | 16.95 | 15.34 | 15.52 | 14.29 | 12.68 | 11.64 | 10.31 | 9.29 | 8.84 | 9.57 | 8.33 | 8.30 | 9.07 | 6.79 |
| 30 | 42.92 | 32.07 | 26.03 | 25.78 | 17.93 | 21.18 | 14.82 | 14.81 | 13.20 | 10.38 | 10.37 | 9.57 | 10.06 | 11.14 | 7.34 | 8.08 | 7.89 | 6.37 | 7.00 | 7.13 |
| 40 | 38.87 | 28.11 | 21.55 | 20.78 | 15.44 | 14.82 | 13.33 | 12.82 | 11.55 | 9.26 | 12.81 | 9.13 | 9.02 | 8.56 | 8.56 | 8.27 | 8.43 | 6.51 | 7.19 | 5.98 |
| 50 | 28.26 | 24.81 | 19.79 | 18.44 | 16.25 | 15.39 | 14.40 | 12.23 | 9.71 | 10.05 | 9.91 | 7.86 | 9.37 | 8.64 | 9.63 | 6.74 | 7.10 | 7.23 | 6.11 | 5.38 |
| 60 | 26.05 | 22.47 | 19.38 | 16.83 | 14.03 | 13.50 | 11.19 | 10.70 | 10.86 | 8.80 | 9.88 | 9.54 | 7.88 | 6.84 | 5.93 | 6.97 | 7.58 | 7.84 | 6.62 | 4.57 |
| 70 | 20.21 | 18.32 | 14.56 | 13.78 | 14.48 | 12.19 | 10.22 | 10.15 | 8.95 | 7.93 | 7.88 | 7.48 | 7.53 | 5.79 | 7.67 | 7.12 | 6.13 | 6.05 | 4.50 | 5.42 |
| 80 | 18.73 | 16.20 | 16.92 | 14.03 | 10.41 | 10.17 | 8.79 | 7.96 | 10.60 | 8.24 | 9.34 | 6.65 | 8.23 | 6.88 | 6.14 | 4.94 | 4.89 | 6.91 | 4.35 | 5.01 |
| 90 | 15.90 | 13.50 | 13.71 | 12.51 | 9.33 | 10.43 | 10.02 | 11.50 | 8.80 | 8.42 | 6.31 | 6.61 | 5.62 | 5.11 | 5.77 | 5.78 | 4.36 | 6.28 | 4.35 | 5.51 |
| 100 | 15.52 | 13.30 | 14.13 | 9.33 | 11.15 | 9.13 | 9.85 | 8.45 | 7.53 | 6.23 | 5.99 | 7.45 | 5.54 | 4.87 | 5.55 | 4.41 | 4.20 | 4.94 | 5.77 | 3.66 |
| 110 | 12.05 | 13.74 | 12.86 | 9.58 | 9.39 | 8.33 | 8.94 | 6.42 | 8.33 | 6.91 | 8.19 | 5.26 | 5.82 | 6.77 | 6.43 | 5.26 | 3.94 | 5.08 | 6.09 | 5.51 |
| 120 | 13.69 | 12.34 | 10.69 | 10.43 | 10.48 | 7.33 | 7.60 | 8.60 | 6.01 | 6.86 | 7.89 | 6.78 | 4.87 | 4.39 | 6.54 | 6.04 | 5.30 | 5.77 | 5.57 | 4.12 |
| 130 | 11.45 | 9.94 | 10.87 | 9.45 | 7.13 | 8.46 | 6.68 | 6.78 | 6.49 | 5.03 | 4.85 | 4.52 | 6.30 | 5.10 | 5.17 | 3.74 | 3.60 | 3.54 | 3.66 | 4.19 |
| 140 | 10.57 | 8.26 | 9.72 | 8.36 | 8.62 | 7.70 | 6.71 | 6.83 | 6.01 | 6.82 | 5.56 | 6.65 | 6.45 | 5.09 | 6.05 | 5.63 | 4.51 | 4.30 | 3.89 | 3.49 |
| 150 | 10.13 | 10.59 | 8.20 | 8.53 | 7.54 | 6.26 | 5.77 | 6.43 | 6.93 | 6.59 | 6.98 | 5.13 | 5.06 | 5.76 | 3.64 | 5.47 | 3.31 | 4.23 | 5.18 | 3.08 |
| 160 | 8.28 | 9.58 | 8.06 | 7.28 | 7.99 | 6.88 | 6.05 | 7.30 | 6.21 | 5.48 | 7.41 | 5.68 | 5.27 | 5.62 | 3.54 | 5.48 | 3.28 | 4.06 | 3.02 | 3.96 |
| 170 | 9.19 | 7.77 | 6.65 | 7.98 | 7.59 | 7.16 | 5.01 | 7.28 | 6.32 | 4.14 | 6.29 | 6.13 | 5.38 | 4.24 | 5.82 | 5.23 | 3.31 | 5.39 | 3.03 | 3.78 |
| 180 | 9.88 | 7.68 | 8.91 | 7.66 | 6.36 | 5.04 | 6.65 | 4.53 | 5.18 | 3.98 | 3.78 | 5.03 | 4.53 | 5.42 | 3.35 | 3.15 | 4.41 | 4.95 | 3.74 | 2.77 |
| 190 | 7.01 | 8.05 | 6.31 | 5.50 | 5.01 | 7.55 | 4.65 | 4.24 | 3.95 | 3.89 | 4.87 | 5.72 | 3.72 | 5.47 | 4.10 | 4.05 | 5.12 | 3.22 | 5.01 | 3.64 |
| 200 | 8.72 | 7.92 | 5.30 | 7.15 | 4.74 | 5.73 | 5.26 | 6.82 | 3.79 | 4.90 | 5.90 | 4.45 | 3.25 | 5.07 | 4.27 | 2.88 | 3.07 | 4.81 | 3.67 | 3.49 |

Table A2 shows the standard deviation of the predictive error for each of the topologies that are tested.

**Table A2.** Standard deviation of the predictive error (%) for each topology as a function of the amount of nodes in the hidden layers.

| | Nodes in the First Hidden Layer | | | | | | | | | | | | | | | | | | | |
|---|---|---|---|---|---|---|---|---|---|---|---|---|---|---|---|---|---|---|---|---|
| Nodes in the second hidden layer | 10 | 20 | 30 | 40 | 50 | 60 | 70 | 80 | 90 | 100 | 110 | 120 | 130 | 140 | 150 | 160 | 170 | 180 | 190 | 200 |
| 10 | 58.96 | 28.75 | **61.09** | 41.95 | 13.42 | 23.40 | 6.31 | 9.94 | 14.81 | 13.57 | 13.25 | 13.00 | 12.76 | 3.60 | 3.21 | 5.28 | 5.72 | 5.05 | 4.87 | 4.62 |
| 20 | 19.33 | 47.56 | 13.39 | 16.32 | 8.62 | 18.35 | 6.16 | 5.14 | 7.12 | 11.86 | 4.15 | 5.40 | 8.93 | 8.26 | 3.32 | 2.75 | 5.08 | 8.61 | 2.39 | 3.99 |
| 30 | 22.01 | 16.41 | 26.93 | 10.95 | 22.21 | 17.56 | 8.28 | 4.40 | 12.88 | 4.23 | 10.87 | 4.80 | 8.39 | 8.28 | 8.64 | 2.87 | 6.43 | 7.93 | 2.21 | 2.53 |
| 40 | 17.07 | 31.23 | 24.15 | 20.37 | 8.87 | 7.79 | 6.70 | 6.03 | 3.40 | 3.28 | 3.88 | 4.19 | 11.49 | 4.30 | 2.86 | 7.13 | 2.30 | 3.65 | 3.23 | 7.03 |
| 50 | 9.52 | 11.56 | 9.35 | 7.92 | 6.87 | 6.95 | 6.25 | 5.54 | 5.34 | 9.62 | 4.03 | 7.33 | 8.89 | 3.87 | 6.29 | 6.88 | 3.23 | 2.39 | 3.70 | 4.72 |
| 60 | 22.65 | 10.43 | 7.95 | 4.66 | 4.46 | 12.15 | 12.45 | 9.26 | 11.20 | 3.04 | 2.95 | 3.99 | 8.77 | 8.28 | 2.87 | 2.18 | 7.99 | 3.06 | 3.33 | 2.24 |
| 70 | 22.05 | 18.31 | 4.78 | 13.17 | 12.53 | 4.21 | 4.64 | 10.87 | 5.03 | 4.91 | 2.94 | 3.79 | 2.70 | 6.73 | 3.78 | 5.19 | 7.10 | 1.97 | 6.82 | 4.71 |
| 80 | 17.48 | 21.58 | 4.26 | 6.41 | 6.29 | 3.91 | 5.55 | 5.11 | 3.36 | 3.97 | 2.53 | 4.49 | 1.90 | 2.50 | 2.32 | 1.78 | 5.58 | 1.82 | 2.17 | 1.97 |
| 90 | 17.69 | 16.30 | 4.24 | 12.62 | 9.19 | 5.24 | 10.07 | 2.52 | 4.04 | 2.64 | 7.91 | 3.56 | 1.94 | 2.66 | 2.29 | 2.82 | 2.84 | 6.09 | 1.79 | 5.04 |
| 100 | 4.50 | 12.01 | 4.13 | 12.25 | 11.17 | 7.81 | 4.86 | 8.29 | 7.72 | 3.51 | 6.66 | 3.55 | 7.05 | 2.39 | 1.64 | 1.47 | 2.65 | 3.20 | 3.82 | 1.23 |
| 110 | 6.31 | 11.91 | 9.64 | 4.32 | 5.36 | 4.71 | 2.59 | 6.47 | 6.06 | 7.08 | 2.10 | 3.62 | 2.27 | 2.59 | 2.23 | 2.13 | 2.02 | 4.48 | 1.26 | 3.52 |
| 120 | 5.57 | 11.89 | 2.95 | 8.35 | 7.60 | 2.90 | 3.01 | 2.70 | 6.68 | 1.82 | 3.50 | 4.76 | 5.99 | 4.37 | 2.14 | 5.54 | 4.87 | 2.35 | 4.48 | 1.12 |
| 130 | 3.80 | 9.36 | 8.35 | 9.85 | 7.08 | 3.39 | 3.90 | 6.64 | 5.35 | 2.09 | 1.59 | 5.57 | 1.44 | 1.84 | 2.71 | 1.92 | 2.39 | 1.19 | 1.79 | 1.70 |
| 140 | 11.68 | 2.80 | 4.81 | 8.22 | 2.50 | 3.56 | 3.93 | 3.52 | 2.45 | 5.78 | 5.55 | 4.94 | 4.24 | 3.92 | 2.51 | 1.89 | 3.45 | 4.52 | 4.18 | 1.53 |
| 150 | 10.64 | 3.94 | 3.58 | 7.68 | 7.39 | 2.30 | 2.57 | 6.58 | 3.49 | 6.99 | 7.12 | 2.16 | 2.55 | 4.93 | 1.68 | 1.18 | 4.50 | 2.60 | 4.16 | 2.29 |
| 160 | 3.19 | 9.65 | 7.71 | 6.33 | 8.70 | 5.79 | 3.60 | 2.45 | 1.63 | 2.27 | 2.08 | 1.52 | 3.94 | 4.45 | 1.37 | 1.44 | 1.59 | **1.05** | 1.70 | 2.45 |
| 170 | 5.24 | 9.33 | 3.48 | 3.42 | 3.28 | 7.52 | 3.63 | 2.18 | 3.25 | 4.41 | 3.99 | 3.95 | 2.81 | 2.44 | 3.32 | 3.50 | 5.10 | 2.95 | 1.34 | 2.77 |
| 180 | 10.02 | 3.72 | 2.33 | 4.27 | 6.64 | 5.74 | 5.60 | 2.10 | 6.98 | 5.18 | 1.97 | 1.55 | 1.77 | 5.35 | 1.63 | 2.55 | 4.05 | 1.94 | 1.59 | 1.90 |
| 190 | 2.29 | 2.30 | 3.78 | 5.29 | 2.60 | 7.36 | 2.21 | 2.18 | 2.72 | 5.01 | 4.60 | 5.89 | 1.78 | 2.59 | 1.07 | 4.19 | 2.14 | 4.17 | 4.71 | 1.38 |
| 200 | 2.83 | 2.29 | 5.43 | 7.18 | 6.58 | 2.21 | 1.55 | 3.09 | 2.05 | 1.21 | 2.77 | 5.38 | 5.40 | 1.64 | 2.58 | 2.01 | 1.48 | 2.00 | 2.62 | 1.74 |

## References


1. Merayo, D.; Rodríguez-Prieto, A.; Camacho, A.M. Prediction of the Bilinear Stress-Strain Curve of Aluminum Alloys Using Artificial Intelligence and Big Data. *Metals* **2020**, *10*, 904. [CrossRef]
2. Alam, T.; Ansari, A.H. Review on Aluminium and Its Alloys for automotive applications. *Int. J. Adv. Technol. Eng. Sci.* **2017**, *5*, 278–294.
3. Merayo, D.; Rodríguez-Prieto, A.; Camacho, A.M. Prediction of mechanical properties by artificial neural networks to characterize the plastic behavior of aluminum alloys. *Materials* **2020**, *13*, 5227. [CrossRef]
4. Ashkenazi, D. How aluminum changed the world: A metallurgical revolution through technological and cultural perspectives. *Technol. Forecast. Soc. Chang.* **2019**, *143*, 101–113. [CrossRef]
5. Hahn, G.; Rosenfield, A. Metallurgical factors affecting fracture toughness of aluminum alloys. *Metall. Trans. A* **1975**, *6*, 653–668. [CrossRef]
6. Yogo, Y.; Sawamura, M.; Iwata, N.; Yukawa, N. Stress-strain curve measurements of aluminum alloy and carbon steel by unconstrained-type high-pressure torsion testing. *Mater. Des.* **2017**, *122*, 226–235. [CrossRef]
7. Kaufman, J.G. *Introduction to Aluminum Alloys and Tempers*; ASM international: Russell Township, Geauga County, OH, USA, 2000.
8. Kamaya, M.; Kawakubo, M. A procedure for determining the true stress–strain curve over a large range of strains using digital image correlation and finite element analysis. *Mech. Mater.* **2011**, *43*, 243–253. [CrossRef]
9. Hu, J.; Marciniak, Z.; Duncan, J. *Mechanics of Sheet Metal Forming*; Elsevier: Amsterdam, The Netherlands, 2002.

10. Hosford, W.F.; Caddell, R.M. *Metal Forming: Mechanics and Metallurgy*; Cambridge University Press: Cambridge, UK, 2011.
11. Zhao, Y.; Song, B.; Pei, J.; Jia, C.; Li, B.; Linlin, G. Effect of deformation speed on the microstructure and mechanical properties of AA6063 during continuous extrusion process. *J. Mater. Process. Technol.* **2013**, *213*, 1855–1863. [CrossRef]
12. Ilyas, M.; Hussain, G.; Rashid, H.; Alkahtani, M. Influence of Forming Parameters on the Mechanical Behavior of a Thin Aluminum Sheet Processed through Single Point Incremental Forming. *Metals* **2020**, *10*, 1461. [CrossRef]
13. Merayo, D.; Rodríguez-Prieto, A.; Camacho, A.M. Prediction of Physical and Mechanical Properties for Metallic Materials Selection Using Big Data and Artificial Neural Networks. *IEEE Access* **2020**, *8*, 13444–13456. [CrossRef]
14. Ghosh, I.; Das, S.K.; Chakraborty, N. An artificial neural network model to characterize porosity defects during solidification of A356 aluminum alloy. *Neural Comput. Appl.* **2014**, *25*, 653–662. [CrossRef]
15. Zhao, D.; Ren, D.; Zhao, K.; Pan, S.; Guo, X. Effect of welding parameters on tensile strength of ultrasonic spot welded joints of aluminum to steel–By experimentation and artificial neural network. *J. Manuf. Process.* **2017**, *30*, 63–74. [CrossRef]
16. Haghdadi, N.; Zarei-Hanzaki, A.; Khalesian, A.; Abedi, H. Artificial neural network modeling to predict the hot deformation behavior of an A356 aluminum alloy. *Mater. Des.* **2013**, *49*, 386–391. [CrossRef]
17. Huang, J.; Liew, J.; Ademiloye, A.; Liew, K. Artificial intelligence in materials modeling and design. *Arch. Comput. Methods Eng.* **2020**, *28*, 3399–3413. [CrossRef]
18. Jordan, M.I.; Mitchell, T.M. Machine learning: Trends, perspectives, and prospects. *Science* **2015**, *349*, 255–260. [CrossRef]
19. Merayo, D.; Rodriguez-Prieto, A.; Camacho, A.M. Comparative analysis of artificial intelligence techniques for material selection applied to manufacturing in Industry 4.0. *Procedia Manuf.* **2019**, *41*, 42–49. [CrossRef]
20. Jackson, P.C. *Introduction to Artificial Intelligence*; Courier Dover Publications: Mineola, NY, USA, 2019.
21. Chen, C.T.; Gu, G.X. Machine learning for composite materials. *MRS Commun.* **2019**, *9*, 556–566. [CrossRef]
22. Joshi, P. *Artificial Intelligence with Python*; Packt Publishing Ltd.: Birmingham, UK, 2017.
23. Dimiduk, D.M.; Holm, E.A.; Niezgoda, S.R. Perspectives on the impact of machine learning, deep learning, and artificial intelligence on materials, processes, and structures engineering. *Integr. Mater. Manuf. Innov.* **2018**, *7*, 157–172. [CrossRef]
24. Balachandran, P.V.; Xue, D.; Theiler, J.; Hogden, J.; Gubernatis, J.E.; Lookman, T. Importance of feature selection in machine learning and adaptive design for materials. In *Materials Discovery and Design*; Springer: Berlin/Heidelberg, Germany, 2018; pp. 59–79.
25. Feng, S.; Zhou, H.; Dong, H. Using deep neural network with small dataset to predict material defects. *Mater. Des.* **2019**, *162*, 300–310. [CrossRef]
26. Mueller, T.; Kusne, A.G.; Ramprasad, R. Machine learning in materials science: Recent progress and emerging applications. *Rev. Comput. Chem.* **2016**, *29*, 186–273.
27. Lefik, M.; Boso, D.; Schrefler, B. Artificial neural networks in numerical modelling of composites. *Comput. Methods Appl. Mech. Eng.* **2009**, *198*, 1785–1804. [CrossRef]
28. Sha, W.; Edwards, K. The use of artificial neural networks in materials science based research. *Mater. Des.* **2007**, *28*, 1747–1752. [CrossRef]
29. Davis, J.R. *Alloying: Understanding the Basics*; ASM International: Russell Township, Geauga County, OH, USA, 2001.
30. Naik, D.L.; Sajid, H.U.; Kiran, R. Texture-based metallurgical phase identification in structural steels: A supervised machine learning approach. *Metals* **2019**, *9*, 546. [CrossRef]
31. Honysz, R. Modeling the Chemical Composition of Ferritic Stainless Steels with the Use of Artificial Neural Networks. *Metals* **2021**, *11*, 724. [CrossRef]
32. Qian, L.; Winfree, E.; Bruck, J. Neural network computation with DNA strand displacement cascades. *Nature* **2011**, *475*, 368–372. [CrossRef]
33. Schmidhuber, J. Deep learning in neural networks: An overview. *Neural Netw.* **2015**, *61*, 85–117. [CrossRef] [PubMed]
34. The Aluminum Association. *Designations and Chemical Composition Limits for Aluminum Alloys in the Form of Castings and Ingot*; The Aluminum Association Inc.: Arlington, VA, USA, 2006.
35. Matmatch GmbH. Matmatch. 2021. Available online: https://matmatch.com/ (accessed on 15 June 2021).
36. Kingma, D.; Ba, J. Adam: A method for stochastic optimization. In Proceedings of the 3rd International Conference for Learning Representations (ICLR 15), San Diego, CA, USA, 7–9 May 2015.
37. Hornik, K. Approximation capabilities of multilayer feedforward networks. *Neural Netw.* **1991**, *4*, 251–257. [CrossRef]

MDPI
St. Alban-Anlage 66
4052 Basel
Switzerland
Tel. +41 61 683 77 34
Fax +41 61 302 89 18
www.mdpi.com

*Metals* Editorial Office
E-mail: metals@mdpi.com
www.mdpi.com/journal/metals